Math at Hand

A Mathematics Handbook

Acknowledgments

We gratefully acknowledge the following teachers and mathematics supervisors who helped make *Math at Hand* a reality.

Sandra P. Alley
Elementary Mathematics Coordinator
Virginia Beach City Public Schools
Virginia Beach, VA

Robert Fair
District Math Coordinator K-12
Cherry Creek Schools
Greenwood Village, CO

Thérèse R. Horn
Third Grade Teacher
Charter Oak School
West Hartford, CT

Matt Larson
K-12 Math Consultant
Lincoln Public Schools
Lincoln, NE

Marilyn LeRud
Retired K-8 Teacher
Tucson Unified School District
Tucson, AZ

Carole M. McKittrick
Grade 5 Teacher
Great Falls Public Schools
Great Falls, MT

Lance Menster
Math Specialist
Alief ISD
Houston, TX

Marni Napierala
Grade 5 Teacher
Great Falls Public Schools
Great Falls, MT

Roberta M. Treinavicz
Instructional Resource Specialist
Hancock School
Brockton, MA

Senior Consultant:
Dr. Marsha W. Lilly
Secondary Mathematics Coordinator
Alief ISD
Alief, TX

Writing/Editorial: Carol DeBold; Justine Dunn; Edward Manfre; Ann Petroni-McMullen, Kane Publishing Services, Inc.; Susan Rogalski
Design Management: Richard Spencer
Production Management: Sandra Easton
Design and Production: Bill SMITH STUDIO
Marketing: Lisa Bingen
Illustration credits: See 548

Printed in the United States of America

International Standard Book Number: 0-669-46807-X (hardcover)
3 4 5 6 7 8 9 0 RRDC 04 03 02 01

International Standard Book Number: 0-669-46922-X (softcover)
4 5 6 7 8 9 0 RRDC 04 03 02 01

Table of Contents

iii

Almanac 419

Yellow Pages 507

Index 539

How This Book Is Organized

vii

Math at Hand is a resource book. That means you're not expected to read it from cover to cover. Instead, you'll want to keep it handy for those times when you're not clear about a math topic and need a place to look up definitions, procedures, explanations, and rules.

Because this is a resource book and because there may be more than one topic on a page, we have given each topic an item number 326. So, when you are looking up a specific topic, look for its item number.

item numbers — 326–327

item number — 326

Computing with Measures — sub-section

— short table of contents

When you add 2 feet and 24 inches, you don't get 26 feet and you don't get 26 inches. But, what do you get? There are rules that help you compute with units that are not the same.

item number — 327

Changing from One Unit to Another — topic

If you want to add 2 feet and 24 inches, one thing you could do is change 24 inches to feet and then add.

When you change from one unit of measure to another, you need to know the relationship between the two units of measure. To change inches to feet or feet to inches, you need to know that 1 foot = 12 inches. The tables of measures in the Almanac can give you that information.

EXAMPLE 1: Change 6 yards to feet. — example

more help — MORE HELP See 149, 485–487

Customary Units of Length		Metric Units of Length	
inch (in.)		millimeter (mm)	1 mm = 0.001
foot (ft)	1 ft = 12 in.	centimeter (cm)	1 cm = 0.01
yard (yd)	1 yd = 3 ft	meter (m)	
mile (mi)	1 mi = 5280 ft	kilometer (km)	1 km = 1000 m

Because feet are smaller than yards, you can expect that you will have more feet than you had yards. Your answer must be more than 6, so multiply 6 by 3.

answer — 6 yards = 18 feet

A good way to get started in this book is to thumb through the pages. Find these parts:

- **Table of Contents**
  This lists the major sections and sub-sections of the book.

- **Sections and Sub-Sections**
  Each section of the handbook has a short table of contents so you know what is in the section. Sections have several subsections and each of these also has its own short table of contents. Notice the color bars across the tops of the pages. Each section has a different color to make it easy to find.

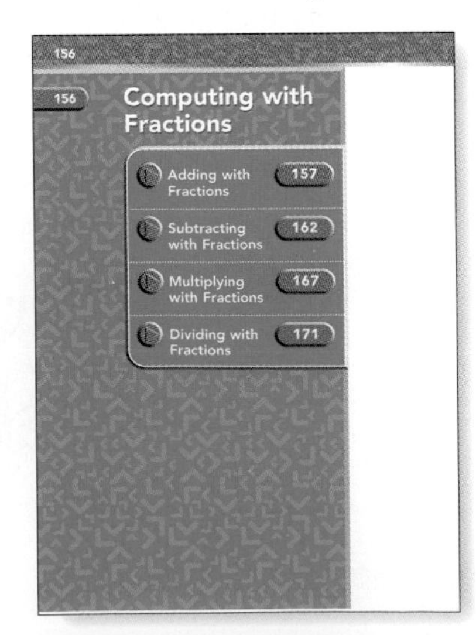

156 Computing with Fractions

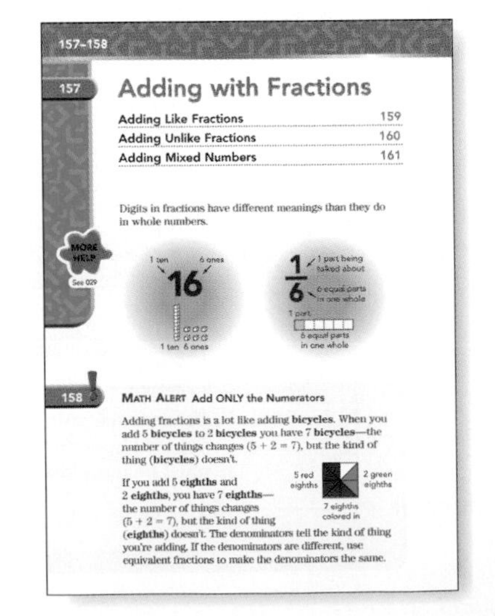

157 Adding with Fractions

Digits in fractions have different meanings than they do in whole numbers.

158 MATH ALERT Add ONLY the Numerators

Adding fractions is a lot like adding **bicycles**. When you add 5 **bicycles** to 2 **bicycles** you have 7 **bicycles**—the number of things changes (5 + 2 = 7), but the kind of thing (**bicycles**) doesn't.

If you add 5 **eighths** and 2 **eighths**, you have 7 **eighths**—the number of things changes (5 + 2 = 7), but the kind of thing (**eighths**) doesn't. The denominators tell the kind of thing you're adding. If the denominators are different, use equivalent fractions to make the denominators the same.

- **Almanac**
  This includes some very helpful tables and lists. It also has hints on how to study, take a test, and use a calculator. Check out all the Almanac entries—you will want to refer to them often.

Almanac 419

This almanac includes very useful tables and lists. It has tips on how to take notes in class and how to study for and take tests. It shows how to use a calculator and even has some computer math tips.

- **Yellow Pages**
  This part of the handbook has three glossaries. The Glossary of Mathematical Formulas is the place you look if you forget a formula. There's also a Glossary of Mathematical Symbols. In the Glossary of Mathematical Terms, you will find math terms that your teacher and textbook use, and terms your parents use.

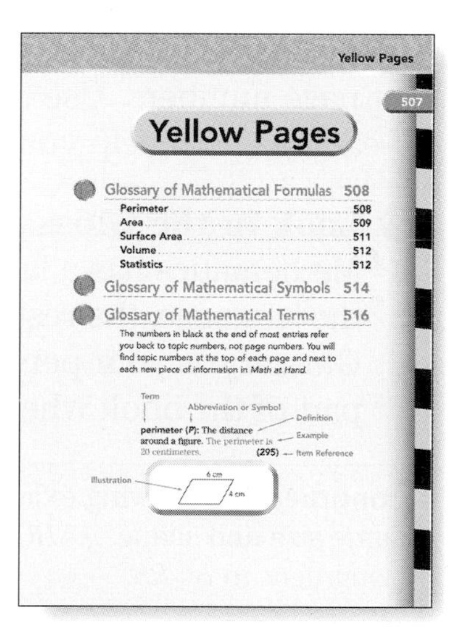

Yellow Pages

The numbers in black at the end of most entries refer you back to topic numbers, not page numbers. You will find topic numbers at the top of each page and next to each new piece of information in *Math at Hand*.

- **Index**
  This is at the very end of the book.

How to Use This Book

There are three ways to find information about the topics in which you are interested:

1 Look in the Index

We listed topics in the Index using any word we thought you might use to describe the topic. For example, you will find "volume of prisms" under both "Volume" and "Prisms."

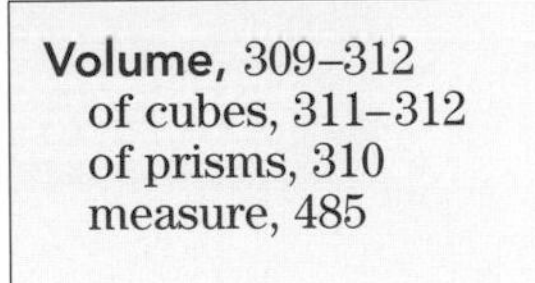
Volume, 309–312
of cubes, 311–312
of prisms, 310
measure, 485

Prisms, 383–384
drawing, 475
nets of, 384
surface area of, 306–308
volume of, 309–311

Remember that you are being directed to item numbers, not page numbers. Use the item numbers located at the top of each page to help you find the topic you are looking for.

2 Look in the Glossary

Mathematics has a language all its own. Once you learn the language, the rest is much easier. Think of this Glossary as your personal interpreter and turn to this part of the book whenever you see an unfamiliar word.

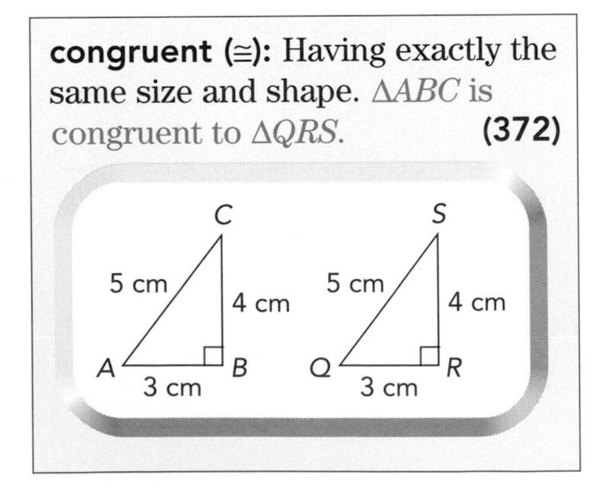
congruent (≅): Having exactly the same size and shape. ΔABC is congruent to ΔQRS. **(372)**

Most Glossary entries will give you an item number to refer to if you want more information about the topic.

3 Look in the Table of Contents

All the major topics covered in this book are listed in the Table of Contents. If you're looking for a general topic, like Plane Figures, the Table of Contents is a quick way to find it.

Section
(If you want to browse through lots of related topics, start here.)

Sub-section
(If you want to browse, but narrow your search, start here.)

Item number
(Don't forget, *Math at Hand* numbers items, not pages!)

001

Numeration

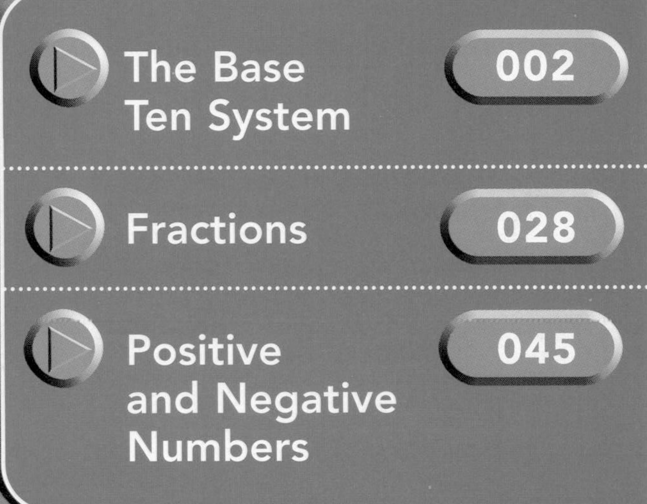

Numbers are everywhere.

002

The Base Ten System

The number system we use for everyday life is based on ten. We use ten symbols—0, 1, 2, 3, 4, 5, 6, 7, 8, 9—called **digits**. Add some punctuation (a comma, a decimal point, and a few others), and you can write numbers for any situation.

With just a handful of symbols, you can write quantities larger than the number of shells in the sea and smaller than the width of one hair on your head.

Imagine if you had to learn a different symbol for every number! Our system lets us write all the numbers you can think of with just a few symbols.

003 Whole Numbers

What are **whole numbers**? They are 0, 1, 2, 3, 4, 5, 6, and so on. If a number has a decimal part, a part that's a fraction, or a negative sign, it's not a whole number.

Whole Numbers

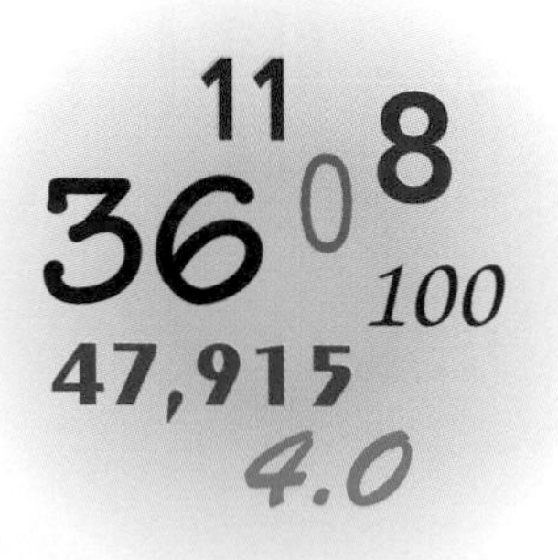

Not Whole Numbers

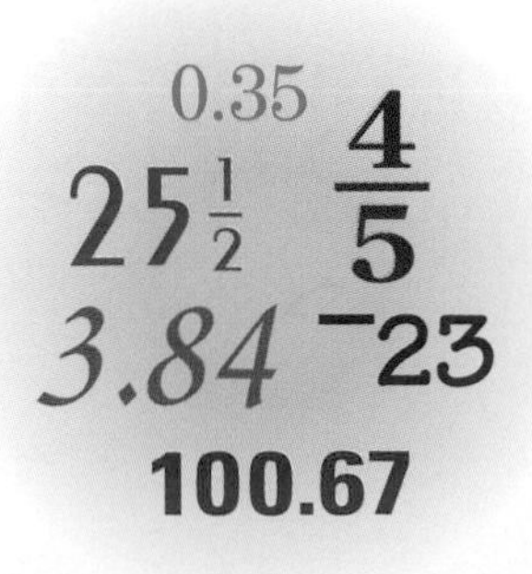

WhOle numbers include O!

Whole Numbers: Place Value

Our number system is based on a simple pattern of tens. Each place has ten times the value of the place to its right.

Place value tells you how much each digit stands for.

Thousands	Hundreds	Tens	Ones
1 thousand is 10 times 1 hundred	1 hundred is 10 times 1 ten	1 ten is 10 times 1 one	

1 thousand + 1 hundred + 1 ten + 1 = 1111

Sometimes four-digit numbers are written without a comma.

2437 is the same as 2,437.

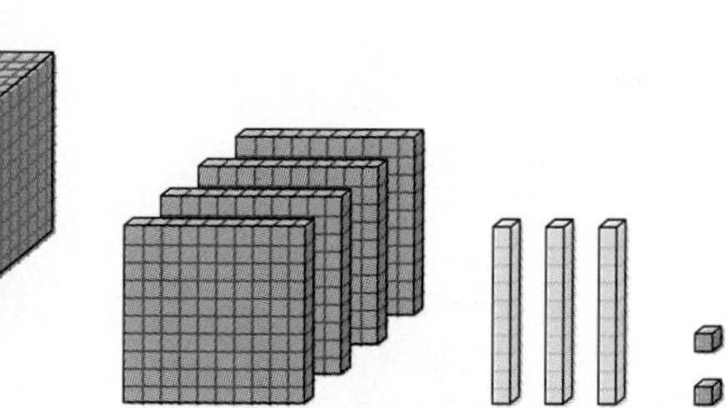

2437 = 2,437

005

Reading and Writing Large Numbers

We arrange numbers into groups of three places called **periods**. The places within periods repeat (hundreds, tens, ones; hundreds, tens, ones; and so on). In the U.S., we usually use commas to separate the periods.

EXAMPLE 1: What is the value of the digit **3** in 905,346,521?

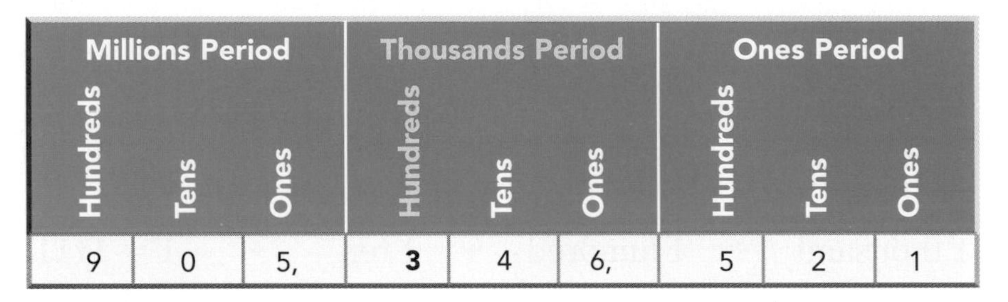

Millions Period			Thousands Period			Ones Period		
Hundreds	Tens	Ones	Hundreds	Tens	Ones	Hundreds	Tens	Ones
9	0	5,	**3**	4	6,	5	2	1

The digit 3 is in the hundred thousands place. Its value is 3 hundred thousand, or 300,000.

EXAMPLE 2: How would you read 905,346,521?

To read a whole number:

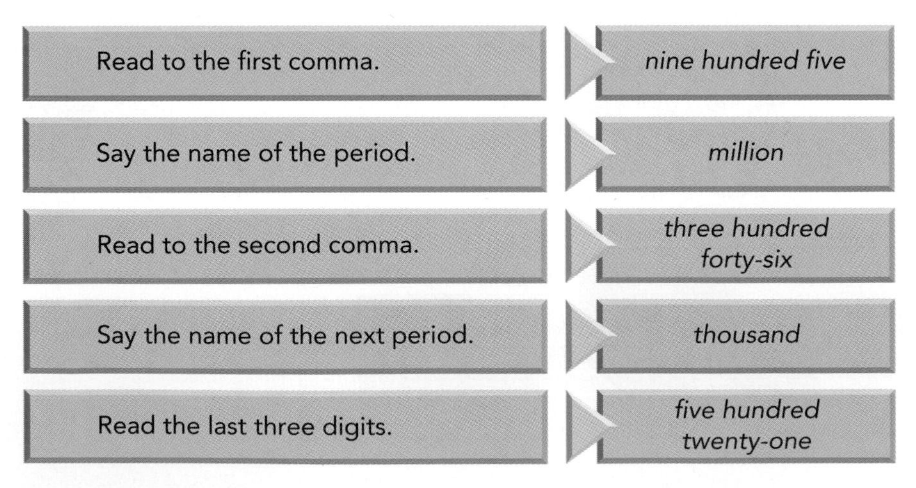

You don't need to say the name of the ones period.

Say: *nine hundred five million, three hundred forty-six thousand, five hundred twenty-one*

Ways of Writing Whole Numbers

006

There are many ways to write the same whole number.

Form	Example
Standard	546,872
Word	five hundred forty-six thousand, eight hundred seventy-two
Expanded	$(5 \times 100{,}000) + (4 \times 10{,}000) + (6 \times 1000) + (8 \times 100) + (7 \times 10) + (2 \times 1)$
Exponential	$(5 \times 10^5) + (4 \times 10^4) + (6 \times 10^3) + (8 \times 10^2) + (7 \times 10^1) + (2 \times 10^0)$

Powers of 10

007

When you multiply 10s together, the product is called a **power of 10**. The numbers 10, 100, 1000, and 10,000 are powers of 10. You can use an **exponent** to show a power of 10. The exponent gives the number of times that 10 is used as a factor.

$10^2 = 10 \times 10$

MORE HELP

See 051, 065

EXAMPLE: There are more than 100,000,000,000 stars in our galaxy, the Milky Way. How can you rewrite this number in a shorter way? *(Source: World Book Encyclopedia)*

Here are some ways to express a large number.

Form	Example
Standard	100,000,000,000
Word	one hundred billion
Factor	$10 \times 10 \times 10 \times 10 \times 10 \times 10 \times 10 \times 10 \times 10 \times 10 \times 10$
Exponential	Write: 10^{11} Say: *ten to the eleventh power,* or just *ten to the eleventh*

There are more than 10^{11} stars in our galaxy!

008 Comparing Whole Numbers

How can you compare two whole numbers?

EXAMPLE 1: Compare 76 and 67.

Write: Say:

$67 < 76$ ***67 is less than 76***

$76 > 67$ ***76 is greater than 67***

Remember, the mouth of the symbol ($>$ or $<$) always opens to the greater number.

You can use what you know about place value to compare two numbers.

EXAMPLE 2: Blanca Peak is 14,345 feet above sea level. Crestone Peak is 14,294 feet above sea level. Which peak is taller? *(Source: U.S. Geological Survey)*

❶ Line up the place values by lining up the ones.	❷ Begin at the left. Find the first place where the digits are different.	❸ Compare the value of the digits.
14,345 14,294	14,345 14,294 ↑↑ same ↑ different	300 > 200 So, 14,345 > 14,294.

Write: $14{,}345 > 14{,}294$ or $14{,}294 < 14{,}345$

Either way, Blanca Peak is taller.

009

MATH ALERT Lining Up by Place Value

Be sure to line up digits with the same place value when comparing or computing.

EXAMPLE: You score 108,464 points. Your friend scores 97,996 points. The higher score wins. Who wins?

Lined up incorrectly	**Lined up at the ones place**
108,464	108,464
97,996	97,996

When one whole number has more digits than another, it is greater.

Write: 108,464 > 97,996 or 97,996 < 108,464

Either way, you win.

Ordering Whole Numbers

If you know how to compare two whole numbers, you also know how to put a group of numbers in order.

MORE HELP

See 008–009

EXAMPLE: Order these numbers from the greatest number of girls to the least.

Girls Playing Youth Soccer	
Age Group	**Number of Girls Playing**
7–8	45,181
9–10	46,758
11–12	39,939

(Source: American Youth Soccer Organization)

① Line up the numbers at the ones place.	② Begin to compare at the left.	③ Continue. Find the first place where the digits are different.
45,181 46,758 39,939	45,181 40,000 > 30,000 46,758 So, 39,939 is the 39,939 least.	45,181 6000 > 5000 46,758 So, 46,758 > 45,181.

Here are the numbers from greatest to least.
46,758 > 45,181 > 39,939

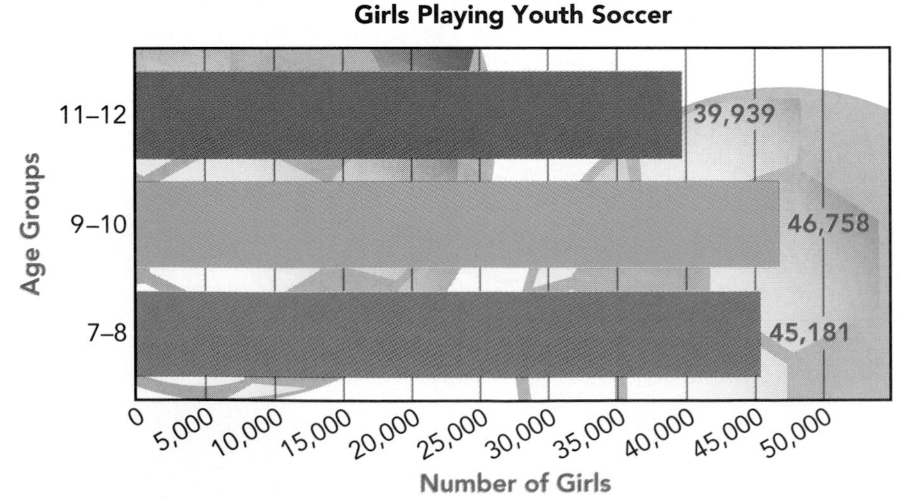

Decimals 011

Decimal numbers are numbers that are written using place value. We use a **decimal point** to separate the whole-number places from the places less than one.

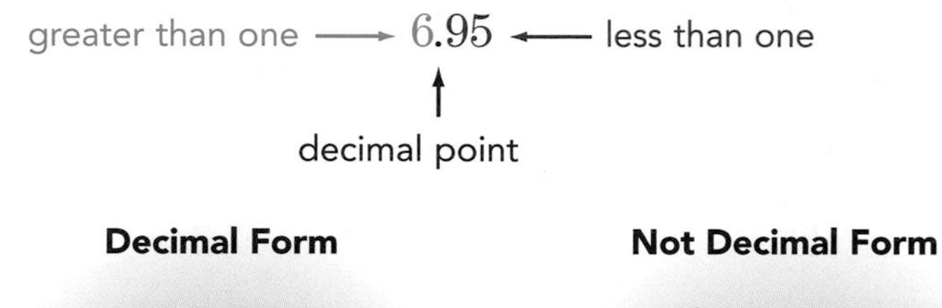

See 019, 043

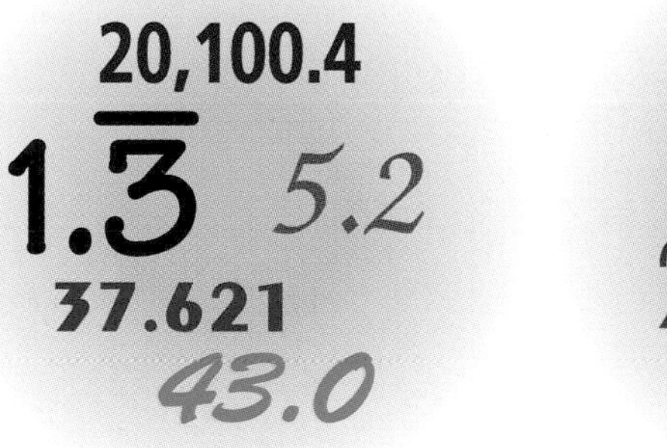

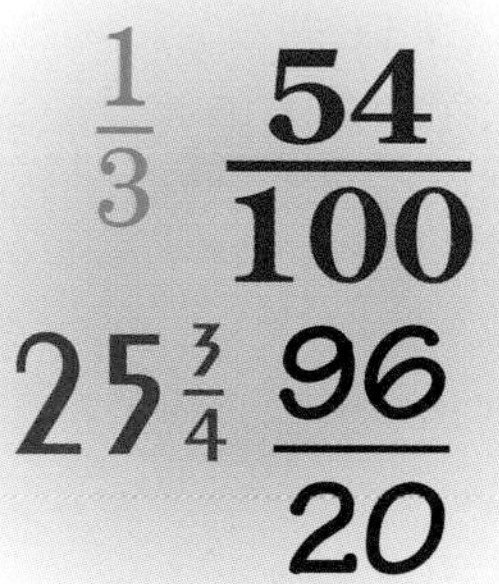

Any fraction can be written in decimal form. See 043 to find out how.

012 Decimals: Place Value

See 006

ONE WAY You can think of money to help you understand decimals and their place values.

1 dollar + 0.1 dollar + 0.01 dollar = $1.11

ANOTHER WAY Decimals follow the place-value pattern. Each place has ten times the value of the place to its right.

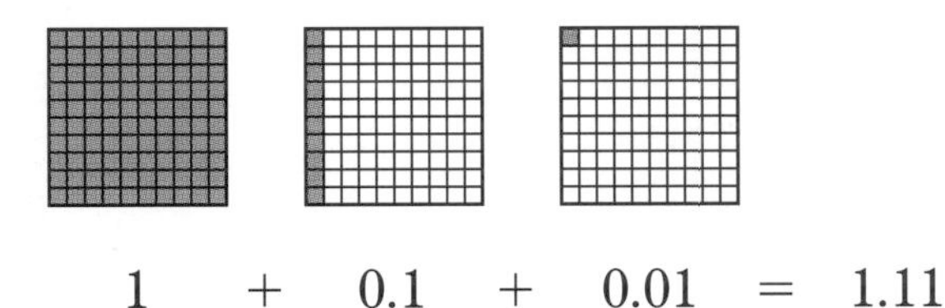

1 + 0.1 + 0.01 = 1.11

EXAMPLE: What is the value of the digit **4** in 12.5**4**?

ONE WAY

Tens	Ones		Tenths	Hundredths
1	2	.	5	**4**

The digit 4 is in the hundredths place. It has a value of 0.04, or 4 hundredths.

ANOTHER WAY You can also use money to help find the value.

Since cents are hundredths of dollars, the **4** in 12.54 has a value of 4 hundredths.

Reading and Writing Decimals

EXAMPLE: Alyssa Kiel and Molly Brammer competed in the 1997–1998 Postal Long Distance Meet in the 1000-yard freestyle race for 10-year-old girls. Alyssa's time was 37.26 seconds faster than Molly's. How would you read 37.26?

(Source: www.usswim.org)

Tens	Ones		Tenths	Hundredths
3	7	.	2	6

The name of every decimal place ends with *ths.*

To read a decimal:

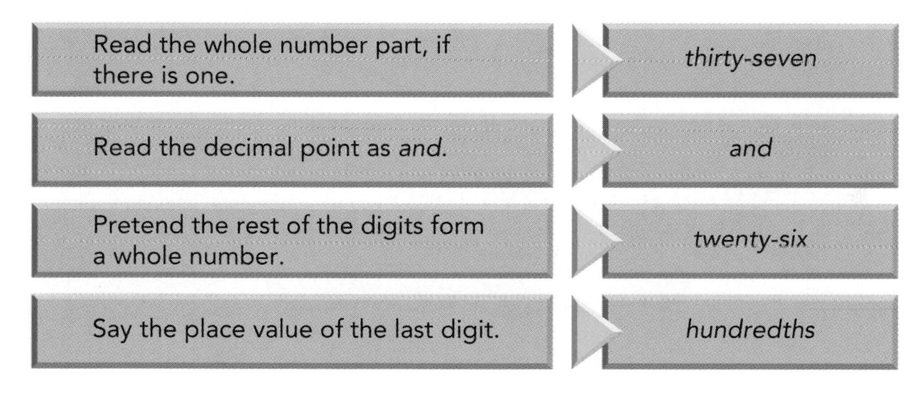

Say: *thirty-seven and twenty-six hundredths*

There are many ways to write the same decimal.

Form	Example
Standard	37.26
Word	thirty-seven and twenty-six hundredths
Expanded	$(3 \times 10) + (7 \times 1) + (2 \times 0.1) + (6 \times 0.01)$

014

MATH ALERT Interpreting *and* in Decimal Numbers

When you read a number, do NOT say the word *and* in any old place. If you do, you'll have trouble when you need to read the decimal point as *and*.

Sometimes, pretty funny misunderstandings can happen.

EXAMPLE: Which is more likely to weigh two hundred and twenty-five thousandths pounds: a baby elephant or a baby hamster?

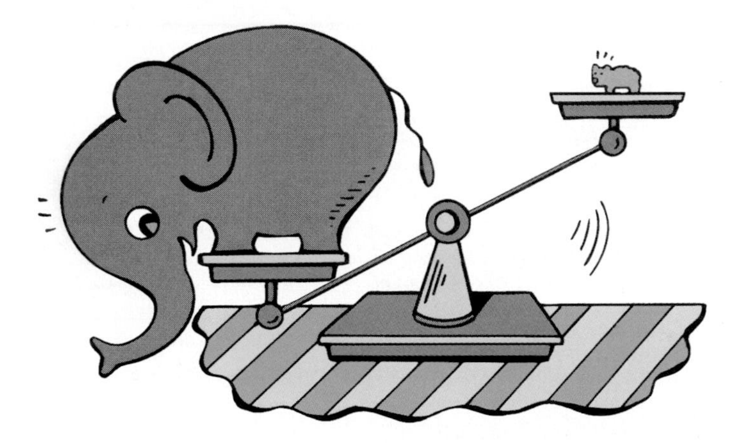

Two hundred and twenty-five thousandths

IS:	IS NOT:
200.025	0.225

If you mean 0.225, say: two hundred twenty-five thousandths.

A baby elephant could weigh two hundred and twenty-five thousandths pounds.

Say *and* only when you come to the decimal point.

Equivalent Decimals

015

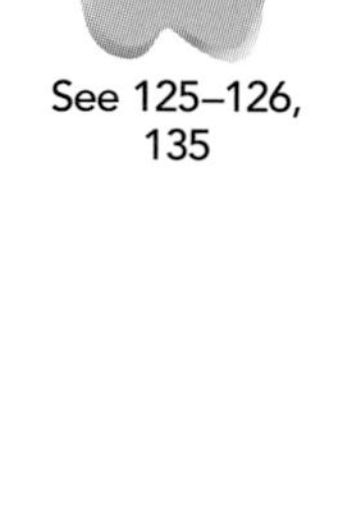

See 125–126, 135

Decimals that name the same amount are called **equivalent decimals**.

0.4 (four tenths) of the square is red.

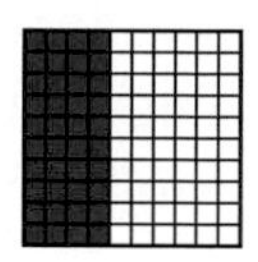

0.40 (forty hundredths) of the square is red.

This number line shows 0.4

This number line shows 0.40

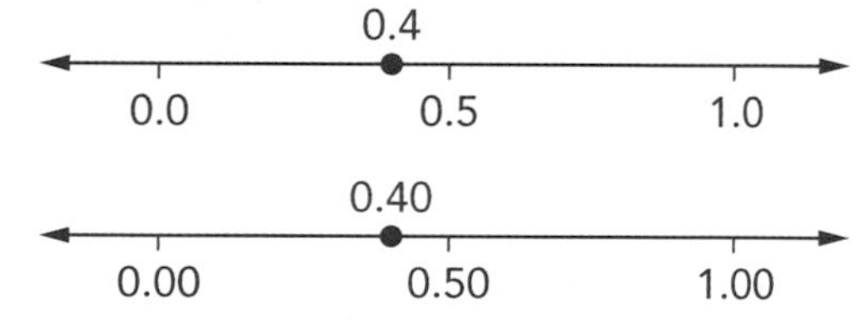

Write: 0.4 = 0.40
Say: *four tenths equals forty hundredths*

Shortcut

A shortcut for writing equivalent decimals is to write zeros in the places to the right of a decimal.

0.3 = 0.30 = 0.300 = 0.3000

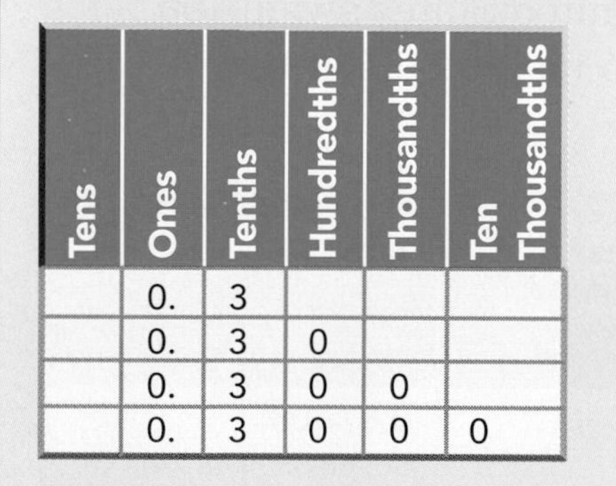

Tens	Ones	Tenths	Hundredths	Thousandths	Ten Thousandths
	0.	3			
	0.	3	0		
	0.	3	0	0	
	0.	3	0	0	0

Don't try this with whole numbers! For whole numbers, if you write zeros at the end, the place values change, so the number changes. 3 does not equal 30!

Writing zeros at the right-hand end of a mixed decimal number does not change its value.

016

Comparing Decimals

Comparing decimals is like comparing whole numbers.

EXAMPLE: The Castillo de San Marcos National Monument in Florida is on 20.51 acres. The Gettysburg National Cemetery in Pennsylvania is on 20.58 acres. Which national landmark is on more land? *(Source: Information Please Almanac)*

❶ Line up the decimal points.	❷ Begin at the left. Find the first place where the digits are different.	❸ Compare the values of the digits.
20.51 20.58	20.51 20.58	$0.08 > 0.01$ So, $20.58 > 20.51$.

The Gettysburg National Cemetery is on more land.

017

MATH ALERT Sometimes Greater Isn't Better!

Be very careful when comparing times and scores. In golf, the lower score wins. In a race, the lower time wins.

EXAMPLE: In 1996, Alexander Popov of Russia swam the Olympic 100-meter freestyle event in 48.74 seconds. In 1988, Matt Biondi of the United States swam 100 meters in 48.63 seconds. Who had the faster time?

(Source: Information Please Almanac)

❶ Line up the decimal points.	❷ Begin at the left. Find the first place where the digits are different.	❸ Compare the values of the digits.
48.74 48.63	48.74 48.63	$0.6 < 0.7$ So, $48.63 < 48.74$.

Matt Biondi had the faster time.

Ordering Decimals

018

If you know how to compare two decimals, you also know how to put a group of decimals in order.

See 010, 016

EXAMPLE: During the softball season, Leah's batting average was 0.322, Emily's average was 0.224, and Erika's average was 0.314. Order the averages from greatest to least.

❶ Line up the decimal points.	❷ Begin at the left. Find the first place where the digits are different.	❸ Compare the remaining digits.
0.322 0.224 0.314	0.322 $\quad 0.3 > 0.2$, so 0.224 $\quad$ 0.224 is the least. 0.314	0.322 $\quad 0.02 > 0.01$ 0.314 $\quad$ So, $0.322 > 0.314$.

Here are the batting averages from greatest to least.
$0.322 > 0.314 > 0.224$

Relating Decimals to Fractions

How are decimals like fractions?

Try saying the decimal aloud. When you do, you'll see that it sounds like a fraction.

See 037

Write: 0.50	Write: $\frac{50}{100}$
Say: *fifty hundredths*	Say: *fifty hundredths*

After you've written a decimal as a fraction, you can simplify it just as you can simplify any fraction.

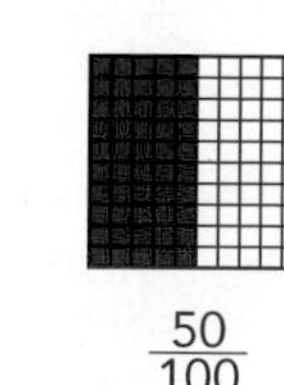
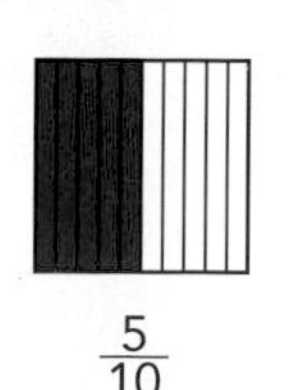
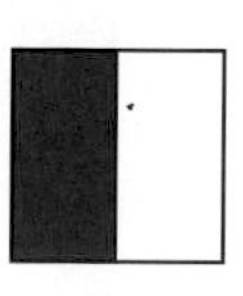

$$0.50 = \frac{50}{100} = \frac{5}{10} = \frac{1}{2}$$

020 Relating Decimals to Percents

The word **percent** means *per hundred*. You can use hundredths or percents to name the same number.

See 178–180, 189–190

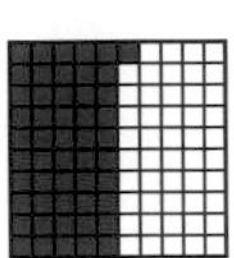

51 of 100 squares are red.

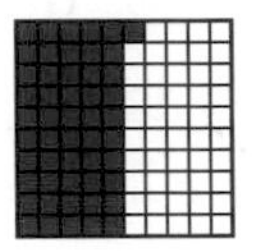

0.51 of the squares are red.

51% of the squares are red.

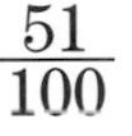

$\frac{51}{100}$ = 0.51 = 51%

The symbol for percent is made up of the digits in 100.

021 **MATH ALERT** Repeating Decimals

Sometimes when you divide, you can keep dividing forever! A **repeating decimal** has one or more digits that repeat in a pattern.

See 148, 153–154

EXAMPLE: Divide 20 by 3.

```
   6.66 . . .  ← The digit 6 repeats forever.
 3)20.00  ← Write the decimal point and zeros.
  −18 00
    2 00
   −1 80
      20
     −18
       2  ← The same remainder repeats forever. So, the decimal 6.66 . . . is a repeating decimal.
```

You can use a bar to show the digits that repeat.

$6.66\ldots = 6.\overline{6}$

$20 \div 3 = 6.\overline{6}$

The three dots ". . ." mean the decimal continues without end.

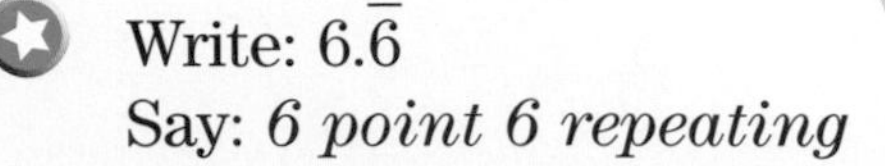

Write: $6.\overline{6}$
Say: *6 point 6 repeating*

Non-Repeating, Non-Terminating Decimals

022

Some decimals continue without ending, but do not have a repeating pattern. These decimals are known as **non-repeating, non-terminating** decimals.

See 069

- The diagonal of a square with a side 1 unit long is $\sqrt{2} \approx 1.414$.
- The ratio of the circumference of a circle to its diameter is $\pi \approx 3.14$.

Write: $\pi \approx 3.14$
Say: *pi is approximately equal to three and fourteen hundredths*

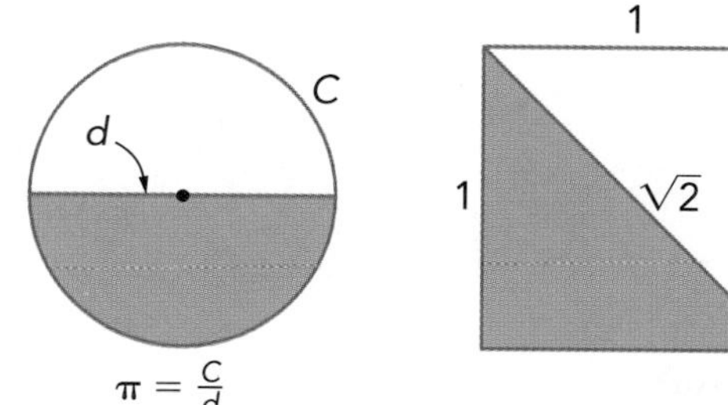

A decimal that doesn't repeat and doesn't end can't be written as a fraction. This kind of decimal is called an irrational number.

023

Money

The Coinage Act of 1792 created the first national money system in the United States. The dollar is the basic unit of U.S. currency.

024

U.S. Coins and Bills

Our coins and bills are based on ones, fives, and tens to make it easier to count money.

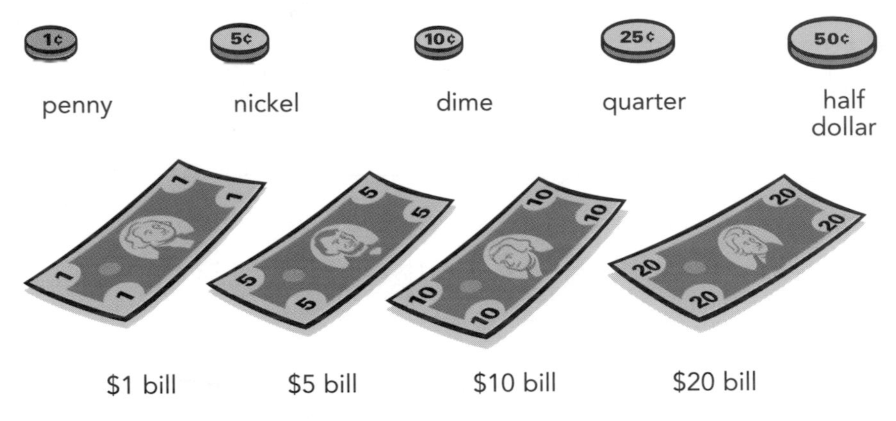

025

Counting Money

The most common way to count money is counting on. Begin with the highest value and count on.

You can also count money by grouping coins and bills.

EXAMPLE: Jacob had four 1-dollar bills, one 5-dollar bill, nine dimes, five pennies, three quarters, and five nickels. How much money was that?

ONE WAY You can sort the bills and coins by value, then add.

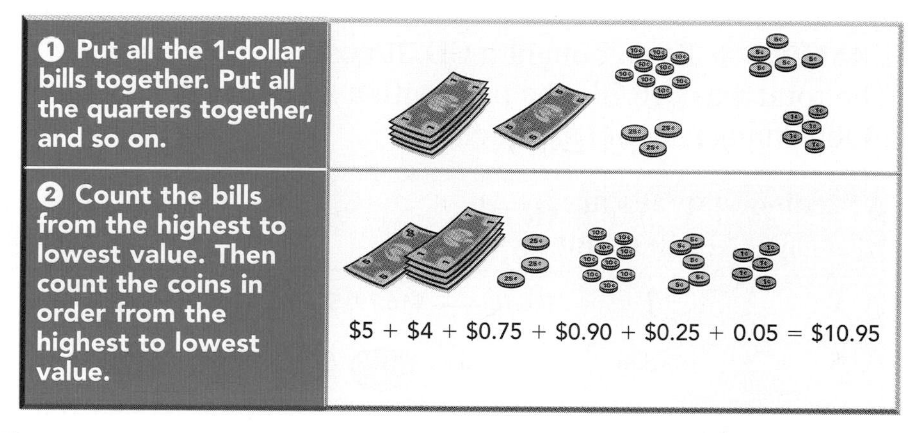

ANOTHER WAY Group the coins into one-dollar amounts, then count.

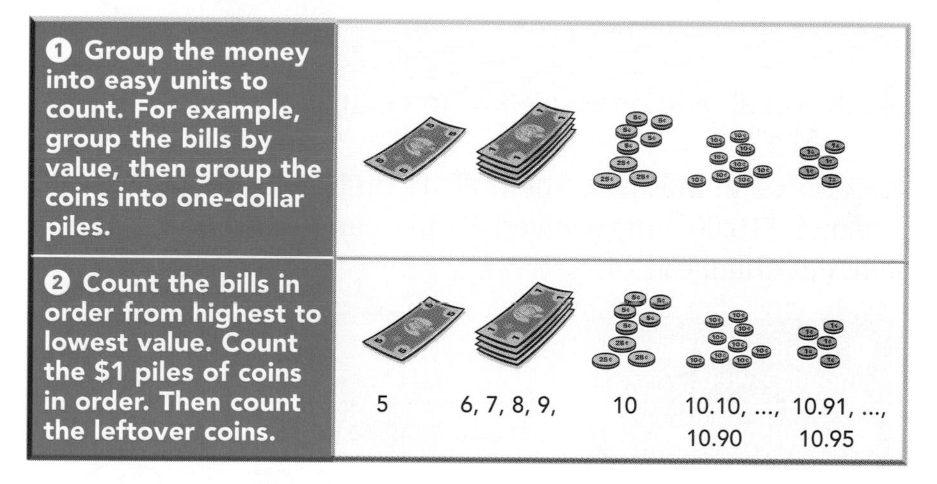

Either way, Jacob had $10.95.

Write: $10.95
Say: *ten dollars and ninety-five cents*

DID YOU KNOW. . .

that the first paper money was made in China? China did not have enough metal to make coins. So, paper money was invented.

026

Making Change

The fastest way to make change is to count up.

See 081

EXAMPLE 1: Kayla bought a CD. It cost $15.98 plus tax. The total was $16.78. She paid with a 20-dollar bill. How much change should she receive?

❶ Begin with the amount owed.	$16.78
❷ Count up from $16.78. Use coins to get to the next dollar.	$16.78 ⟶ $16.79, $16.80, $16.90, $17.00 1¢ 1¢ 10¢ 10¢
❸ Count up with bills to reach $20.00.	$17.00 ⟶ $18.00, $19.00, $20.00

Kayla should receive $3.22 in change.

EXAMPLE 2: Matthew spent $6.47 on lunch. He gave the cashier $10.00 and received $3.43 change. Did he get the correct change?

❶ Begin with the amount owed.	$6.47
❷ Count up from $6.47. Use coins to get to the next dollar.	$6.47 ⟶ $6.48, $6.49, $6.50, $6.75, $7.00 1¢ 1¢ 1¢ 25¢ 25¢
❸ Count up with bills to reach $10.00.	$7.00 ⟶ $8.00, $9.00, $10.00

Matthew should have received $3.53 change. He did not get the correct change.

Exchanging Currency

When you travel to another country, you exchange U.S. dollars for local money—Japanese yen, Mexican pesos, German marks, or whatever currency is used where you are going.

See 035, 142–143, 185

When you exchange money, a currency exchange rate is used to figure out how much foreign money you will get for your dollars. These rates change every day.

This table shows some exchange rates on Monday, October 26, 1998.

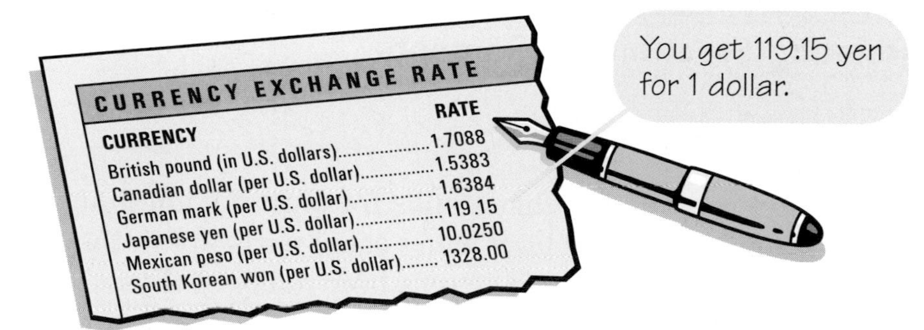

CURRENCY EXCHANGE RATE

CURRENCY	RATE
British pound (in U.S. dollars)	1.7088
Canadian dollar (per U.S. dollar)	1.5383
German mark (per U.S. dollar)	1.6384
Japanese yen (per U.S. dollar)	119.15
Mexican peso (per U.S. dollar)	10.0250
South Korean won (per U.S. dollar)	1328.00

(Source: Bloomberg Financial Markets)

EXAMPLE: Suppose on October 26, 1998, you exchanged $1000 for Japanese yen. How many yen should you have received?

When you use an exchange rate multiply the rate times the number of dollars you will exchange.

rate $\times$ dollars exchanged = yen received

119.15 $\times$ 1000 = 119,150

You should have received 119,150 yen for 1000 U.S. dollars.

028

Fractions

A **fraction** is a number that stands for part of something. The **denominator** tells how many equal parts are in the whole or set. The **numerator** tells how many of those parts you're talking about.

$\frac{3}{8}$ ← numerator (parts you are talking about)
← denominator (equal parts in one whole or set)

Why would you need to use a fraction? There are two main reasons.

CASE 1 You can use a fraction to name a part of one thing.

$\frac{3}{8}$ ← slices of pizza that are gone
← slices of pizza in a whole pizza

$\frac{3}{8}$ of the pizza is gone.

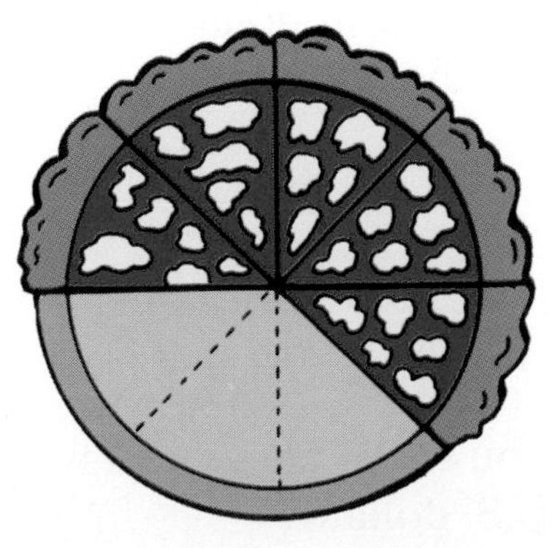

CASE 2 You can use a fraction to name part of a collection of things.

$\frac{3}{8}$ ← dogs in the group
← animals in the group

$\frac{3}{8}$ of the animals in the group are dogs.

Reading and Writing Fractions 029

Reading a fraction is different from reading a whole number. The numerator is easy, just say the number. To read the denominator, use words like *thirds*, *fourths*, *twenty-seconds*, etc.

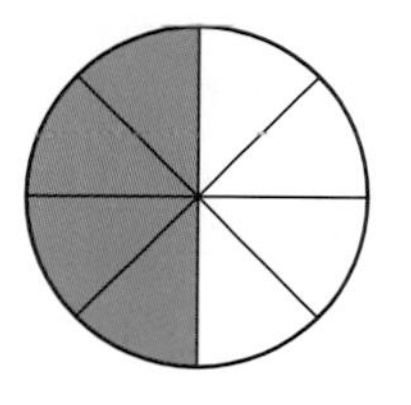

Write: $\frac{4}{8}$

Say: *four eighths*

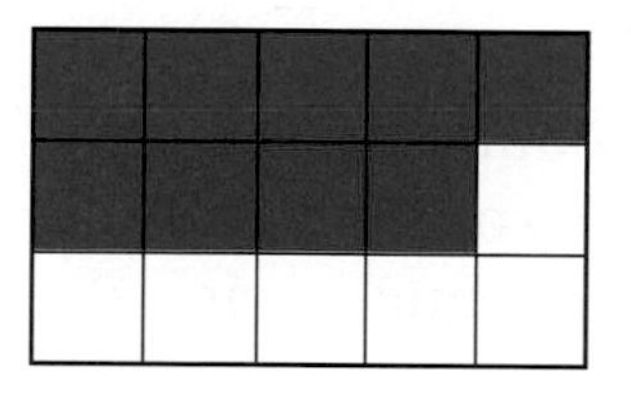

Write: $\frac{9}{15}$

Say: *nine fifteenths*

030 Fraction of a Whole

Sometimes you need to find a fractional part of an area or a thing. These are fractions of a whole.

$\frac{1}{2}$ of a pizza

$1\frac{1}{4}$ acres

$2\frac{3}{8}$ yards of fabric

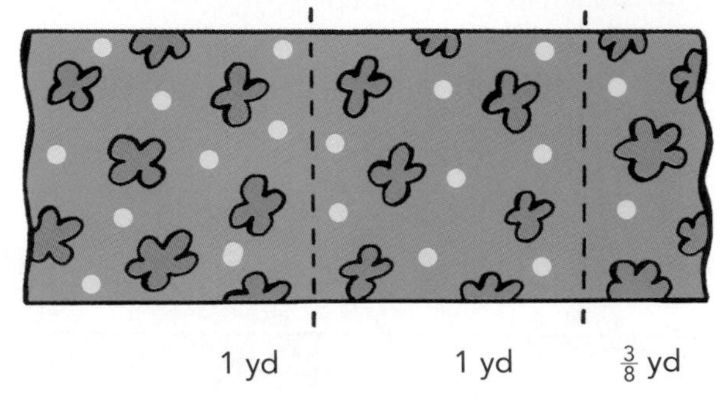

031 Fractions on a Number Line

A number line can be a useful tool when thinking about fractions.

EXAMPLE: Show $\frac{3}{5}$ on a number line.

$\frac{3}{5}$ is less than 1. Draw a number line from 0 to 1.

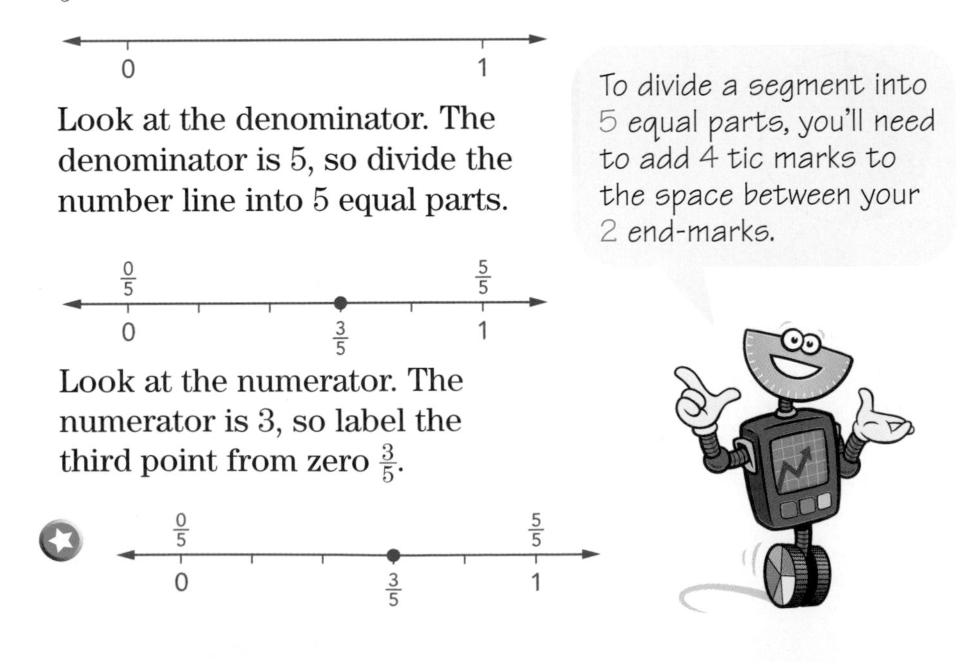

Look at the denominator. The denominator is 5, so divide the number line into 5 equal parts.

Look at the numerator. The numerator is 3, so label the third point from zero $\frac{3}{5}$.

Fractions in Measurement

032

We use fractions often in our systems of measure.

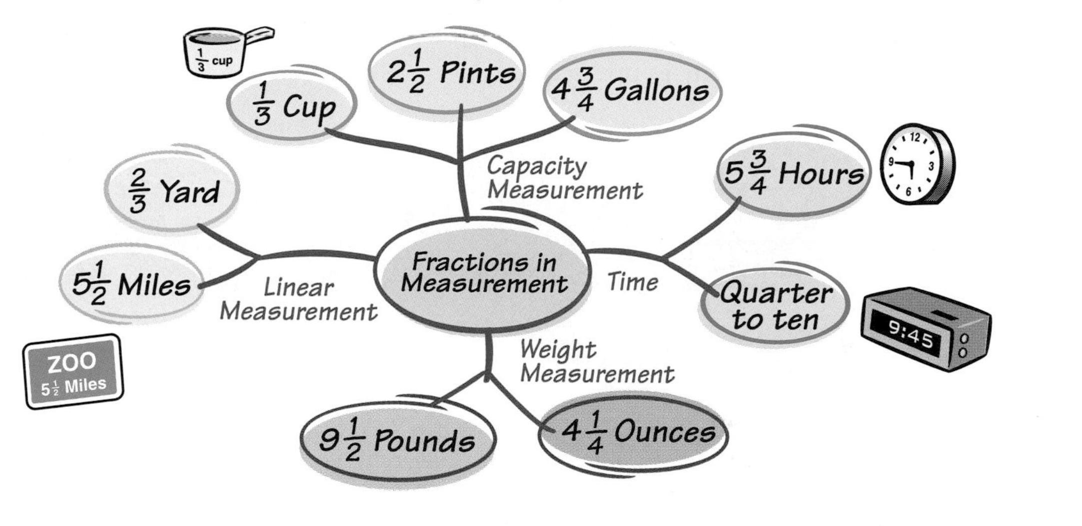

Fraction of a Set

033

Sometimes you need to talk about a fractional part of a group of items. This is a fractional part of a set.

See 035

EXAMPLE: What fraction of the dozen bagels are poppy seed?

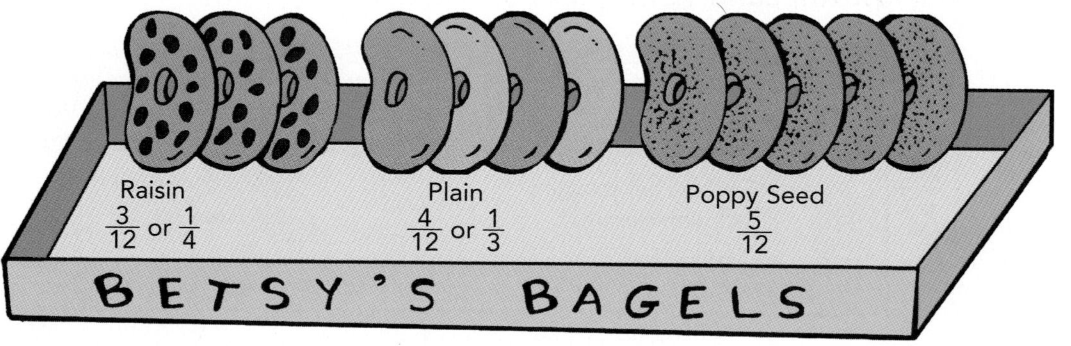

$\frac{5}{12}$ of the dozen bagels are poppy seed.

034 Fractions Greater than One

You can use fractions for numbers that are greater than one. A **mixed number** has a part that is a whole number and a part that is a fraction.

A fraction with a numerator greater than its denominator has a value greater than one. $\frac{7}{4} = 1\frac{3}{4}$ $7 > 4$, so $\frac{7}{4} > 1$.

If the numerator and denominator are equal, the fraction is equal to one. $\frac{4}{4} = \blacksquare$ $4 = 4$, so $\frac{4}{4} = 1$

CASE 1 You can rewrite a fraction greater than 1 as a mixed number or as a whole number.

EXAMPLE 1: Write $\frac{9}{4}$ as a mixed number.

① Divide the numerator by the denominator.		② Use the remainder to write the fraction part of the quotient.	
$\begin{array}{r} 2 \\ 4\overline{)9} \\ -8 \\ \hline 1 \end{array}$	There are enough fourths for two wholes.	$\begin{array}{r} 2\frac{1}{4} \\ 4\overline{)9} \\ -8 \\ \hline 1 \end{array}$	There is one fourth left over.

$\frac{9}{4}$ can be written as $2\frac{1}{4}$.

CASE 2 You can rewrite a mixed number or whole number as a fraction.

EXAMPLE 2: Write $1\frac{1}{2}$ as a fraction.

ONE WAY You can use a diagram to understand the problem.

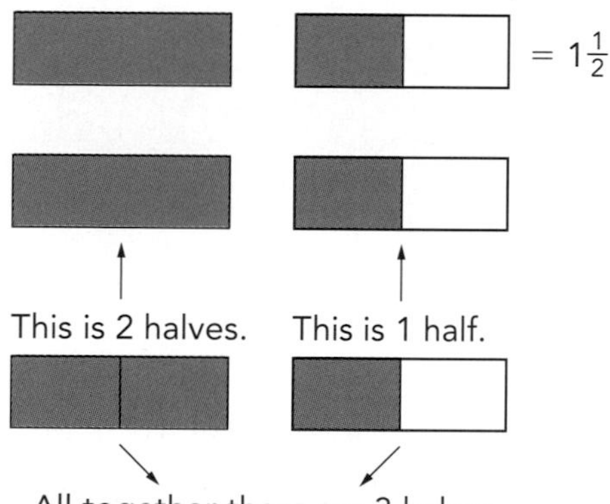

ANOTHER WAY Look at the whole-number part first.

❶ Write the whole number as a fraction.	❷ Add the fractions.
$1\frac{1}{2}$ ↓ $\frac{2}{2}$	$\frac{2}{2} + \frac{1}{2} = \frac{3}{2}$

Multiply the denominator by the whole number, then add the numerator.

$$1\frac{1}{2} = \frac{(2 \times 1) + 1}{2} = \frac{3}{2}$$

A fraction with a numerator greater than (or equal to) its denominator is sometimes called an **improper fraction**. This doesn't mean there's anything wrong with the fraction.

Any way you look at it, $1\frac{1}{2}$ can be rewritten as $\frac{3}{2}$.

035 Equivalent Fractions

Equivalent fractions name the same amount. You can use equivalent fractions to add, subtract, and compare fractions.

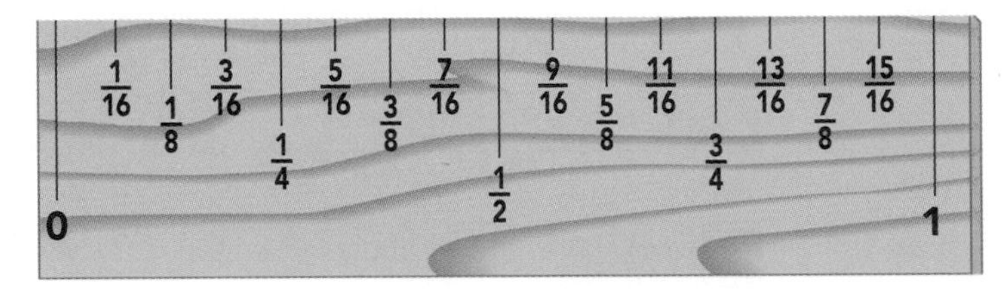

$$\frac{1}{2} = \frac{2}{4} = \frac{4}{8} = \frac{8}{16}$$

See 160, 165, 167–170, 227

To find equivalent fractions, you can multiply or divide the numerator and the denominator by the same non-zero number.

This does not change the value of the fraction, because you are really just multiplying (or dividing) by 1.

$$\frac{1}{2} = \frac{1 \times 4}{2 \times 4} = \frac{4}{8}$$

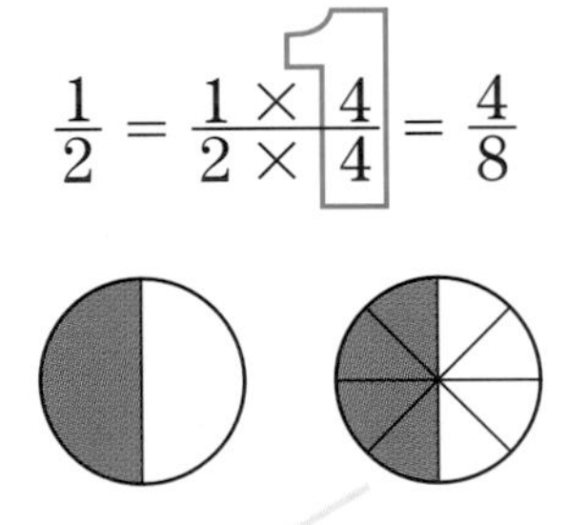

$$\frac{2}{4} = \frac{2 \div 2}{4 \div 2} = \frac{1}{2}$$

Same amount shaded, but 4 times as many parts, so 4 times as many parts shaded.

Same amount shaded, but half as many parts shaded.

Sometimes, you may need to find a missing numerator or a missing denominator in equivalent fractions.

EXAMPLE 1: Find the missing numerator.

$\frac{3}{4} = \frac{\blacksquare}{24}$

Multiplying the numerator and denominator by the same number is just like multiplying by 1. So, find the missing factor, $4 \times ? = 24$ and use that factor to find the missing numerator.

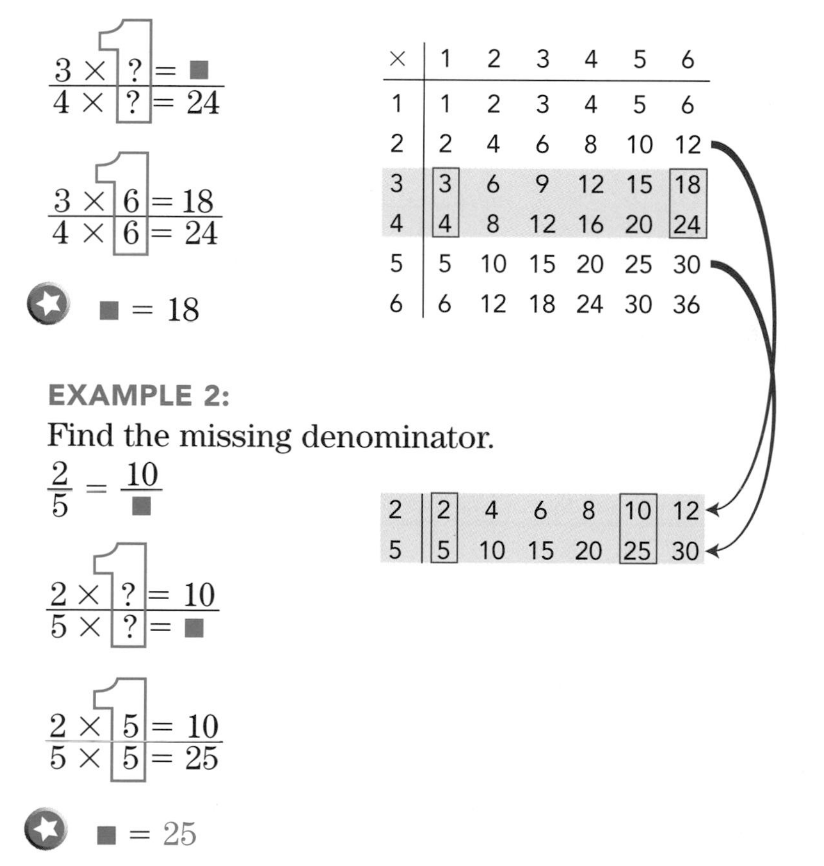

$$\frac{3 \times ?}{4 \times ?} = \frac{\blacksquare}{24}$$

$$\frac{3 \times 6}{4 \times 6} = \frac{18}{24}$$

$\blacksquare = 18$

×	1	2	3	4	5	6
1	1	2	3	4	5	6
2	2	4	6	8	10	12
3	3	6	9	12	15	18
4	4	8	12	16	20	24
5	5	10	15	20	25	30
6	6	12	18	24	30	36

EXAMPLE 2:

Find the missing denominator.

$\frac{2}{5} = \frac{10}{\blacksquare}$

2	2	4	6	8	10	12
5	5	10	15	20	25	30

$$\frac{2 \times ?}{5 \times ?} = \frac{10}{\blacksquare}$$

$$\frac{2 \times 5}{5 \times 5} = \frac{10}{25}$$

$\blacksquare = 25$

036

Least Common Denominator

When fractions have the same denominator, you can say they have a **common denominator**.

See 035, 059, 061

When you add, subtract, or compare fractions, it often helps to find the **least common denominator**. The least common denominator of two or more fractions is the least common multiple (LCM) of the denominators of the fractions.

You can also say that they have **like denominators**.

EXAMPLE: Find equivalent fractions with the least common denominator for $\frac{1}{8}$ and $\frac{7}{12}$.

❶ Find the least common multiple of both denominators.	❷ Rewrite each fraction as an equivalent fraction with the LCM as the denominator.
multiples of 8 → 8, 16, 24, . . .	$\frac{1 \times 3}{8 \times 3} = \frac{3}{24}$
multiples of 12 → 12, 24, . . . The LCM of 8 and 12 is 24.	$\frac{7 \times 2}{12 \times 2} = \frac{14}{24}$

Equivalent fractions with the least common denominator for $\frac{1}{8}$ and $\frac{7}{12}$ are $\frac{3}{24}$ and $\frac{14}{24}$.

1	2	3	4	5	6	7	8	9	10
11	12	13	14	15	16	17	18	19	20
21	22	23	24	25	26	27	28	29	30

The multiples of 8 are highlighted in yellow.
The multiples of 12 are highlighted in pink.
The first yellow and pink is the LCM.

Simplest Form

037

A fraction is in **simplest form** when its numerator and denominator have no common factor other than 1.

MORE HELP

See 057–058, 061

EXAMPLE: Find the simplest form of $\frac{12}{18}$.

You can divide the numerator and denominator by common factors until the only common factor is 1.

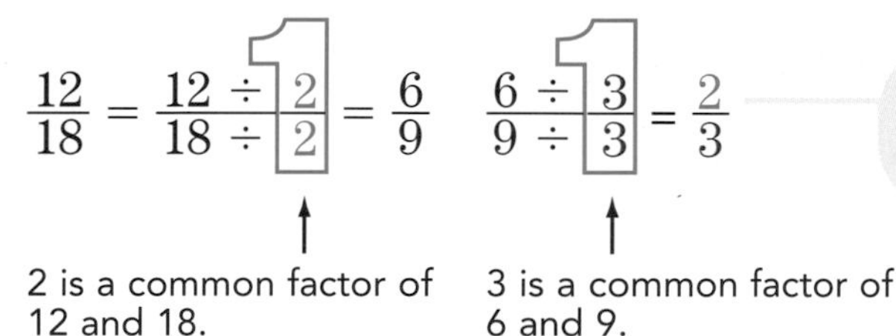

$$\frac{12}{18} = \frac{12 \div 2}{18 \div 2} = \frac{6}{9} \quad \frac{6 \div 3}{9 \div 3} = \frac{2}{3}$$

2 is a common factor of 12 and 18.

3 is a common factor of 6 and 9.

2 and 3 have no common factor greater than 1.

You can divide the numerator and denominator by the greatest common factor (GCF).

Factors of 12: 2, 3, 4, 6
Factors of 18: 2, 3, 6, 9

$$\frac{12}{18} = \frac{12 \div 6}{18 \div 6} = \frac{2}{3}$$

6 is the GCF of 12 and 18.

Either way, $\frac{12}{18}$ in simplest form is $\frac{2}{3}$.

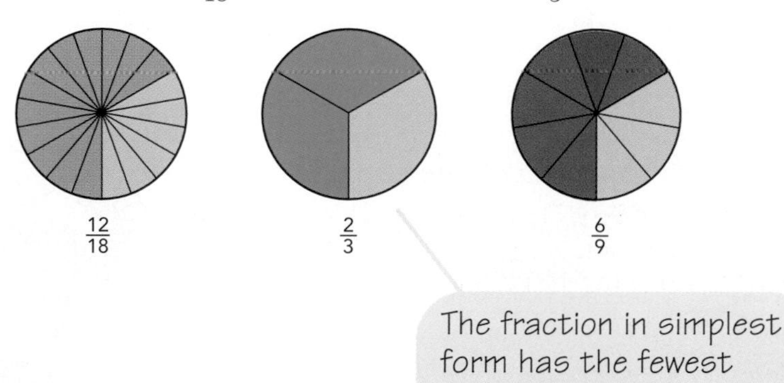

The fraction in simplest form has the fewest possible pieces.

038 Comparing and Ordering Fractions

Sometimes you can compare fractions just by looking at them.

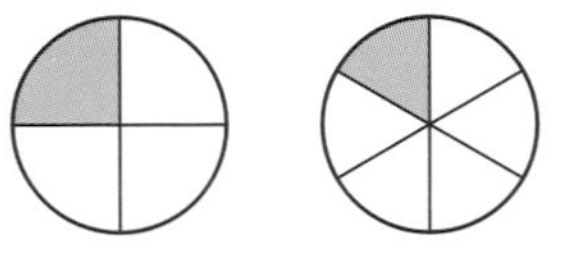

Write: $\frac{1}{4} > \frac{1}{6}$
Say: *one fourth is greater than one sixth*

If you know how to compare two or more fractions, you can also order fractions.

039 Comparing Fractions with Like Denominators

Comparing fractions with the same denominators is a lot like comparing whole numbers.

EXAMPLE: George walks $\frac{5}{10}$ mile to school. Susan walks $\frac{7}{10}$ mile. Who walks a greater distance?

When fractions have the same denominator, compare the numerators.

$7 > 5$, so $\frac{7}{10} > \frac{5}{10}$.

 Susan walks a greater distance.

This makes sense because 7 of something is more than 5 of the same thing.

7 hours > 5 hours
7 tenths > 5 tenths

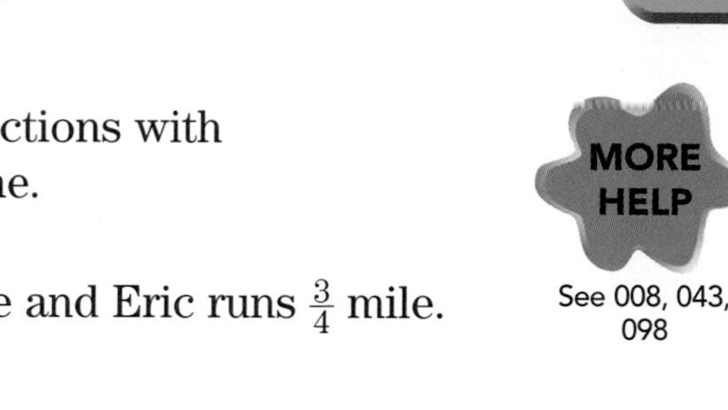

Comparing Fractions with Unlike Denominators

040

Here are two ways to compare fractions with denominators that are not the same.

MORE HELP

See 008, 043, 098

EXAMPLE 1: Adrienne runs $\frac{3}{10}$ mile and Eric runs $\frac{3}{4}$ mile. Who runs farther?

ONE WAY Compare the fractions to your favorite benchmarks.

$\frac{3}{10} < \frac{1}{2}$ because
3 is less than half of 10.

$\frac{3}{4} > \frac{1}{2}$ because
3 is more than half of 4.

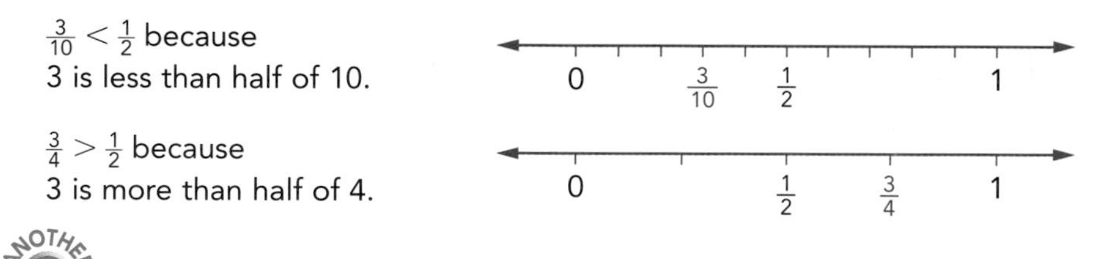

ANOTHER WAY Find equivalent fractions with the same denominator.

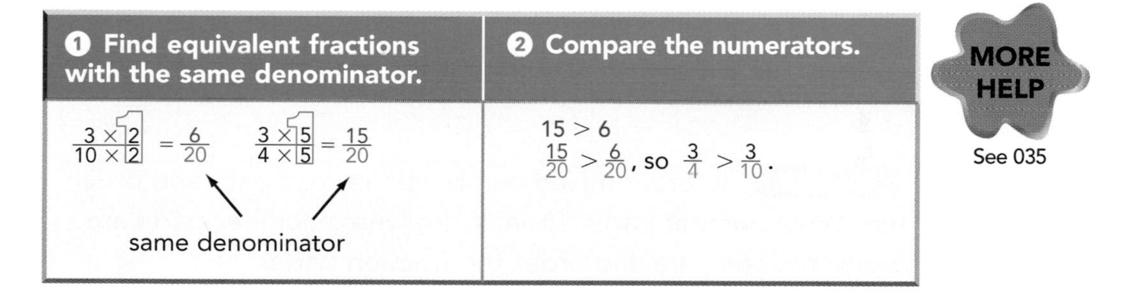

❶ Find equivalent fractions with the same denominator.	❷ Compare the numerators.
$\frac{3 \times 2}{10 \times 2} = \frac{6}{20}$ $\frac{3 \times 5}{4 \times 5} = \frac{15}{20}$ (same denominator)	$15 > 6$ $\frac{15}{20} > \frac{6}{20}$, so $\frac{3}{4} > \frac{3}{10}$.

MORE HELP

See 035

Either way, $\frac{3}{4} > \frac{3}{10}$. Eric runs farther.

MATH ALERT All Halves Are Not Equal

041

When you compare two fractions, make sure the wholes are the same.

What if Ben ordered a small pizza and Spencer ordered a large pizza? Then you couldn't compare fractions of the pizzas because the wholes are not the same.

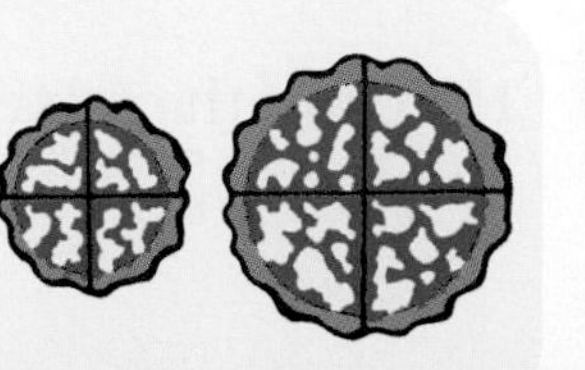

042 Ordering Fractions and Mixed Numbers

CASE 1 If you know how to compare two fractions, you also know how to order a group of fractions.

MORE HELP

See 036

EXAMPLE 1: For a science project on weather, Samantha measured rainfall for three weeks. Which week had the least rain?

① Rewrite the fractions with the same denominators.	② Compare the numerators.
Multiples of 4 ⟶ 4, 8, 12, 16, ... Multiples of 8 ⟶ 8, 16, ... Multiples of 16 ⟶ 16, 32, ... The least common denominator is 16. $\frac{11}{16} = \frac{11}{16}$ $\frac{5}{8} = \frac{10}{16}$ $\frac{3}{4} = \frac{12}{16}$	$12 > 11 > 10$ $\frac{12}{16} > \frac{11}{16} > \frac{10}{16}$, so $\frac{3}{4} > \frac{11}{16} > \frac{5}{8}$.

The least rain fell in the second week.

CASE 2 To order mixed numbers, first compare and order the whole-number parts. Then, if the whole-number parts are the same, compare and order the fraction parts.

EXAMPLE 2: Order from least to greatest: $3\frac{11}{16}$, $2\frac{5}{8}$, $2\frac{3}{4}$.

① Look at the whole-number parts.	② Compare the fraction parts of the remaining numbers.
$3\frac{11}{16}$ $2\frac{5}{8}$ $2\frac{3}{4}$ $3 > 2$, so $3\frac{11}{16}$ is greatest.	$\frac{5}{8} = \frac{5}{8}$ $\frac{3}{4} = \frac{6}{8}$ $\frac{5}{8} < \frac{6}{8}$, so $2\frac{5}{8} < 2\frac{3}{4}$.

Here are the mixed numbers in order from least to greatest: $2\frac{5}{8}$, $2\frac{3}{4}$, $3\frac{11}{16}$.

Relating Fractions to Decimals 043

To write a decimal for any fraction, you can divide the numerator by the denominator.

This works because a fraction is a way of showing division. The line between the numerator and denominator of a fraction means *is divided by.*

See 021, 154–155

EXAMPLE: Write $\frac{1}{8}$ as a decimal.

Think: $\frac{1}{8} = 1 \div 8$

$$\begin{array}{r} 0.125 \\ 8\overline{)1.000} \\ \underline{-800} \\ 200 \\ \underline{-160} \\ 40 \\ \underline{-40} \\ 0 \end{array}$$

$0.125 = \frac{1}{8}$

This is a handy fact to remember! If you know $\frac{1}{8} = 0.125$, you can find decimals for all the other eighths.

$\frac{2}{8} = 2 \times 0.125 = 0.250 = 0.25$

$\frac{3}{8} = 3 \times 0.125 = 0.375$

$\frac{4}{8} = 4 \times 0.125 = 0.500 = 0.5$

$\frac{5}{8} = 5 \times 0.125 = 0.625$

$\frac{6}{8} = 6 \times 0.125 = 0.750 = 0.75$

$\frac{7}{8} = 7 \times 0.125 = 0.875$

$\frac{8}{8} = 8 \times 0.125 = 1.000 = 1$

044 Relating Fractions to Percents

Percents are really just fractions with a denominator of 100. For example, 52% means 52 *per hundred*, or $\frac{52}{100}$.

See 035, 099, 043

EXAMPLE 1: Write $\frac{1}{4}$ as a percent.

To write $\frac{1}{4}$ as a percent, first find the equivalent fraction with a denominator of 100.

$$\frac{1}{4} = \frac{■}{100}$$

$$\frac{1 \times 25}{4 \times 25} = \frac{25}{100}$$

Multiplying the numerator by 25 and the denominator by 25 is just like multiplying the fraction by $\frac{25}{25}$ or 1.

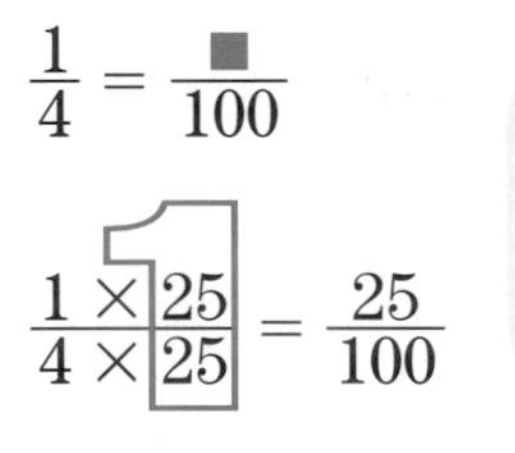

$\frac{1}{4} = 25\%$.

EXAMPLE 2: Write $\frac{2}{5}$ as a percent.

To write $\frac{2}{5}$ as a percent, first find the equivalent fraction with a denominator of 100.

$$\frac{2}{5} = \frac{■}{100}$$

$$\frac{2 \times 20}{5 \times 20} = \frac{40}{100}$$

$\frac{40}{100} = 40\%$

EXAMPLE 3: Write 75% as a fraction.

Since 75% means 75 *per hundred*, $75\% = \frac{75}{100}$

$\frac{75}{100} = \frac{3}{4}$ so in simplest form, $75\% = \frac{3}{4}$.

Positive and Negative Numbers

Integers 046

Positive numbers are numbers that are greater than zero. **Negative numbers** are numbers that are less than zero.

Zero is neither positive nor negative. You can show positive and negative numbers on a number line.

negative zero positive

$^{-}5$ $^{-}4$ $^{-}3$ $^{-}2$ $^{-}1$ 0 1 2 3 4 5

Opposite numbers are the same distance from zero in the opposite direction. Every whole number, fraction, and decimal has an opposite.

2 and $^{-}2$ are opposites.

4.5 and $^{-}4.5$ are opposites.

$3\frac{1}{5}$ and $^{-}3\frac{1}{5}$ are opposites.

Zero is its own opposite.

$^{-}4.5$ $^{-}3\frac{1}{5}$ $^{-}2$ 2 $3\frac{1}{5}$ 4.5

$^{-}5$ $^{-}4$ $^{-}3$ $^{-}2$ $^{-}1$ 0 1 2 3 4 5

Opposites have a sum of zero!

$2 + {^{-}2} = 0$

$0 + 0 = 0$

046 Integers

Integers are the set of whole numbers and their opposites:

. . . $^{-}5, {}^{-}4, {}^{-}3, {}^{-}2, {}^{-}1, 0, 1, 2, 3, 4, 5$ . . .

You may have seen integers used like this.

110
100
90
80
70
60
50
40
30
20
10
0
¯10
¯20
°F

047 Comparing Integers

See 008

You can use a number line to compare any two integers.

EXAMPLE 1: Which is larger, $^{-}2$ or 1?

A negative number is less than a positive number.

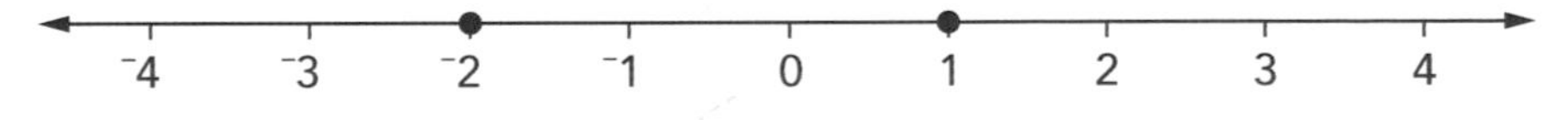

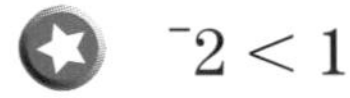
$^{-}2 < 1$

That makes sense because a mountain top 1 mile *above* sea level is higher than an ocean floor 2 miles *below* sea level.

EXAMPLE 2: Which is greater, $^{-}1$ or $^{-}4$?

The negative number closer to 0 is greater.

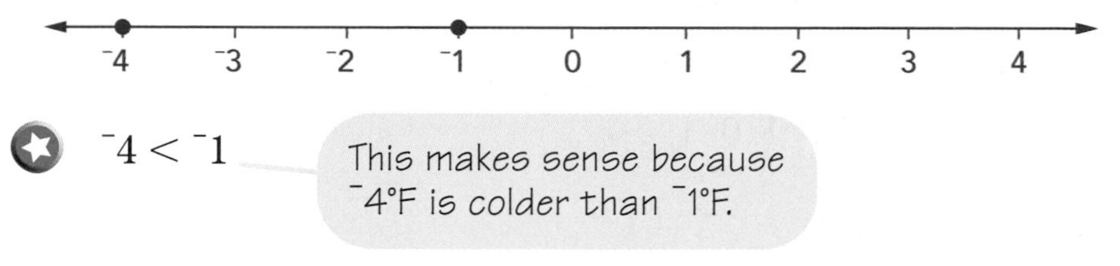

$^{-}4 < {^{-}1}$

This makes sense because $^{-}4$°F is colder than $^{-}1$°F.

Ordering Integers

If you know how to compare integers, you also know how to order a group of integers.

See 010

EXAMPLE: For a science project, you record the outside temperature at 7:00 A.M. every day for a week. What were the daily temperatures, from coldest to warmest?

MORNING TEMPERATURE (7:00 A.M.)

Time	Temperature
JAN. 5	-12°F
JAN. 6	-2°F
JAN. 7	9°F
JAN. 8	5°F
JAN. 9	-3°F

To solve the problem, you can use a number line. Numbers increase as you go from left to right on the number line.

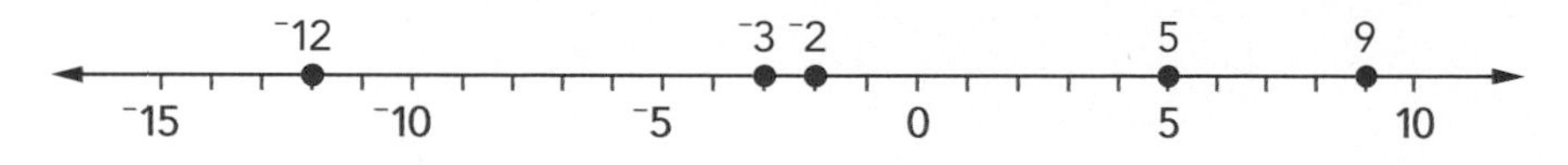

Here are the temperatures from coldest to warmest: $^{-}12$°F, $^{-}3$°F, $^{-}2$°F, 5°F, 9°F.

Number Theory

Number theory is a part of mathematics where you look for number patterns and look at how numbers work together.

Number theory may sound as if it's only for mathematicians. But, some of the things you learn can save you time when you're making calculations.

050

Factors and Multiples

What's your favorite number?

Think of two numbers that can be multiplied to make your number. Your number is a **multiple** of the two numbers you just thought of. The two numbers are **factors** of your number.

If you are good at skip-counting, you already know a lot about factors and multiples. They're handy to know about because they make it easier to compute with fractions and other numbers

051

Factors

If you can find factors of a number, you can do these things, too.

- You can find the greatest common factor of two numbers.
- You can write fractions in simplest form.
- You can tell whether one number can be divided by another.

Pick a whole number. 18

Now find two whole numbers that can be multiplied to get your number.

$3 \times 6 = 18$
$1 \times 18 = 18$ or
$2 \times 9 = 18$

One pair of factors of 18 is 3 and 6. Another pair of factors of 18 is 1 and 18. Another is 2 and 9. You can say that 1, 2, 3, 6, 9, and 18 are **factors** of 18.

EXAMPLE: Find all the factors of 36.

Think of all the pairs of numbers that you can multiply to get 36.

When the factors in a pair are the same or close together, you're done.

$36 = 1 \times 36$
$36 = 2 \times 18$
$36 = 3 \times 12$
$36 = 4 \times 9$
$36 = 6 \times 6$

1, 2, 3, 4, 6, 9, 12, 18, 36
36
36
36
36

The factors of 36 are 1, 2, 3, 4, 6, 9, 12, 18, and 36.

Factorials

052

Take any whole number. 4

Multiply all the counting numbers from that number down to 1.

$4 \times 3 \times 2 \times 1$

This product is called a **factorial**.

Write: 4!
Say: *four factorial*

EXAMPLE: Find 10!

$10! = 10 \times 9 \times 8 \times 7 \times 6 \times 5 \times 4 \times 3 \times 2 \times 1$

$10! = 3{,}628{,}800$

053 Prime Numbers

If you're talking about whole numbers greater than zero, a **prime number** has exactly two different factors, one and itself.

$13 = 1 \times 13$

Since there are no other whole-number factors of 13, 13 is a prime number.

All the prime numbers from 1 to 100 are in color.

If you don't believe they are prime numbers, just try to find more than two whole-number factors for any of them. Bet you can't!

1	2	3	4	5	6	7	8	9	10
11	12	13	14	15	16	17	18	19	20
21	22	23	24	25	26	27	28	29	30
31	32	33	34	35	36	37	38	39	40
41	42	43	44	45	46	47	48	49	50
51	52	53	54	55	56	57	58	59	60
61	62	63	64	65	66	67	68	69	70
71	72	73	74	75	76	77	78	79	80
81	82	83	84	85	86	87	88	89	90
91	92	93	94	95	96	97	98	99	100

Did You Know. . .

that every even number greater than 2 can be written as the sum of two prime numbers?

$4 = 2 + 2$ $\quad$ $26 = 13 + 13$

$10 = 3 + 7$ $\quad$ $48 = 11 + 37$

This is called **Goldbach's conjecture**, named after the mathematician Christian Goldbach.

MATH ALERT One Is Not Prime

054

Some people think that one is a prime number, but it is not. It is really a lonely number because it is neither prime nor composite. It's not prime because it does not have *exactly* two different factors. It's not composite because it does not have *more than* two factors.

MORE HELP
See 055

Composite Numbers

055

.Think about whole numbers greater than zero. Every one of these numbers, except 1, is either a composite number or a prime number. In the prime number table in 053, the numbers in black from 4 through 100 are **composite** numbers. Each has more than two different factors.

MORE HELP
See 053

EXAMPLE: Tell whether each number is composite or prime: 16, 24, 89.

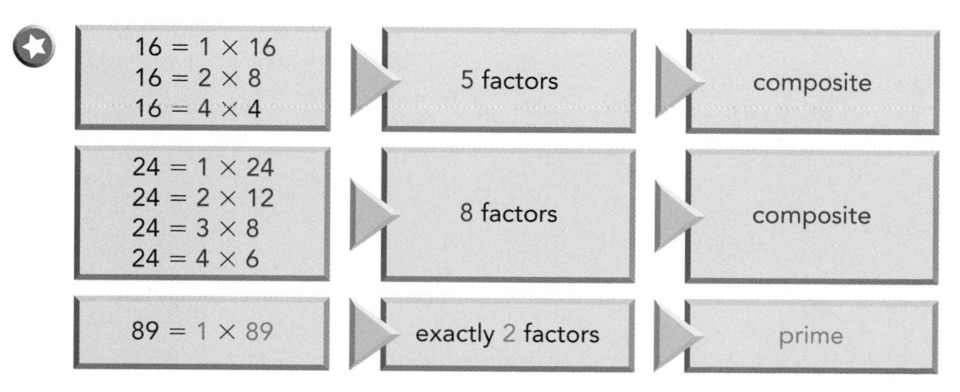

056

Prime Factorization

A composite number can be written as the product of prime numbers. This is called the **prime factorization** of the number.

ONE WAY You can find the prime factorization by making a **factor tree**. First, find any pair of factors. Then find pairs of factors for the factors. Keep this up until you can't do it any more.

EXAMPLE 1: Find the prime factorization of 12.

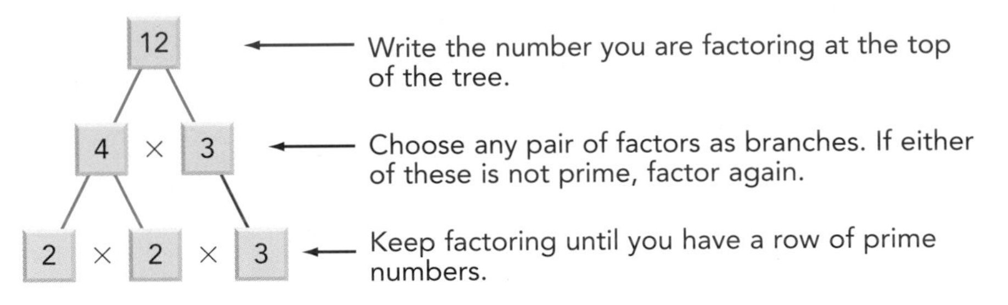

See 065–066

Find the prime factorization of 12 in a different way.

12 — Write the number you are factoring at the top of the tree.

2 × 6 — Choose any pair of factors as branches. If either of these is not prime, factor again.

2 × 2 × 3 — Keep factoring until you have a row of prime numbers.

The prime factorization of 12 is $2 \times 2 \times 3$, or $2^2 \times 3$.

ANOTHER WAY You can find the prime factorization by using division.

MORE HELP
See 144

EXAMPLE 2: Find the prime factorization of 112.

2. Keep dividing each quotient by a prime number until the quotient is 1.

1. Divide 112 by a prime number.

$$\begin{array}{r} 1 \\ 7\overline{)7} \\ 2\overline{)14} \\ 2\overline{)28} \\ 2\overline{)56} \\ 2\overline{)112} \end{array}$$

If you try a prime number and you get a remainder, that means your dividend is not divisible by that prime. Try a different prime number.

The prime factorization of 112 is the product of the divisors, $2 \times 2 \times 2 \times 2 \times 7$, or $2^4 \times 7$.

EXAMPLE 3: Find the prime factorization of 87.

See 054, 062

2. Divide each quotient by a prime number until the quotient is 1.

1. Divide 87 by a prime number.

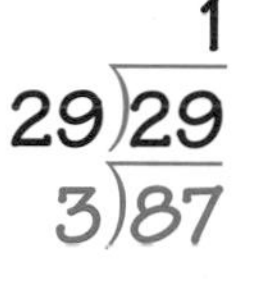

$$\begin{array}{r} 1 \\ 29\overline{)29} \\ 3\overline{)87} \end{array}$$

The prime factorization of 87 is the product of the divisors, $3 \times 29 = 87$.

Remember: One is not included in the prime factorization because one is not prime.

057 Common Factors

A group of two or more whole numbers may have some factors that are the same. These factors are called **common factors.**

EXAMPLE 1: Find the common factors of 16 and 24.

$16 = 1 \times 16$
$16 = 2 \times 8$
$16 = 4 \times 4$

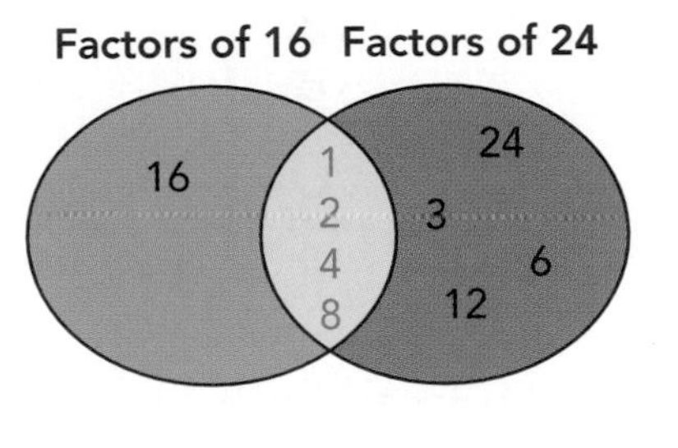

$24 = 1 \times 24$
$24 = 2 \times 12$
$24 = 3 \times 8$
$24 = 4 \times 6$

Common factors are factors of both numbers.

The common factors of 16 and 24 are 1, 2, 4, and 8.

EXAMPLE 2: Find the common factors of 9, 18, and 81.

Factors of 9: 1, 3, 9
Factors of 18: 1, 2, 3, 6, 9, 18
Factors of 81: 1, 3, 9, 27, 81

The common factors of 9, 18, and 81 are 1, 3, and 9.

Greatest Common Factor

To find common factors of a group of numbers, list all the factors of each number. The factors that appear in all the lists are **common factors.**

Find the greatest of those **common factors.** This is the **greatest common factor (GCF)** of the numbers.

EXAMPLE 1: Find the GCF of 6 and 12.

$6 = 1 \times 6$
$6 = 2 \times 3$

$12 = 1 \times 12$
$12 = 2 \times 6$
$12 = 3 \times 4$

Factors of 6 Factors of 12
1
2
3
6
4
12

The greatest number that appears in *both* lists is 6.

The greatest common factor (GCF) of 6 and 12 is 6.

EXAMPLE 2: Find the GCF of 14 and 21.

Factors of 14: 1, 2, 7, 14
Factors of 21: 1, 3, 7, 21

The greatest number that is in both lists is 7.

The greatest common factor (GCF) of 14 and 21 is 7.

EXAMPLE 3: Find the GCF of 18 and 25.

Factors of 18: 1, 2, 3, 6, 9, 18
Factors of 25: 1, 5, 25

The greatest common factor (GCF) of 18 and 25 is 1.

Two numbers that have 1 as the greatest common factor are **relatively prime.** 18 and 25 are relatively prime.

059 Multiples

When you skip-count, you are saying multiples of a number. **Multiples** of any number are the products of that number and any whole number.

EXAMPLE: Find 8 multiples of 5 starting with 5.

ONE WAY Skip count by 5s starting with 5.
5, 10, 15, 20, 25, 30, 35, and 40

ANOTHER WAY Multiply to find multiples.

First Factor	×	Second Factor	=	Multiple
5	×	1	=	5
5	×	2	=	10
5	×	3	=	15
5	×	4	=	20
5	×	5	=	25
5	×	6	=	30
5	×	7	=	35
5	×	8	=	40

Either way, the first 8 multiples of 5 are 5, 10, 15, 20, 25, 30, 35, and 40.

060 MATH ALERT Zero Is a Multiple but Not a Factor of Every Number

- You can multiply any number by zero. When you do, the product is zero. So, $5 \times 0 = 0$ and $500 \times 0 = 0$. You could say that zero is a multiple of any number. Usually, when we list multiples of a number, we do not list zero.
- Since $0 \times$ any number $= 0$, zero is *not* a factor of any number except itself.

Least Common Multiple

To make a list of multiples, multiply your number by whole numbers. To find **common multiples** of several numbers, find multiples that appear in all the lists.

Find the smallest of those common multiples (other than zero). This is the **least common multiple (LCM)** of the numbers.

EXAMPLE 1: Find the LCM of 6 and 9.

In this table, all multiples of 6 are in red squares. All multiples of 9 are in blue circles.

1	2	3	4	5	6	7	8	9	10
11	12	13	14	15	16	17	18	19	20
21	22	23	24	25	26	27	28	29	30
31	32	33	34	35	36	37	38	39	40
41	42	43	44	45	46	47	48	49	50
51	52	53	54	55	56	57	58	59	60
61	62	63	64	65	66	67	68	69	70
71	72	73	74	75	76	77	78	79	80
81	82	83	84	85	86	87	88	89	90
91	92	93	94	95	96	97	98	99	100

The first number that has both a red square and a blue circle is 18.

The least common multiple (LCM) of 6 and 9 is 18.

EXAMPLE 2: Find the LCM of 3 and 12.

Some multiples of 3: 3, 6, 9, 12, 15, . . .
Some multiples of 12: 12, . . .

You can stop finding multiples when you find the first match—that's the least common multiple.

When one of the numbers is a multiple of the other, the greater number will be the least common multiple.

The least common multiple (LCM) of 3 and 12 is 12.

062 Divisibility

One whole number is **divisible** by another whole number if the remainder is zero when you divide.

Mathematicians have discovered patterns that make it easier to tell if one number is divisible by another.

Try testing these rules with your own numbers.

Divisor	Rule	Test with 324
2	The ones digit is 0, 2, 4, 6, or 8. (The number is even.)	324: the ones digit is 4 Four is an even number. So, 324 is divisible by 2.
3	The sum of the digits is divisible by 3.	324: 3 + 2 + 4 = 9 Nine is divisible by 3. So, 324 is divisible by 3.
4	The number formed by the last two digits is divisible by 4.	324: the last two digits are 24 Twenty-four is divisible by 4. So, 324 is divisible by 4.
5	The ones digit is 0 or 5.	324: the ones digit is not 0 or 5 So, 324 is not divisible by 5.
6	The number is divisible by 2 and by 3.	324: the ones digit is even and the sum of the digits is 9 So, 324 is divisible by 2 and by 3. So, 324 is divisible by 6.
9	The sum of the digits is divisible by 9.	324: 3 + 2 + 4 = 9 Nine is divisible by 9. So, 324 is divisible by 9.
10	The ones digit is 0.	324: the ones digit is not 0 So, 324 is not divisible by 10.

Divisibility tests for 7 and 8 are not as simple as the tests for the other numbers from 1 through 10. Just go ahead and do the division.

EXAMPLE 1: Can 6 people share equally 354 CDs?

To answer this question, think, *Is 354 divisible by 6?* A number that is divisible by 6 must be divisible by 2 and by 3.

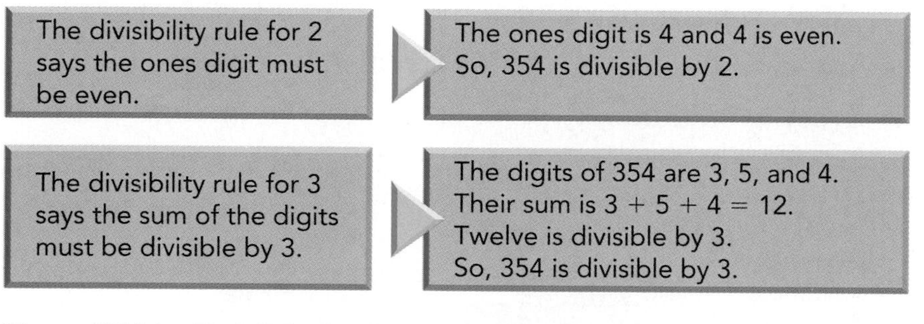

Since 354 is divisible by 2 and by 3, it must be divisible by 6.

Yes, 6 people can share equally 354 CDs.

EXAMPLE 2: There will be 127 people at a party. Each table can seat 5 people. Will all the tables be full?

To answer this question, think, *Is 127 divisible by 5?* Look at the divisibility rule for 5.

A number is divisible by 5 if its ones digit is 0 or 5. The ones digit is 7. So, 127 is not divisible by 5.

The tables will not all be full.

063 Even and Odd Numbers

If a whole number is divisible by 2, it is an **even number**.

- Any even number of things can be divided into two equal-size groups.
- Any even number of things can also be put into pairs.
- Every even number has 0, 2, 4, 6, or 8 in its ones place.

The 12 flowers can be divided into two equal-size groups.
There is an even number of flowers.

The 12 flowers are all in pairs. There is an even number of flowers.

Odd numbers are *not* divisible by 2.

- When you try to put an odd number of things into two equal-size groups, there is always 1 thing left over.
- When you try to put an odd number of things into pairs, there is always 1 thing left over.
- Every odd number has 1, 3, 5, 7, or 9 in its ones place.

The 11 flowers cannot be divided into two equal-size groups. There is an odd number of flowers.

You can use even and odd number patterns to check your computation.

- The sum of two even numbers is always even.
- The sum of two odd numbers is always even.
- The sum of an even and an odd number is always odd.

EXAMPLE: Amy added 469 and 357 and got 827. Could that be right?

Since 469 is odd and 357 is odd, the sum must be even.

The sum cannot be 827, which is odd.

064

Powers and Roots

Mathematicians are always looking for simpler ways to write and think about things. Just as we use the shortcut 7×8 to write $8 + 8 + 8 + 8 + 8 + 8 + 8$, we can use a shortcut to write $6 \times 6 \times 6 \times 6 \times 6 \times 6 \times 6$.

065

Positive Exponents

Suppose you want to multiply by the same factor more than once. You can use exponents to show what you mean.

In **exponential form**, the **base** is the repeated factor. The **exponent** is the number of times the factor is repeated.

EXAMPLE 1: Use exponential form to show 3 used as a factor 5 times.

$3 \times 3 \times 3 \times 3 \times 3 = 3^5$ ← **exponent** (5); **base** (3)

A **power** of a number tells how many times that number is used as a factor.

Write: 3^5
Say: *three to the fifth power* or *three to the fifth*

Did You Know. . .

that the largest number you can write with two 9s is 9^9?

EXAMPLE 2: Pick which prize you would you rather have.

- $100 a day for 30 days
- 2¢ on Day 1, 4¢ on Day 2, 8¢ on Day 3, and so on for 30 days

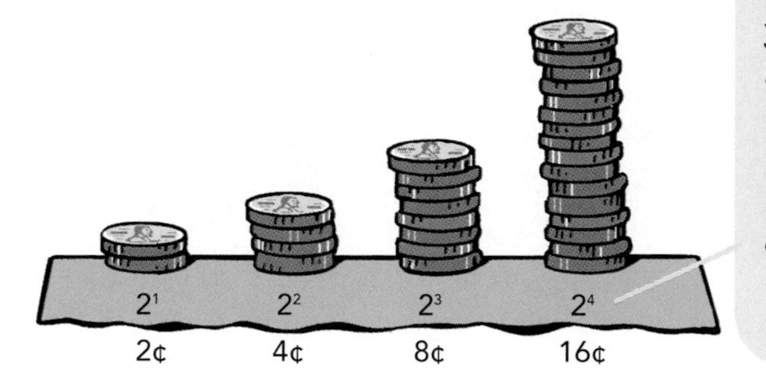

Since the amount in the pattern starts at 2¢ and doubles each day, you can use a base of 2 and an exponent to show the day. This may not seem like much money, but the amount grows fast. Try figuring it out day by day for a month and you'll see.

This table shows how to write and read positive exponents.

Repeated Factors	Write	Say	Standard Form
2×2	2^2	*two to the second power or two squared*	4
$2 \times 2 \times 2$	2^3	*two to the third power or two cubed*	8
$2 \times 2 \times 2 \times 2$	2^4	*two to the fourth power*	16
$2 \times 2 \times 3 \times 3 \times 3$	$2^2 \times 3^3$	*two squared times three cubed*	108

See 067–069

MATH ALERT Zero as an Exponent

Any number to the zero power is one. So, $4^0 = 1$ and $62^0 = 1$.

067 Squares and Square Roots

Look at these pictures. What shape is each design?

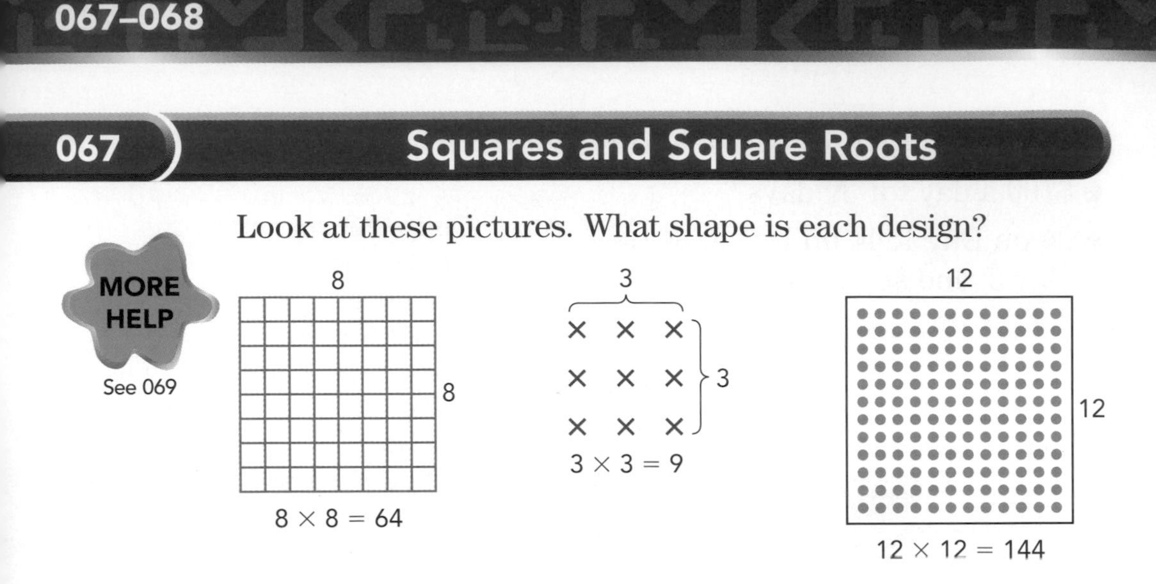

See 069

Do you see why we say that the product of a number and itself is the **square** of the number? The square of 8 is 64.

Write: $8^2 = 64$
Say: *eight squared is 64* or *64 is the square of 8*

When we look for a number that we can multiply by itself to get another number, we are looking for the **square root**. The square root of 64 is 8.

Write: $\sqrt{64} = 8$
Say: *the square root of 64 is 8*

068 MATH ALERT Square Roots Can Be Negative

See 210

You know that if you multiply two positive numbers, you get a positive product. Two negative factors also have a positive product.

$6 \times 6 = 36$ $\qquad$ ${}^-6 \times {}^-6 = 36$

That means ${}^-6$ is also a square root of 36. However, $\sqrt{}$ means the principal square root, which is the positive root. For example, $\sqrt{25} = 5$ and $\sqrt{100} = 10$.

Using a Square Root Table

069

The table shows squares and square roots for the numbers 1 through 10.

To find the square of a number, look for the number in the first column (n). Then move to the right along the row to the column labeled n^2.

To find the square root of a number, look for the number in the first column (n). Then move to the right along the row to the column labeled $\sqrt{n}$.

n	n^2	$\sqrt{n}$ rounded to the nearest thousandth
1	1	1
2	4	1.414
3	9	1.732
4	16	2
5	25	2.236
6	36	2.449
7	49	2.646
8	64	2.828
9	81	3
10	100	3.162

MORE HELP

See 012–013, 096

Perfect Powers

070

If a number is a power of a whole number, it is a **perfect power**. Since 36 is 6^2, 36 is a perfect square.

6

6

If a whole number is a power of a number between whole numbers, it is not a perfect power. Since 37 is between 6^2 (36) and 7^2 (49), it is not a perfect square.

Mental Math and Estimation

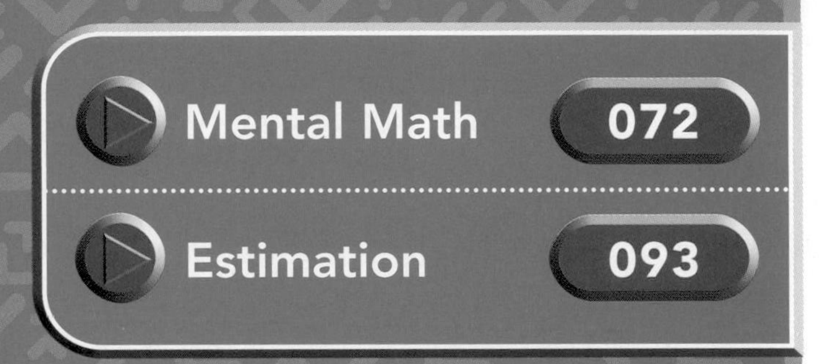

What is mental math?
It is doing math mentally—in other words, in your head.

Estimation is a kind of mental math. When you estimate, you are looking for a number close to the exact answer. You do this when you don't need an exact answer.

When you do other mental math, your answer is the exact answer.

072

Mental Math

A calculator is a great tool, but many times you don't need it to solve a problem.

You don't need a calculator to add 10 + 10, subtract 101 − 1, or multiply 3000 × 20. You can compute in your head. That's **mental math**.

073 Mental Addition

Every day, you probably add using mental math.

Sometimes, you add mentally because the numbers are easy. At other times, paper or a calculator might not be available.

999 + 1 = ■
Why would it be faster to add this mentally than to use a calculator?

Making Tens to Add

You can add using mental math by grouping numbers to make ten.

	Milk Orders Room 222 Week of May 10					
	Mon.	Tues.	Wed.	Thurs.	Fri.	Total
Regular	22	18	0	9	0	
Skim	4	3	7	6	8	
Chocolate	4	6	9	8	2	

EXAMPLE 1: How many cartons of skim milk did the students order that week?

See 221

To solve the problem, you can add.

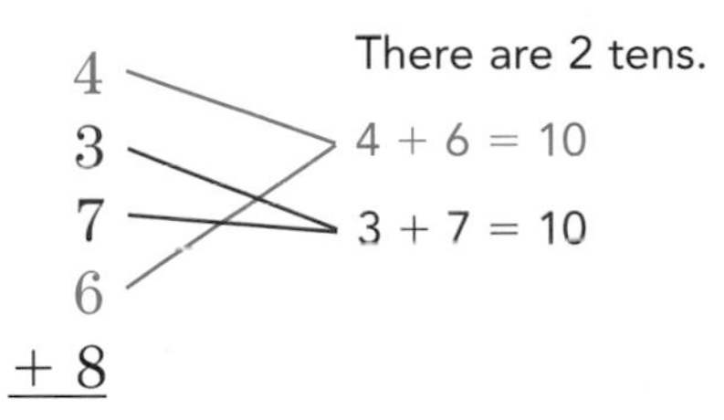

2 tens + 8 ⟶ 20 + 8 = 28

The students ordered 28 cartons of skim milk.

EXAMPLE 2: How many cartons of regular milk did the students order that week?

	Tens	Ones	
There are 3 tens.	2	2	2 + 8 = 10
	1	8	(Here's another ten.)
	+	9	

3 tens + 1 ten + 9 ⟶ 30 + 10 + 9 = 49

The students ordered 49 cartons of regular milk.

075 Adding 9

It's easy to add 10 to a number. What is 27 + 10?

Now, what is 27 + 9? It takes some time to add these numbers mentally if you think of it this way:

A much easier way to add 9 is to think of 9 as 10 − 1.

You just add 10 and take away 1.

Give and Take to Add

076

Sometimes you can take from one number and give the same amount to another to make it easier to add.

See 075

EXAMPLE: Evan buys these two computer games. How much money does he spend for both items?

To solve the problem, you can add 19 and 75.

ONE WAY You can make the second addend an even ten.

It's easier to add 80 than 75. Take 5 from 19 and use it to make 75 into 80.

$$\begin{array}{ccc} 19 & + & 75 \\ \downarrow -5 & & \downarrow +5 \\ 14 & + & 80 = 94 \end{array}$$

Since $19 + 75 = 14 + 80$, then $19 + 75 = 94$.

ANOTHER WAY You can make the first addend an even ten.

It's easier to add 20 than 19. Take 1 from 75 and use it to make 19 into 20.

$$\begin{array}{ccc} 19 & + & 75 \\ \downarrow +1 & & \downarrow -1 \\ 20 & + & 74 = 94 \end{array}$$

Since $19 + 75 = 20 + 74$, then $19 + 75 = 94$.

Either way, Evan spends $94.

Grouping in Column Addition

Suppose you want to add a few numbers in your head. You can group them in the way that makes it easiest for you to add.

MORE HELP

See 105, 217, 221

EXAMPLE: You're traveling through Arizona. You want to take the scenic route marked on the map. How many miles will you travel?

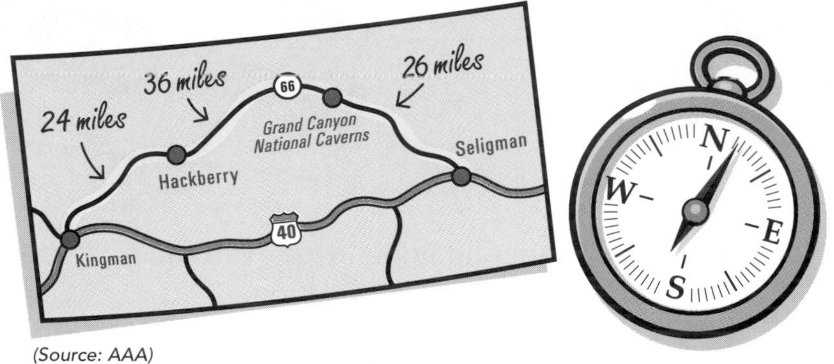

(Source: AAA)

Group the numbers in the way that makes adding easiest.

You can add 26 and 24 first.

$$\begin{array}{r} 26 \\ 36 \\ +\ 24 \\ \hline \end{array} \qquad \begin{array}{r} 36 \\ 26 \\ +\ 24 \\ \hline \end{array} \longrightarrow \begin{array}{r} 36 \\ \\ +\ 50 \\ \hline 86 \end{array}$$

Just because 26 and 36 are the first two addends, you don't need to start with them. If you think 26 and 24 are easier to add, add them first!

You can add 36 and 24 first.

$$\begin{array}{r} 26 \\ 36 \\ +\ 24 \\ \hline \end{array} \longrightarrow \begin{array}{r} 26 \\ +\ 60 \\ \hline 86 \end{array}$$

Either way, you will travel 86 miles.

Mental Subtraction 078

You subtract using mental math all the time. When you think about how much money you will have left after you buy something, you use mental subtraction.

Subtracting 9 079

It's easy to subtract 10 from a number. What is 24 − 10?

Now, what is 24 − 9? It may be hard to subtract mentally if you think of it this way:

A much easier way to subtract 9 is to subtract 10, then add 1.

080 Subtracting in Parts

You can subtract in your head by subtracting a number in parts. Break up the number you are subtracting into parts that are easy to subtract.

See 105

EXAMPLE: You have 64¢. You want to buy an apple for 26¢. How much money will you have left?

To solve the problem, you can subtract 26 from 64.

ONE WAY

Think of 26 as 24 + 2, then subtract 24 and 2 separately.

64 − 24 = 40

↓

40 − 2 = 38

Break up 26 twice.

64 − 20 = 44 Think of 26 as 20 + 6.

↓

44 − 4 = 40 Think of 6 as 4 + 2.

↓

40 − 2 = 38

Think about the coins.

	64¢
	Take away 2 dimes. 44¢ left
	Take away 4 pennies. 40¢ left
40¢ − 2¢ = 38¢	Think about taking away 2¢ more. 38¢ left

No matter which method you use, you will have 38¢ left.

Counting Up to Subtract

Sometimes it's easier to count up from the number you're subtracting to the number you started with.

EXAMPLE: Your sister rented a boring movie that is 126 minutes long. You have already sat through 60 minutes of the movie. How much longer will the movie run?

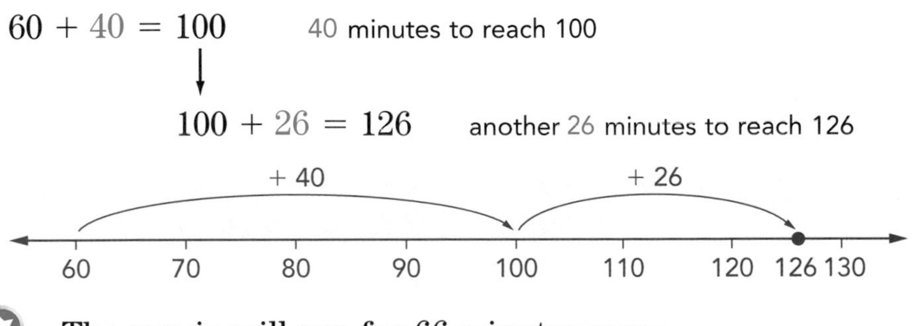

The movie will run for 66 minutes more.

082 Finding Easier Numbers to Subtract

The pattern below shows a great way to make mental subtraction easier.

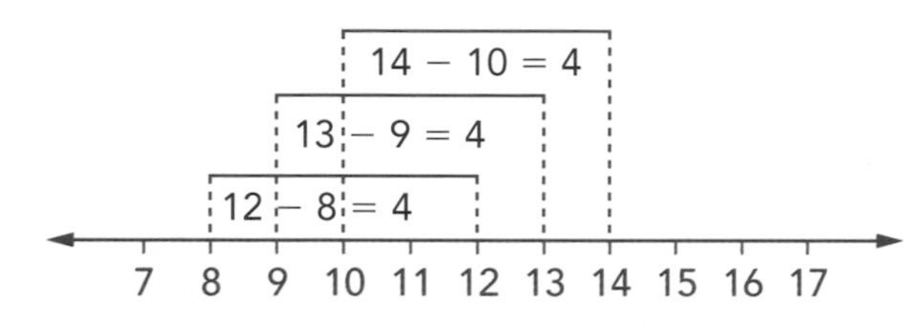

That makes sense! If you move the left and right endpoints of the bar the same distance in the same direction, the bar will stay the same length!

EXAMPLE: The Drama Club printed 150 copies of the program for their play. They handed out 98 copies of the program for the first show. How many programs do they have left for the second show?

To solve this problem, you can subtract 98 from 150.

$$150 - 98$$

$+2$ $\quad$ $+2$

$$152 - 100 = 52$$

It may be hard to subtract 98 in your head. If you add 2 to each number, you can subtract 100. This is just like sliding the bar 2 units to the right.

 There are only 52 programs left for the second show.

083 Mental Multiplication

You often multiply mentally. When you decide how many quarters you will get by putting $10 into a change machine, you might multiply 4 by 10 mentally.

Multiplying with Easy Numbers

Multiplying by 2 To multiply by 2, you can add a number to itself.

See 224–226

EXAMPLE 1: Multiply. $2 \times 44 = ■$

$44 + 44 = 88$

$2 \times 44 = 88$

Multiplying by 5 To multiply by 5, think of 5 as $10 \div 2$. Multiply by 10, then divide by 2.

EXAMPLE 2: Multiply. $5 \times 12 = ■$

12 12 12 12 12 — 5×12 is half of 10×12.

$10 \times 12 = 120$ — 12 12 12 12 12

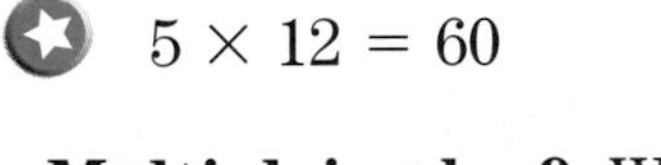

$5 \times 12 = 60$

Multiplying by 9 When you multiply by 9, think of 9 as $10 - 1$. Multiply by 10, then subtract the 1 times the original factor.

EXAMPLE 3: Multiply. $9 \times 15 = ■$

15 15 15 15 15 15 15 15 15 — 9×15

15 15 15 15 15 15 15 15 15 15 — 10×15

$9 \times 15 = (10 \times 15) - (1 \times 15)$

$150 - 15 = 135$

$9 \times 15 = 135$

085

Multiplying by 10, 100, and 1000

When you multiply by powers of 10 (like 10, 100, or 1000), you can multiply in your head. Look for patterns in this table.

Number	Number × 10	Number × 100	Number × 1000
6	60	600	6000
64	640	6400	64,000
648	6480	64,800	648,000
6.486	64.86	648.6	6486
64.86	648.6	6486	64,860

SHORTCUT

- To multiply a whole number by 10, tack on 1 zero at the right.
  8 × 10 = 80 567 × 10 = 5670
- To multiply a whole number by 100, tack on 2 zeros at the right.
  8 × 100 = 800 567 × 100 = 56,700
- To multiply a whole number by 1000, tack on 3 zeros at the right.
  8 × 1000 = 8000 567 × 1000 = 567,000
- To multiply a decimal number by 10, move the decimal point 1 place to the right.
  6.782 × 10 = 67.82 0.8 × 10 = 8
- To multiply a decimal number by 100, move the decimal point 2 places to the right.
  6.782 × 100 = 678.2
  0.8 × 100 = 80

Sometimes you have to tack on zeros in order to move the decimal point.

- To multiply a decimal number by 1000, move the decimal point 3 places to the right.
  6.782 × 1000 = 6782 0.8 x 1000 = 800

Multiplying with Multiples of 10

When one or both factors are multiples of 10 (say 40, 400, or 4000), you can multiply in your head. Look at these patterns to find out how.

See 059–061

$8 \times 3 = 24$	$5 \times 4 = 20$
$8 \times 30 = 240$	$5 \times 40 = 200$
$8 \times 300 = 2400$	$5 \times 400 = 2000$
$8 \times 3000 = 24{,}000$	$5 \times 4000 = 20{,}000$
$80 \times 300 = 24{,}000$	$50 \times 400 = 20{,}000$
$800 \times 300 = 240{,}000$	$500 \times 400 = 200{,}000$

That makes sense, because
8 × 3 tens = 24 tens and
8 × 3 hundreds = 24 hundreds.

SHORTCUT

To multiply by multiples of 10, first find the product of the front non-zero digits. Then tack on the zeros from both factors.

EXAMPLE: Souvenir hats come 200 to a box. The stadium orders 60 boxes. How many hats are in 60 boxes?

You can multiply 200 by 60 in your head.

$60 \times 200 = \blacksquare$

Think:

$6 \times 2 = 12$	6 and 2 are the front non-zero digits.
60×200	60 has 1 zero. 200 has 2 zeros.
12 000 = 12,000	Tack on 3 zeros to 12.

There are 12,000 hats in 60 boxes.

MORE
HELP

See 136

087

Repeated Doubling

Sometimes you can find a product in your head by doubling.

EXAMPLE: Each van at Concord Academy carries 12 passengers. How many passengers can 2 vans carry? How many can 4 vans carry? How many can 8 vans carry?

2×12 is $12 + 12$.
$12 + 12 = 24$

 2 vans can carry 2×12, or 24 passengers.

4×12 is double 2×12.
$24 + 24 = 48$

4 vans can carry 4×12, or 48 passengers.

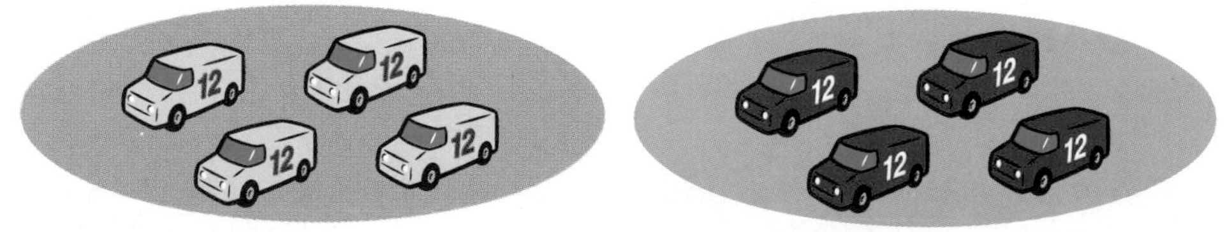

8×12 is double 4×12.
$48 + 48 = 96$

 8 of the Concord Academy vans can carry 96 passengers.

Breaking Apart and Putting Together Numbers to Multiply

088

One way to multiply in your head is to break the factors into parts and multiply the parts.

See 224–225

EXAMPLE: The school store ordered 15 boxes of pencils. There are 12 pencils in each box. How many pencils did the store order?

To solve the problem, you can multiply 12 by 15. Think of 15 as 10 + 5.

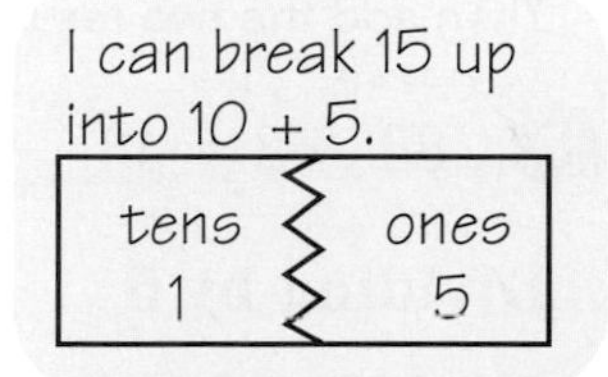

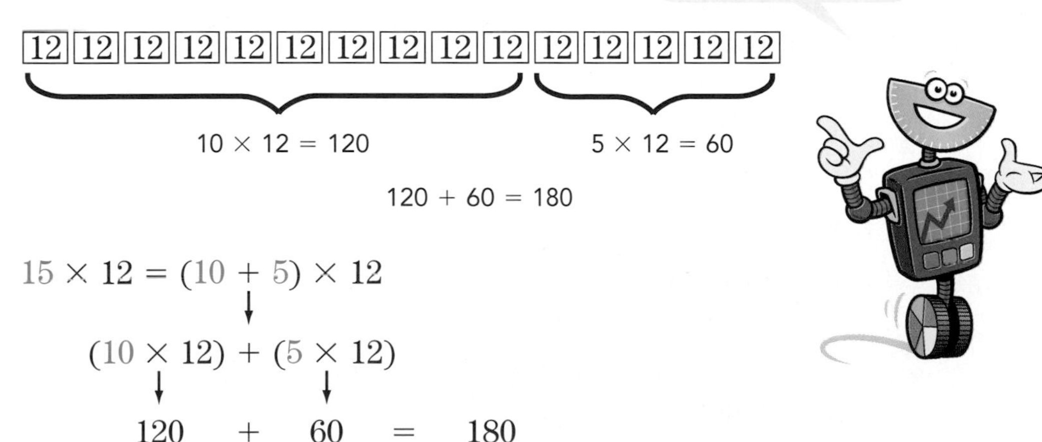

$15 \times 12 = (10 + 5) \times 12$

$(10 \times 12) + (5 \times 12)$

$120 + 60 = 180$

The store ordered 180 pencils.

089 Mental Division

You probably divide in your head all the time. If you and some friends are forming teams, you might divide in your head to see how many players will be on each team or how many teams you will make.

090 Dividing by 2 or 5

Some numbers are easy to divide by in your head.

MORE HELP

See 224–226

Dividing by 2

EXAMPLE 1: Divide. $86 \div 2 = ■$
Think of 86 as 80 + 6.

Divide 80 by 2. → $80 \div 2 = 40$

Divide 6 by 2. → $6 \div 2 = 3$

Then add the two results. → $40 + 3 = 43$

 $86 \div 2 = 43$

Dividing by 5

EXAMPLE 2: Divide. $80 \div 5 = ■$
Think of 5 as $10 \div 2$.

ONE WAY If you divide by 10 first, your quotient will be half what it should be.

Divide by 10. → $80 \div 10 = 8$

Then multiply by 2. → $8 \times 2 = 16$

If you multiply by 2 first, then you can divide by 10 to get the quotient.

Multiply by 2. → $80 \times 2 = 160$

Then divide by 10. → $160 \div 10 = 16$

 Either way, $80 \div 5 = 16$.

Dividing by 10, 100, and 1000

091

When you divide by powers of 10 (like 10, 100, or 1000) you can divide in your head. Look for patterns in this table.

Number	Number ÷ 10	Number ÷ 100	Number ÷ 1000
4000	400	40	4
43,000	4300	430	43
56,700	5670	567	56.7
82	8.2	0.82	0.082
8.36	0.836	0.0836	0.00836

See 015

SHORTCUT

See 144, 153–154

- Sometimes you can divide by 10, 100, or 1000 just by crossing off zeros. Cross off the same number of zeros in the dividend and the divisor.

  $600 \div 10 \longrightarrow 60\not{0} \div 1\not{0} \longrightarrow 60 \div 1 = 60$

  $45{,}000 \div 100 \longrightarrow 45{,}0\not{0}\not{0} \div 1\not{0}\not{0} \longrightarrow 450 \div 1 = 450$

- Sometimes you need to move the decimal point in the dividend.

  To divide by 10, move the decimal point in the dividend 1 place to the left.

  $567 \div 10 = 56.7$ $8.9 \div 10 = 0.89$

  To divide by 100, move the decimal point in the dividend 2 places to the left.

  $567 \div 100 = 5.67$ $8.9 \div 100 = 0.089$

  Sometimes you need to tack on zeros to give yourself enough places.

  To divide by 1000, move the decimal point in the dividend 3 places to the left.

  $567 \div 1000 = 0.567$ $8.9 \div 1000 = 0.0089$

092

Dividing with Multiples of 10

When you're dividing with tens or multiples of 10, you can often divide in your head.

Look at this pattern. It shows that a basic fact (like $24 \div 4 = 6$) can help you divide with multiples of 10.

$$240{,}000 \div 4 = 60{,}000$$
$$240{,}000 \div 40 = 6000$$
$$240{,}000 \div 400 = 600$$
$$240{,}000 \div 4000 = 60$$
$$240{,}000 \div 40{,}000 = 6$$

SHORTCUT

Divide the basic fact. Then count the zeros on the end of the dividend. Subtract the number of zeros on the end of the divisor. The difference tells you how many zeros to tack onto the end of the quotient.

EXAMPLE: If you and 29 friends won the $24,000,000 lottery, how much would each of you win?

To solve the problem, you can divide:
24,000,000 ÷ 30 = ■

dividend	÷	divisor	=	quotient
24,000,000	÷	30	=	800,000
↓		↓		↓
6 zeros	−	1 zero	=	5 zeros

24 ÷ 3 = 8

Each of you would win $800,000!

I just cross off the same number of zeros in the dividend and the divisor.
24,000,00~~0~~ ÷ 3~~0~~ = 800,000

Estimation

093

Most people use **estimation** every day when they don't need an exact answer. For example, shoppers estimate which product is the best buy, carpenters estimate how much a job will cost, and students estimate how much time they will need in the morning to get ready for school.

Rounding

094

Rounding is one way to estimate. When you round a number, you change it to a nearby number that may have one or more nice, "round" zeros at the end.

095 Rounding Whole Numbers

Rounded numbers give a rough idea of an amount.

You can round numbers by using a number line.

See 004

EXAMPLE 1: Round 273 to the nearest hundred.

200 is the closest hundred below 273.
300 is the closest hundred above 273.

273

200 210 220 230 240 250 260 270 280 290 300

273 is closer to 300 than to 200. So, 273 rounds up to 300.

ANOTHER WAY You can also round numbers by using place value.

EXAMPLE 2: Round 361 to the nearest hundred.

- Find the hundreds place. 361
- Look at the digit one place to its right. 361

If the digit is 5 or greater, round up. $6 > 5$ Round up.
If the digit is less than 5, round down.

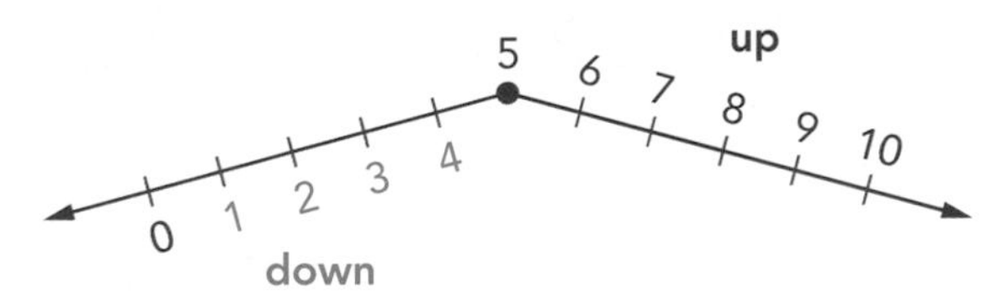

To the nearest hundred, 361 rounds up to 400.

EXAMPLE 3: Round 361 to the nearest ten.

- Find the tens place. 361
- Look at the digit one place to its right. 361

  If the digit is 5 or greater, round up.
  If the digit is less than 5, round down. $1 < 5$ Round down.

To the nearest ten, 361 rounds down to 360.

Rounding Decimals

096

You can round decimals just as you round whole numbers.

See 012, 095

EXAMPLE 1: On June 23, 1998 one British pound was worth 1.664100 U.S. dollars. Round the number of dollars to the nearest cent. *(Source:www.tradingfloor.net/currency.htm)*

- Find the hundredths place. 1.664100
- Look at the digit one place to its right. 1.664100

  If the digit is 5 or greater, round up.
  If the digit is less than 5, round down. $4 < 5$ Round down.

To the nearest cent, $1.664100 rounds down to $1.66.

EXAMPLE 2: A centimeter is approximately equal to 0.3937 inch. Round this number to the nearest tenth.

After you round a decimal number, you drop the digits to the right of the place you are rounding to.

- Find the tenths place. 0.3937
- Look at the digit one place to its right. 0.3937

  If the digit is 5 or greater, round up. $9 > 5$ Round up.
  If the digit is less than 5, round down.

To the nearest tenth, 0.3937 rounds up to 0.4. A centimeter is about four tenths of an inch.

097 Estimation Benchmarks

Some numbers are easier to work with and to picture than others. We often compare other numbers to these *easy* numbers. We call them **benchmarks.**

098 Benchmark Fractions

MORE HELP

See 035, 038

The most common benchmarks for comparing fractions are 0, $\frac{1}{2}$, and 1. Here is a way to decide whether a fraction is closer to 0, $\frac{1}{2}$, or 1.

- If the numerator is *about half of the denominator*, the fraction is close to $\frac{1}{2}$. $\frac{3}{8}$ is close to $\frac{1}{2}$.

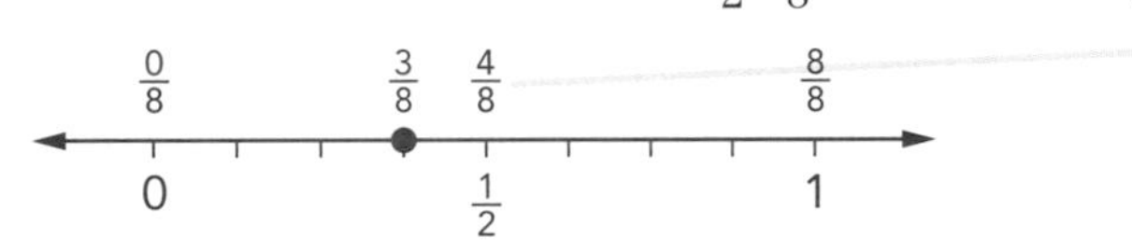

$\frac{1}{2}$ of 8 is 4.
$\frac{1}{2} = \frac{4}{8}$
$\frac{3}{8}$ is close to $\frac{4}{8}$.

- If the numerator is *much less than half of the denominator*, the fraction is close to 0. $\frac{1}{10}$ is close to 0.

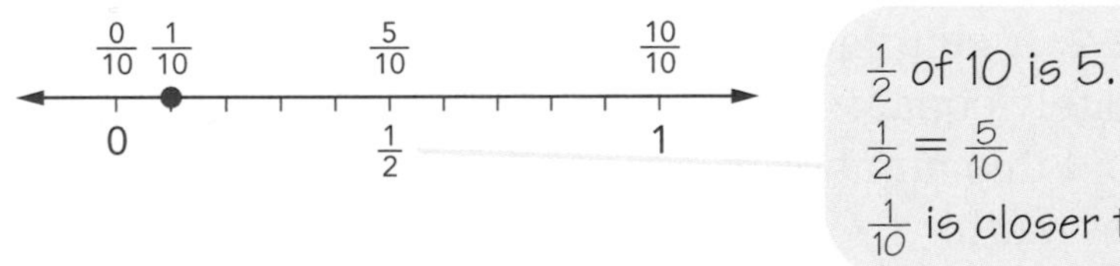

$\frac{1}{2}$ of 10 is 5.
$\frac{1}{2} = \frac{5}{10}$
$\frac{1}{10}$ is closer to $\frac{0}{10}$ than to $\frac{5}{10}$.

- If the numerator is *much more than half of the denominator*, the fraction is close to 1. $\frac{15}{16}$ is close to 1.

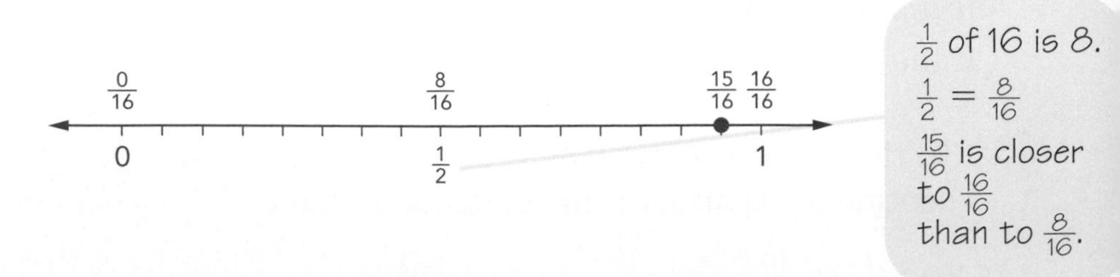

Benchmark Percents

099

When working with percents, certain percents are helpful reference points. These *easy* percents are known as **benchmark percents**.

MORE HELP

See 189–193

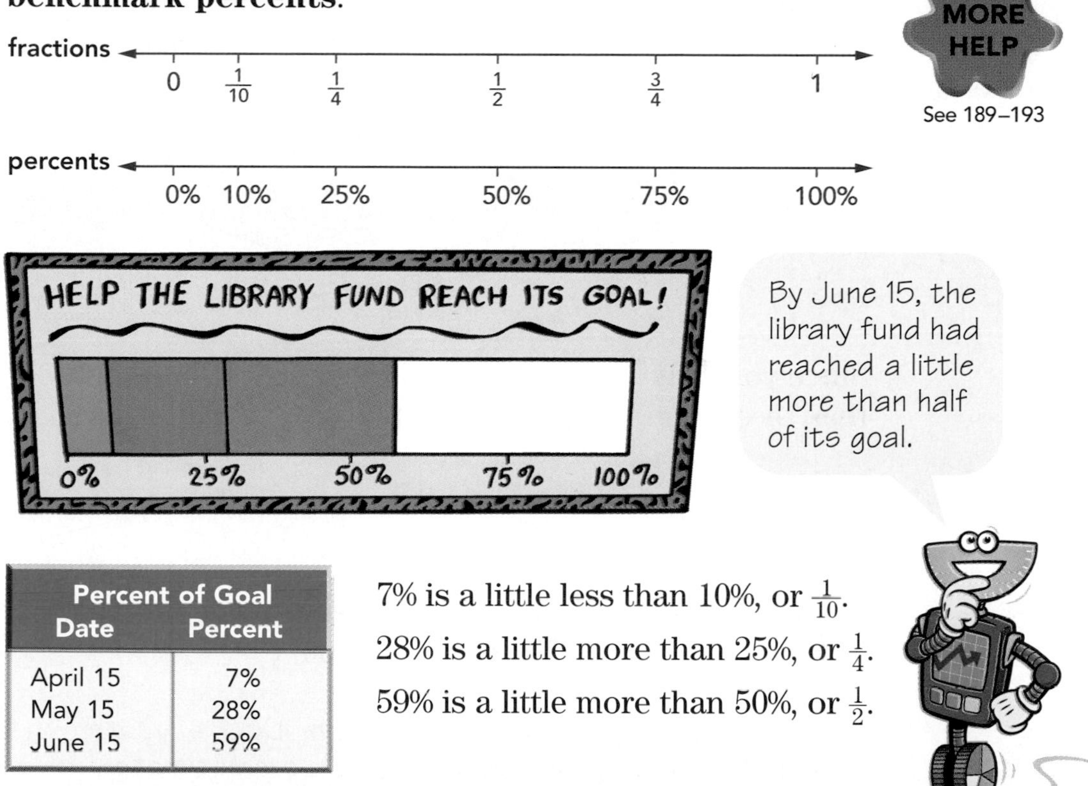

Percent of Goal	
Date	**Percent**
April 15	7%
May 15	28%
June 15	59%

7% is a little less than 10%, or $\frac{1}{10}$.

28% is a little more than 25%, or $\frac{1}{4}$.

59% is a little more than 50%, or $\frac{1}{2}$.

Estimating Sums and Differences

100

The **sum** is the total you get when you add two or more numbers. The **difference** is the number you get when you subtract two numbers. Sometimes, you don't need an exact sum or difference. In these cases you can estimate. You can also use estimates to check computation, especially when you use a calculator.

101 Rounding to Estimate Sums and Differences

You can use rounding to estimate sums and differences.

See 095–096

EXAMPLE 1: Is \$1000 enough to buy a modem and a printer? Do you need an exact sum or an estimate?

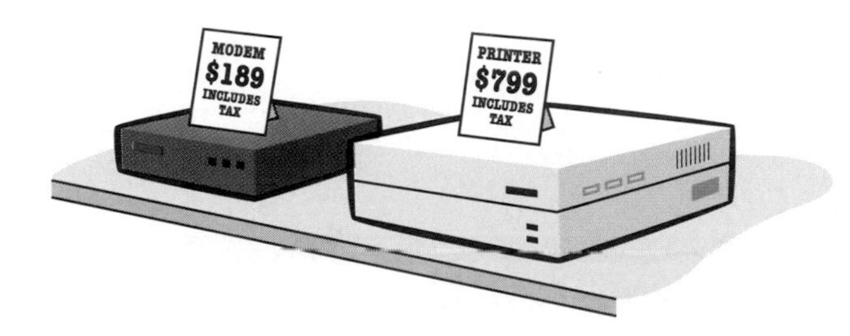

Since you only need to know whether 189 + 799 is less than or equal to 1000, an estimate is enough.

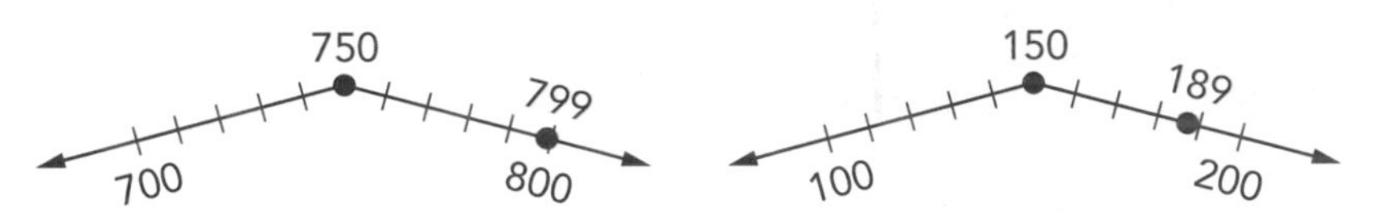

799 rounds up to 800.

189 rounds up to 200.

$800 + 200 = 1000$

$800 > 799$ and $200 > 189$. So, 1000 is greater than the actual sum. 1000 is an **overestimate.**

Since \$1000 is an overestimate, you know \$1000 is enough money.

EXAMPLE 2: You have \$7.00. If you buy a CD for \$5.19, can you pay the \$2.50 bus fare to get home?

Since you only need to know whether \$7.00 − \$5.19 is greater than or equal to \$2.50, an estimate is enough.

\$5.19 rounds down to \$5.00.

\$7.00 − \$5.00 = \$2.00

\$5 < \$5.19
So, \$2 is more than the actual difference. It is an overestimate.

You don't have enough money for bus fare.

Using Benchmarks to Estimate Sums and Differences of Fractions

102

You can use what you know about benchmark fractions to estimate sums and differences of fractions and mixed numbers.

See 034, 098, 161

EXAMPLE 1: Mariah rode her bike $2\frac{8}{10}$ miles to school and then $1\frac{1}{5}$ miles to Rebecca's house. About how many miles did she ride altogether?

Since you only need to know *about* how many miles she rode, you can estimate the sum of $2\frac{8}{10}$ and $1\frac{1}{5}$.

❶ Use benchmark fractions to round each addend.			❷ Add the rounded numbers.
$2\frac{8}{10}$ ⟶	3	$\frac{8}{10}$ is close to 1. So, $2\frac{8}{10}$ is close to 3.	$3 + 1 = 4$
$1\frac{1}{5}$ ⟶	1	$\frac{1}{5}$ is close to 0. So, $1\frac{1}{5}$ is close to 1.	

Mariah rode about 4 miles altogether.

EXAMPLE 2: Andrew lives $6\frac{4}{10}$ miles from the library. Sam lives $2\frac{9}{10}$ miles from the library. About how much farther from the library does Andrew live than Sam?

❶ Use benchmark fractions to round each number.		❷ Subtract the rounded numbers.
$6\frac{4}{10}$ ⟶	$6\frac{1}{2}$	$6\frac{1}{2} - 3 = 3\frac{1}{2}$
$2\frac{9}{10}$ ⟶	3	

Andrew lives about $3\frac{1}{2}$ miles farther from the library than Sam does.

Front-End Estimation of Sums and Differences

You can estimate sums or differences using the front digits. This is called **front-end estimation**.

CASE 1 When the addends have the same number of digits, add the front digits.

EXAMPLE 1: Estimate 4239 + 2256 + 1175.

In addition, front-end estimation always gives a sum less than the actual sum. This is called an **underestimate.**

$$\begin{array}{rcr} 4239 & \longrightarrow & 4000 \\ 2256 & \longrightarrow & 2000 \\ \underline{+1175} & \longrightarrow & \underline{+1000} \\ & & 7000 \end{array}$$

The sum is about 7000.

EXAMPLE 2: Estimate 2466 − 1284.

$$\begin{array}{rcr} 2466 & \longrightarrow & 2000 \\ \underline{-1284} & \longrightarrow & \underline{-1000} \\ & & 1000 \end{array}$$

The difference is about 1000.

CASE 2 You can still use front-end estimation when the addends have a different number of digits.

EXAMPLE 3: Estimate 2142 + 726 + 854 + 317.

One of the front digits is in the thousands place. The others are in the hundreds place. To estimate the sum, you can use the thousands *and* hundreds digits.

$$\begin{array}{rcr} 2142 & \longrightarrow & 2100 \\ 726 & \longrightarrow & 700 \\ 854 & \longrightarrow & 800 \\ \underline{+\ 317} & & \underline{+\ 300} \\ & & 3900 \end{array}$$

The sum is at least 3900.

Adjusting Front-End Sums and Differences

104

Adjusting your front-end estimate by looking at the front two digits brings you closer to the exact answer.

Front-End Estimate		Adjusted Estimate		Exact Answer
101,082	100,000	101,082	100,000	101,082
− 78,122	− 70,000	− 78,122	− 78,000	− 78,122
	30,000		22,000	22,960

Many times, a front-end estimate is close enough. If you need a closer estimate, make an adjustment.

Estimating by Adding and Subtracting Compatible Numbers

105

Compatible numbers work well together. Number pairs that are easy to add or subtract are compatible. When estimating, you can replace actual numbers with compatible numbers.

See 101–102

EXAMPLE: Two classes have gym together. One class has 17 students. The other class has 26 students. About how many students are in the gym class? About how many more students are in the bigger class?

To solve the problems, you can estimate by choosing compatible numbers.

$17 + 26$
↓
$15 + 25 = 40$

Since $15 + 25 = 40$, $17 + 26$ is a bit more than 40.

$26 - 17$
↓
$25 - 15 = 10$

Since $25 - 15 = 10$, $27 - 16$ is a bit more than 10.

There are about 40 students in gym class. The bigger class has about 10 more students.

106 Estimating Products

A **product** is the result of multiplication. Sometimes when you multiply, you don't need an exact product. In these cases you can estimate products. You can also use estimates to check exact answers you've found, especially when you use a calculator

107 Rounding to Estimate Products

Rounding can be helpful when you only need to estimate a product.

See 095–096

EXAMPLE 1: Approximately what is the net weight of 66 boxes of Bonnie's Blueberry Bars?

66 rounds up to 70
10.73 rounds up to 11

$$\begin{array}{r} 70 \\ \times 11 \\ \hline 770 \end{array}$$

Since both numbers were rounded up, the estimated product will be an **overestimate**.

The net weight of the breakfast bars is less than 770 ounces.

EXAMPLE 2: About how much will the 66 boxes cost?

\$2.29 rounds down to \$2.00
66 rounds up to 70

$$\begin{array}{r} \$2.00 \\ \times \quad 70 \\ \hline \$140.00 \end{array}$$

Since one number was rounded down, and one number was rounded up, you can't be sure if you have an overestimate or an underestimate.

The 66 boxes will cost about $140.00.

Using Benchmarks to Estimate Products of Fractions

You can use what you know about benchmark fractions to estimate products of fractions and mixed numbers.

See 034, 098, 170, 301

EXAMPLE 1: About how many square feet of carpet would cover a floor $9\frac{10}{12}$ feet long and $15\frac{1}{12}$ feet wide?

Since you only need to know *about* how many square feet of carpet you need, you can estimate the product of $9\frac{10}{12}$ and $15\frac{1}{12}$.

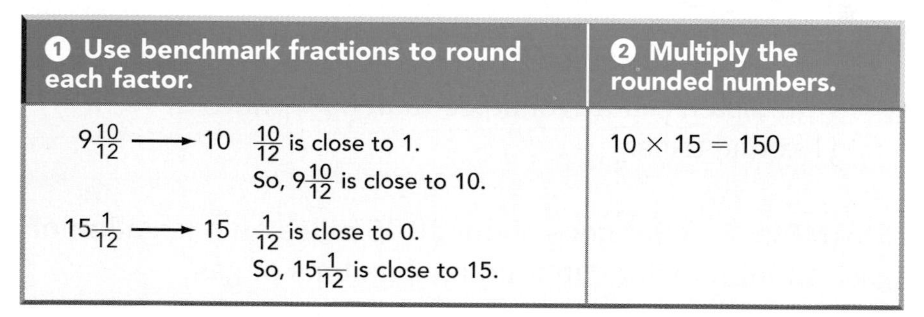

① Use benchmark fractions to round each factor.	② Multiply the rounded numbers.
$9\frac{10}{12} \longrightarrow 10$ $\frac{10}{12}$ is close to 1. So, $9\frac{10}{12}$ is close to 10. $15\frac{1}{12} \longrightarrow 15$ $\frac{1}{12}$ is close to 0. So, $15\frac{1}{12}$ is close to 15.	$10 \times 15 = 150$

About 150 square feet of carpet would cover the floor.

EXAMPLE 2: Estimate the product of $\frac{12}{15}$ and $36\frac{9}{10}$.

① Use benchmark fractions to round each factor.	② Multiply the rounded numbers.
$\frac{12}{15} \times 36\frac{9}{10}$ ↓ ↓ 1 37	$1 \times 37 = 37$

The product is about 37.

109 Front-End Estimation of Products

To estimate products, you can multiply the front digits.

EXAMPLE 1: Your older sister attends a college 336 miles away. Will she travel at least 5000 miles if she makes the trip 22 times (11 round trips) in one year?

Since you only need to know if she will travel at least 5000 miles, you can estimate 22×336.

$$\begin{array}{r} 336 \\ \times\ 22 \\ \hline \end{array} \longrightarrow \begin{array}{r} 300 \\ \times\ 20 \\ \hline 6000 \end{array}$$

The exact product is greater than 6000.

Your sister will travel more than 5000 miles in 11 round trips.

EXAMPLE 2: A car goes about 19.45 miles on one gallon of gas. About how far can it go on 9 gallons of gas?

$$\begin{array}{r} 19.45 \\ \times\quad 9 \\ \hline \end{array} \longrightarrow \begin{array}{r} 20 \\ \times\ 9 \\ \hline 180 \end{array}$$

The exact gas mileage is less than 20 miles per gallon.

The car can travel a little less than 180 miles.

110 Adjusting Front-End Estimates of Products

Adjusting your front-end estimate by looking at more digits brings you closer to the exact answer.

Front-End Estimate	Adjusted Estimate	Exact Answer
$\begin{array}{r} 35{,}486 \\ \times\quad 121 \\ \hline \end{array}$ $\begin{array}{r} 30{,}000 \\ \times\quad 100 \\ \hline 3{,}000{,}000 \end{array}$	$\begin{array}{r} 35{,}486 \\ \times\quad 121 \\ \hline \end{array}$ $\begin{array}{r} 35{,}000 \\ \times\quad 100 \\ \hline 3{,}500{,}000 \end{array}$	$\begin{array}{r} 35{,}486 \\ \times\quad 121 \\ \hline 4{,}293{,}806 \end{array}$

Many times, a front-end estimate is close enough. If you need a closer estimate, make an adjustment.

Multiplying Compatible Numbers

111

Compatible numbers are numbers that work well together. In multiplication, they are number pairs that are easy to multiply.

To estimate products, replace factors with compatible numbers.

MORE HELP
See 007, 106–108

EXAMPLE 1: The Pottery Club meets for 9 weeks in the fall. If 36 students each create a different piece of pottery each week, how many pieces of pottery will have been created at the end of 9 weeks?

Any factor is compatible with 10.

$$\begin{array}{r} 36 \\ \times\ 9 \\ \hline \end{array} \longrightarrow \begin{array}{r} 36 \\ \times\ 10 \\ \hline 360 \end{array}$$

9 × 36 is close to 10 × 36.

 About 360 pieces of pottery will have been created.

EXAMPLE 2: Use compatible numbers to estimate 8 × 27.

8 × 27 is close to 8 × 25.

Since 8 × 25 = 200,
8 × 27 is a little more than 200.

When I multiply by 25,
I think about quarters.
Eight quarters are the
same as $2.00.
So, 8 × 25 = 200.

112 Estimating Quotients

When you divide, sometimes you only need to know *about* how many things are in each group or *about* how many groups can be formed. So, sometimes you can estimate quotients. You can also use estimates to check your computation.

113 Estimating Quotients Using Compatible Numbers

Compatible numbers are numbers that work well together. In division, they are number pairs that are easy to divide. They are often the numbers from a basic fact.

You can use compatible numbers to estimate quotients by replacing the actual numbers with compatible numbers.

EXAMPLE 1: Ryan can run a mile in 8 minutes. If he runs for 31 minutes, about how many miles will he run?

You can estimate $31 \div 8$.
Find compatible numbers close to 31 and 8.

$31 \div 8$
↓ ↓
$32 \div 8 = 4$

Since $32 > 31$, $32 \div 8$ will be greater than the quotient of $31 \div 8$. So, 4 miles is an **overestimate**.

Ryan will run a little less than 4 miles.

EXAMPLE 2: In 1995, the population of Tokyo was about 8,112,000. Tokyo's area is 223 square miles. In 1995, about how many people were there for each square mile?

(Source: Information Please Almanac)

MORE HELP
See 089–092

You can estimate: 8,112,000 ÷ 223 ≈ ■.
Find compatible numbers close to 8,112,000 and 223.

$$8{,}112{,}000 \div 223$$
$$\downarrow \qquad \downarrow$$
$$8{,}000{,}000 \div 200 = 40{,}000$$
$$\uparrow \qquad \uparrow \qquad \uparrow$$
6 zeros − 2 zeros = 4 zeros

Remembering basic facts like 8 ÷ 2 = 4 can help you find compatible numbers.

In 1995, Tokyo had about 40,000 people for each square mile.

See 153–155

You can also use compatible numbers to estimate decimal quotients.

EXAMPLE 3: You want to divide 3.4 by 0.6. The calculator display shows 5.6666666.

To check this, you can estimate the quotient and compare. If your estimate is not close, you should compute again.

$$3.4 \div 0.6$$
$$\downarrow \qquad \downarrow$$
$$3.6 \div 0.6 = 6$$

Since 36 and 6 are compatible, 3.6 and 0.6 are compatible.

6 is close to 5.6666666. So, your answer is reasonable.

114 Using Multiples of 10 to Estimate Quotients

MORE HELP
See 085–086

You can use what you know about multiplying with multiples of 10 (like 30, 300, or 3000) to help you estimate quotients.

Multiplication is the opposite of division. For each division equation, there is a multiplication equation that uses the same numbers.

$$12 \div 4 = \blacksquare \longrightarrow \blacksquare \times 4 = 12$$
$$3 \times 4 = 12$$
$$\text{So, } 12 \div 4 = 3$$

EXAMPLE: Estimate $625 \div 7$.

Think: $\blacksquare \times 7$ is about 625.

↑ quotient ↑ divisor ↑ dividend

$7 \times \blacksquare$	Target: 625	High or Low?
7×60	420	too low
7×70	490	close, still too low
7×80	560	closer, STILL too low
7×90	630	VERY close!

The dividend, 625, is very close to 630, so the quotient is close to 90.

Estimating with Percents

115

Percents may name the amount taken off the regular price, or the number of students in a group, or the number of successful free throws. When you compute, you use decimal or fraction equivalents for percents. Sometimes an estimate is enough.

Wilson said that he was pleased by his team's accuracy at the foul line. The team made 75% of their free throws. That is the highest percentage recorded in the league since 1981. Defensively, the

Using Benchmark Percents to Estimate the Percent of a Number

116

Benchmark percents can be helpful if you want to estimate the percent of a number.

See 099, 193

EXAMPLE 1: About how many students said they liked Sci Fi shows?

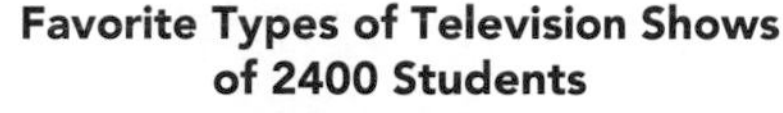

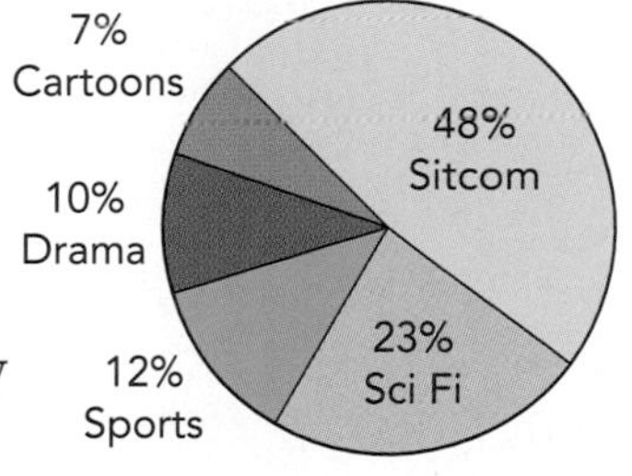

23% of 2400 students said they liked Sci Fi shows. 23% is close to 25% or $\frac{1}{4}$.

$\frac{1}{4} \times 2400 = 600$

About 600 students said they liked Sci Fi shows.

EXAMPLE 2: You see a sweater marked 30% off. The regular price is $28. What is the sale price of the sweater?

You know you will save 30% of $28. 30% of $28 is a little more than 25% of $28 or $\frac{1}{4}$ of $28. $\frac{1}{4} \times 28 = 7$

You know you will save at least $7. So, the sweater will cost less than $21.00.

$28 − $7 = $21

117

Computing with Whole Numbers and Decimals

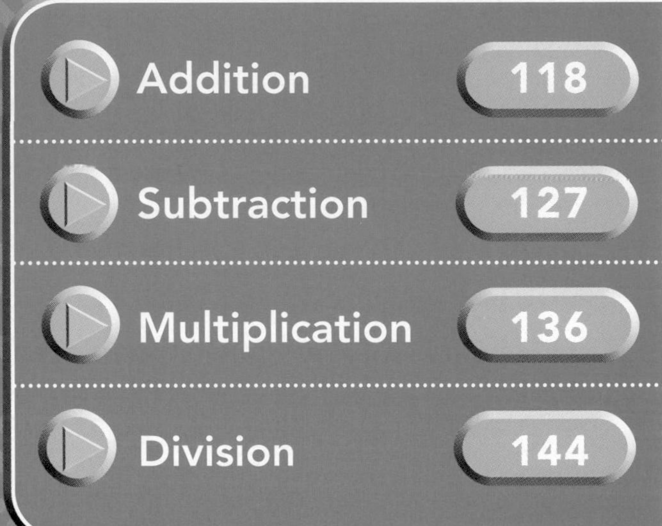

Some people think arithmetic is just a lot of hard work. But when you think about it, adding, subtracting, multiplying, and dividing save us time and make things easier. That's because arithmetic is faster than counting, and there's less chance of making a mistake. If you have a 64-page album filled with 8 photos on each page, it sure is quicker to multiply 64 by 8 than to count every one of the 512 photos.

118

Addition

A **mathematical operation** allows you to do something to one or more numbers which creates an answer. Addition may be the most important mathematical operation.

- Subtraction is the opposite of addition, so you can't subtract unless you can add.
- Multiplication is adding the same thing over and over again, so you can't multiply unless you understand addition.
- Division is the opposite of multiplication, so even division is related to addition!

Terms you may hear used in addition are **addend** and **sum**.

$$\begin{array}{r} 23 \\ +52 \\ \hline 75 \end{array} \begin{array}{l} \leftarrow \text{addend} \\ \leftarrow \text{addend} \\ \leftarrow \text{sum} \end{array}$$

119 Adding with Whole Numbers

Adding whole numbers is one of the basic skills in mathematics. When you **add**, you combine quantities.

If you know how to add ones and tens, you also know how to add hundreds, thousands, hundred thousands, and even billions. That's because of place value. When you line numbers up and add them place by place, what you do in each place is the same.

Adding Without Regrouping

See 006

EXAMPLE: For the final game of the basketball tournament, 115 fans of the home team and 81 fans of the visiting team attended the game. How many fans were there?

To solve the problem, add 115 and 81.

Lining up the digits helps you add ones to ones, tens to tens, and so on.

ONE WAY You can add this way.

```
  H T O
  1 1 5
+   8 1   <- Line up the digits at the ones place.
  -----
  1 9 6   <- Add the ones. There are 6 ones.
```

- Add the tens. There are 9 tens.
- Add the hundreds. There is 1 hundred.

ANOTHER WAY You can also write the addends in expanded form and add them.

$$\begin{array}{rcr} 115 & \longrightarrow & 100 + 10 + 5 \\ \underline{+\ 81} & \longrightarrow & \underline{+\ 80 + 1} \\ & & 100 + 90 + 6 = 196 \end{array}$$

Either way, 196 fans attended the game.

MATH ALERT Regrouping

Sometimes you have too many digits to squeeze into the one-digit space for each place value. Adding 9 tens to 8 tens gives you 17 tens, but the tens place only has room for one digit. When what you do in one place changes the value in another place, you are **regrouping**.

EXAMPLE 1: How can you show 17 tens so that there is just one digit in each place?

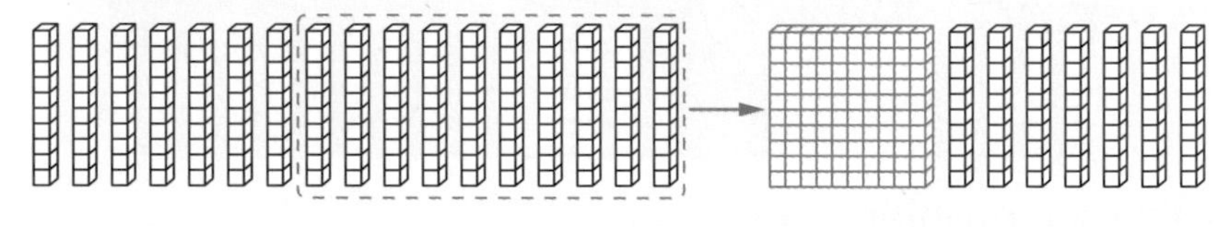

You started with 17 tens.

17 tens is the same as 1 hundred + 7 tens.

You now have 1 hundred + 7 tens + 0 ones, or 170.

EXAMPLE 2: How can you show 9 tens 13 ones so that there is just one digit in each place?

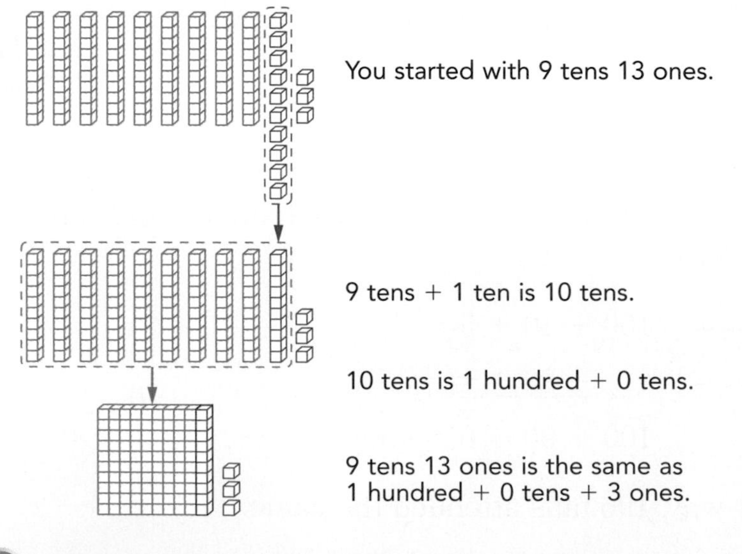

You now have 1 hundred + 0 tens + 3 ones, or 103.

Adding with Regrouping

Sometimes when you add, the sum in a place has two digits or more. You can use place value to regroup the sum so that there is just one digit in each place.

EXAMPLE: The students in your school sold gift wrap to raise money to buy library books. They sold 154 rolls of birthday gift wrap and 238 rolls of holiday gift wrap. How many rolls of gift wrap did they sell?

To solve the problem, add 154 and 238.

ONE WAY You can add this way.

```
 H T O
   1
 1 5 4
+2 3 8
------
 3 9 2
```

Line up the digits at the ones place.

Add the ones. There are 12 ones. Since 12 ones is 1 ten + 2 ones, write 2 in the ones place. Write 1 as a new addend in the tens place.

Add the tens. There are 9 tens.

Add the hundreds. There are 3 hundreds.

ANOTHER WAY You can write the addends in expanded form and add them.

$$\begin{array}{rcr} 154 & \longrightarrow & 100 + 50 + 4 \\ +238 & \longrightarrow & +\ 200 + 30 + 8 \\ \hline & & 300 + 80 + 12 = 392 \end{array}$$

Either way, the students sold 392 rolls of gift wrap.

123 Checking Addition

No matter how good you are at addition, it never hurts to check your answer.

Check your addition by using subtraction. You can do this because subtraction is the opposite, or **inverse**, of addition.

MORE HELP

See 128–134

To check your answer, subtract one addend from the sum. The answer should be the other addend. If it isn't, you should try doing the addition again.

EXAMPLE 1: Does 869 + 435 = 1304? Use subtraction to check.

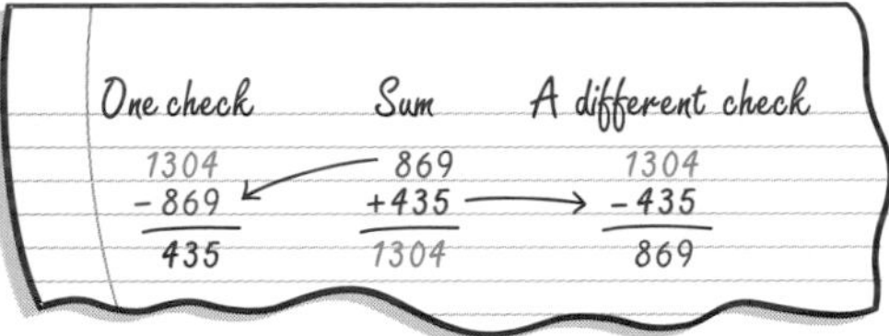

Since completing the subtraction gives the other addend, the addition is correct.

You can also check your addition by adding again in a different order.

MORE HELP

See 217

EXAMPLE 2: Is the sum correct? Add in a different order to check.

1140 does not equal 1240. When adding in a different order gives you a different sum, you should do the addition again.

Column Addition

Sometimes you need to add more than two numbers. You can still add place by place and regroup if you need to.

EXAMPLE: These five counties make up New York City. How many schools are in New York City?

County	Number of Schools, K–12
Bronx	310
Kings	692
New York	432
Queens	421
Richmond	132

(Source: Market Data Retrieval)

See 074, 076–077, 100–101, 103–105, 121

To solve the problem, add.

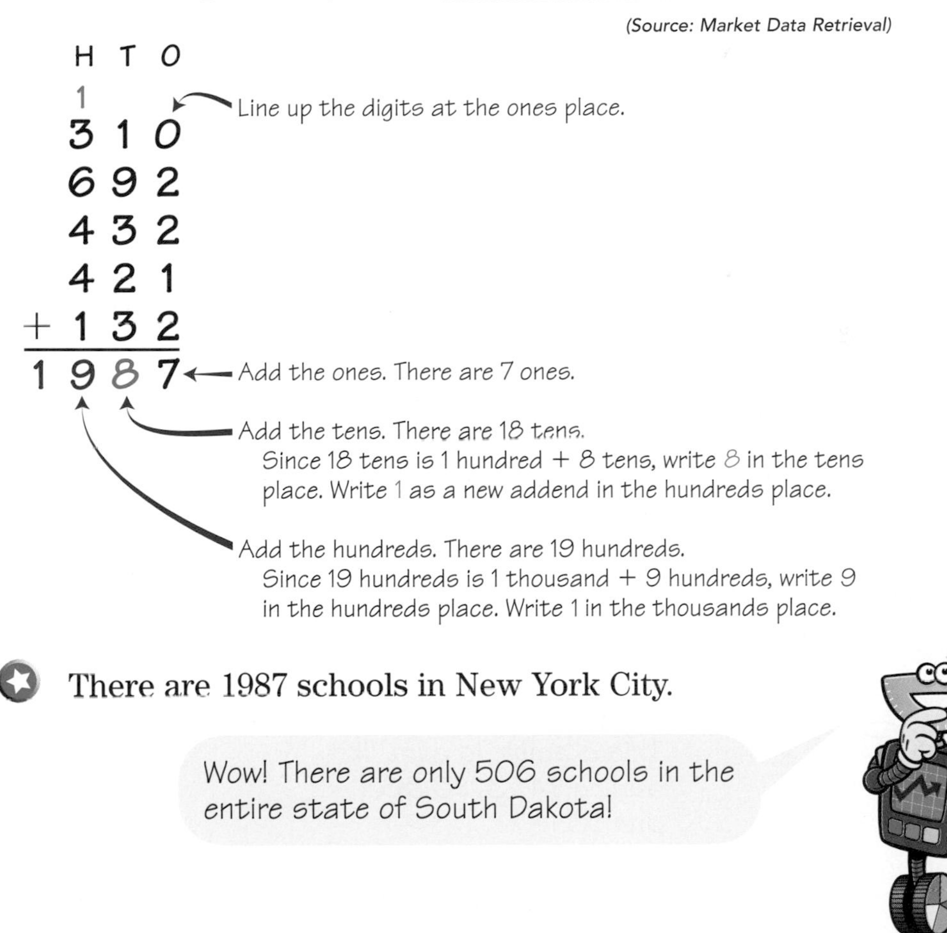

There are 1987 schools in New York City.

125 Adding with Decimals

If you know how to add whole numbers, you also know how to add decimals. You still need to line up the ones digits. This means the decimal points line up, too.

There are a few important things to remember.

- Pay attention to place value.
- Add in each place.
- Regroup when needed.

See 011–014, 119–124

EXAMPLE: Add 4.73 and 0.86 and 3.14.

```
  O  T H
  1  1
  4 . 7 3
  0 . 8 6
+ 3 . 1 4
---------
  8 . 7 3
```

Line up the decimal points. Write the decimal point in the sum.

Add the hundredths. There are 13 hundredths. Since 13 hundredths is 1 tenth + 3 hundredths, write 3 in the hundredths place. Write 1 as a new addend in the tenths place.

Add the tenths. There are 17 tenths. Since 17 tenths is 1 one and 7 tenths, write 7 in the tenths place. Write 1 as a new addend in the ones place.

Add the ones. There are 8 ones.

$4.73 + 0.86 + 3.14 = 8.73$

When you add money, you are really adding decimals. This example is a lot like adding \$4.73, \$0.86, and \$3.14.

It is helpful to estimate the sum to make sure your answer is reasonable. One way to estimate is to round.

See 096, 100–101, 103–105

$$\begin{array}{rcr} 4.73 & \approx & 5 \\ 0.86 & \approx & 1 \\ +\ 3.14 & \approx & +\ 3 \\ \hline 8.73 & & 9 \end{array}$$

Since your sum is close to your estimate, 8.73 is a reasonable answer.

Write: $8.73 \approx 9$
Say: *eight and seventy-three hundredths is approximately* (or *about*) *equal to nine*

MATH ALERT What to Do with a Ragged Right Side

Sometimes you line up the ones and decimal points and the right side of your problem looks ragged. To make it easier to keep track while you compute, give all your decimals the same number of places by tacking on zeros.

See 015

EXAMPLE: Write equivalent decimals for 2, 6.5, and 4.02 so all three numbers have the same number of decimal places.

The number in this group with the greatest number of decimal places is four and two hundredths. You need to write 2 and 6.5 as decimals in hundredths.

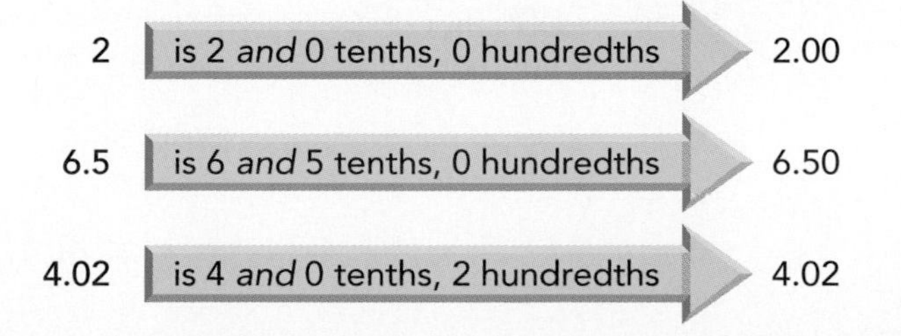

127

Subtraction

When you **subtract**, you are separating quantities. You use subtraction for different reasons.

- If you want to remove one amount from another, you subtract.

  You have \$30. You plan to spend \$18. Subtract 18 from 30 to find what you will have left.

- If you want to compare one amount to another, you subtract.

  Your book bag weighs 14.5 pounds. Kayla's weighs 9.75 pounds. Subtract 9.75 from 14.5 to see how much more weight you carry around than Kayla does.

- If you know part of a quantity and the whole quantity but need to know the other part, you subtract.

  Your public library has 25 books on hiking. There are 21 of these books on the shelf. How many books on hiking are loaned out? Subtract 21 from 25 to find out how many books are loaned out.

Three terms you may hear used in subtraction are **minuend**, **subtrahend**, and **difference**.

$$\begin{array}{r l} 25 & \leftarrow \text{minuend} \\ -\ 21 & \leftarrow \text{subtrahend} \\ \hline 4 & \leftarrow \text{difference} \end{array}$$

Relating Addition and Subtraction 128

Addition is the opposite, or **inverse**, of subtraction. So, that means subtraction is the inverse of addition. An addition table can show you the relationship between addition and subtraction.

To add two numbers on an addition table, choose one number from the beginning of a row and one number from the top of a column. The sum of the numbers will be in the place where the row and column meet.
$1 + 2 = 3$

+	0	1	2	3	4
0	0	1	2	3	4
1	1	2	3	4	5
2	2	3	4	5	6
3	3	4	5	6	7
4	4	5	6	7	8

To subtract, do the reverse. Choose a number in the table. The difference between that number and the row number is the column number.
$3 - 1 - 2$

Or, choose a number in the table. The difference between that number and the column number is the row number.
$3 - 2 = 1$

Subtracting with Whole Numbers 129

If you know how to subtract tens and ones, you know how to subtract the digits in *any* place. That's because of place value. When you line numbers up and subtract them place by place, what you do in each place is the same.

$7 - 3 = 4$	7 ones − 3 ones = 4 ones
$70 - 30 = 40$	7 tens − 3 tens = 4 tens
$700 - 300 = 400$	7 hundreds − 3 hundreds = 4 hundreds
$7000 - 3000 = 4000$	7 thousands − 3 thousands = 4 thousands

130 Subtracting Without Regrouping

To subtract whole numbers, begin at the ones place. Subtract one place at a time. If you are writing the two numbers yourself, line them up at the ones place.

EXAMPLE: Your class raised \$289 for a field trip. The bus will cost \$75. How much money will be left for the rest of the trip?

To solve the problem, subtract 75 from 289.

ONE WAY You can subtract this way.

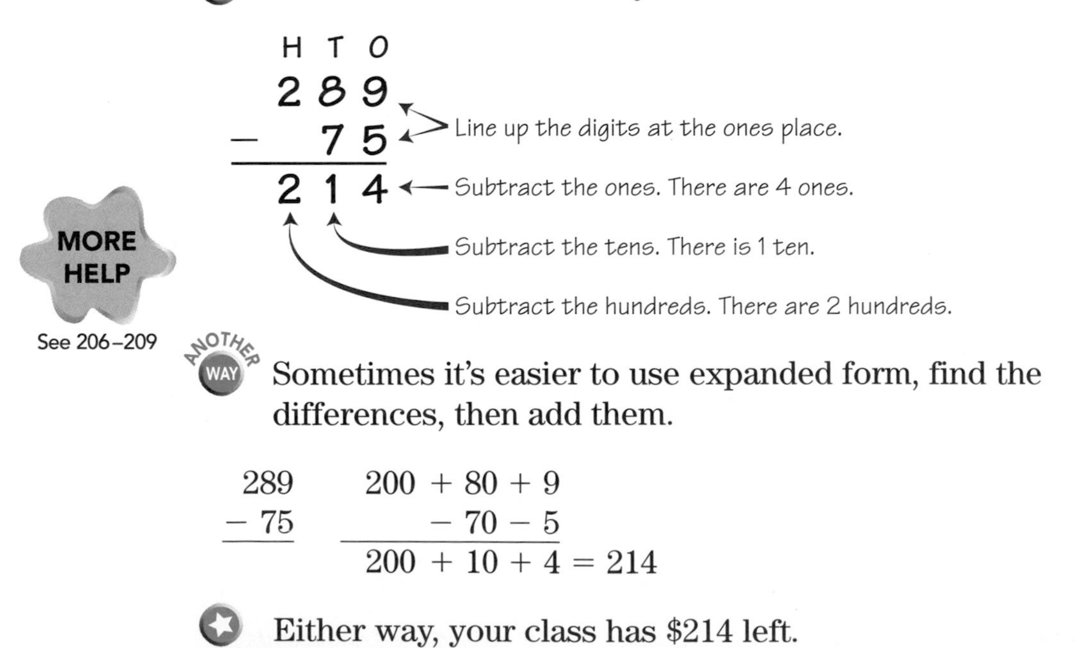

MORE HELP

See 206–209

ANOTHER WAY Sometimes it's easier to use expanded form, find the differences, then add them.

$$\begin{array}{r} 289 \\ -\ 75 \\ \hline \end{array} \qquad \begin{array}{r} 200 + 80 + 9 \\ -\ 70 - 5 \\ \hline 200 + 10 + 4 = 214 \end{array}$$

Either way, your class has \$214 left.

Math Alert Regrouping Numbers to Subtract

You can regroup to help you subtract.

EXAMPLE: Regroup 327 so you can subtract 5 tens.

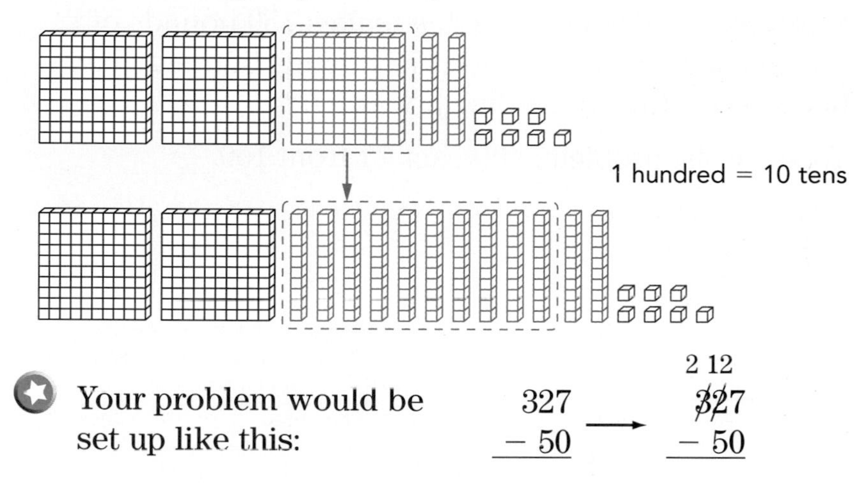

Your problem would be set up like this:

$$\begin{array}{r} 327 \\ -\ 50 \\ \hline \end{array} \longrightarrow \begin{array}{r} {\scriptstyle 2\,12} \\ \not{3}\not{2}7 \\ -\ 50 \\ \hline \end{array}$$

Subtracting with Regrouping

132

Sometimes it looks like you need to subtract a number that's too big. When that happens, you need to regroup.

EXAMPLE: The school year has 180 days. Today is the 129th day of school. How many days are left?

To solve the problem, subtract 129 from 180.

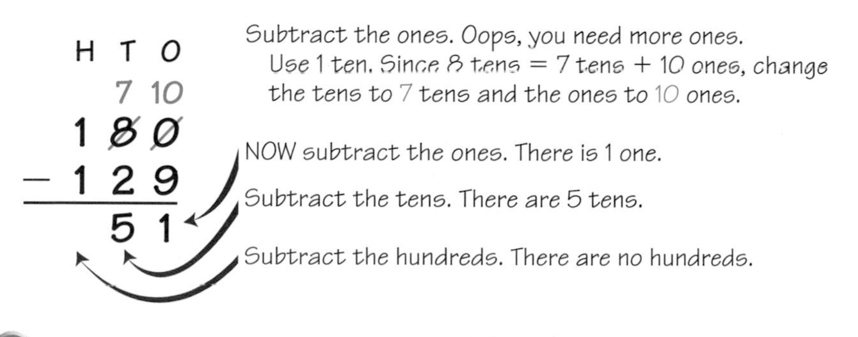

There are 51 school days left.

133 MATH ALERT Tricky Regrouping

CASE 1 Sometimes you have to regroup more than once while you subtract.

EXAMPLE 1: The Pizza Palace makes 150 pounds of dough each day. By 2:00 P.M., 77 pounds of dough had been used. How much dough was left?

To solve the problem, subtract 77 from 150.

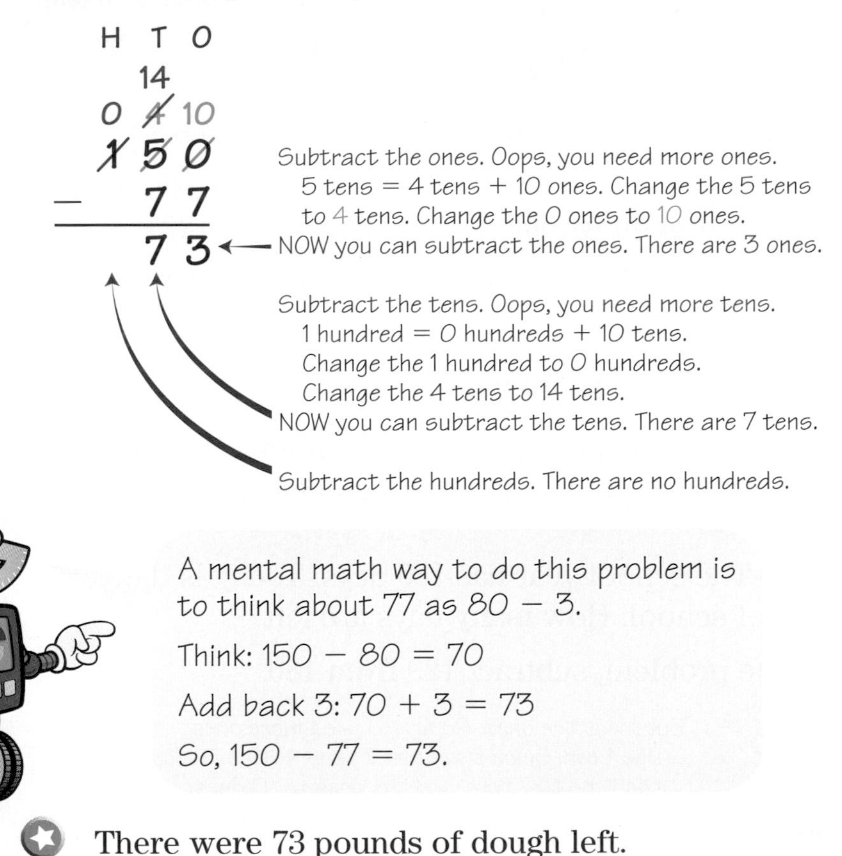

There were 73 pounds of dough left.

CASE 2 Sometimes you have to regroup more than once before you can begin subtracting.

EXAMPLE 2: The Pizza Palace ordered 2000 pizza boxes from the printer. The boxes are shipped flat. So far, your cousin Vince has folded 1287 boxes. How many boxes are left to be folded?

To solve the problem, subtract 1287 from 2000.

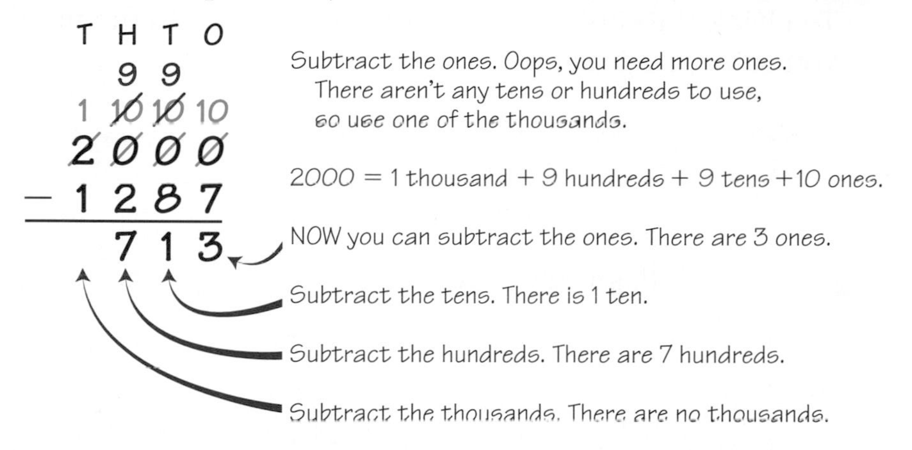

There are 713 pizza boxes left to be folded.

134 Checking Subtraction

It is always a good idea to check your answers.

See 119–123

One way to check your subtraction is to use addition. You can do this because addition is the opposite, or inverse, of subtraction.

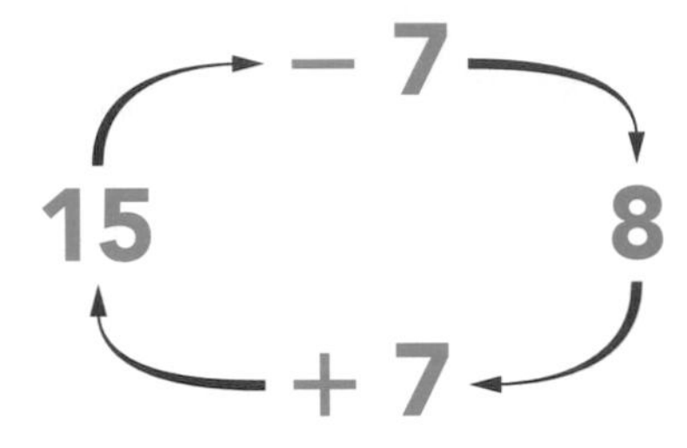

EXAMPLE 1: Does $528 - 371 = 157$? Use addition to check.

To check the subtraction, add the difference to the number you subtracted. Then compare.

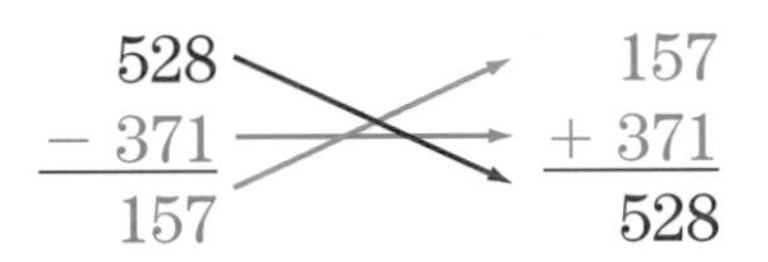

$528 = 528$, so $528 - 371 = 157$.

EXAMPLE 2: Does $5005 - 348 = 3657$? Use addition to check.

To check the subtraction, add the difference to the number you subtracted. Then compare.

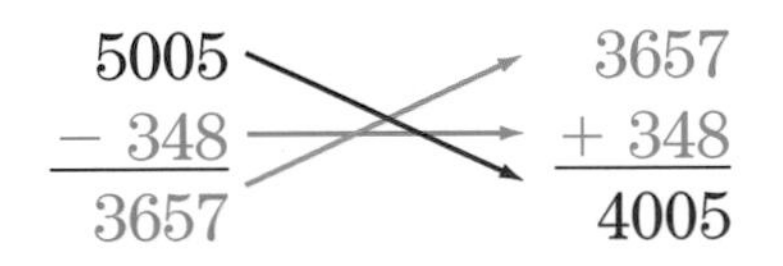

Hey, I can just add from the bottom up!

5005 does not equal 4005,
so $5005 - 348$ does not equal 3657.
You should do the subtraction again.

Subtracting with Decimals 135

If you know how to subtract whole numbers, you also know how to subtract decimals. Be sure to line up the ones digits. This means the decimal points line up, too.

See 012, 013, 126, 129–134

With both decimals and whole numbers, there are a few things to remember.

- Pay attention to place value.
- Subtract in each place.
- Regroup when you have to.

EXAMPLE: Subtract 5.4 from 6.26.

Write both numbers as two-place decimals. Line up the decimal points and tack on zeros in empty places.

O		T	H
5		12	
~~6~~	.	~~2~~	6
− 5	.	4	0
0	.	8	6

Line up the decimal points. Write the decimal point in the difference.

Write both numbers as two-place decimals.

Subtract the hundredths. There are 6 hundredths.

Try to subtract the tenths. You need more tenths. Since 1 one = 10 tenths, change the number of ones from 6 to 5. Change the number of tenths from 2 to 12.

NOW subtract the tenths. There are 8 tenths.

Subtract the ones. There are 0 ones.

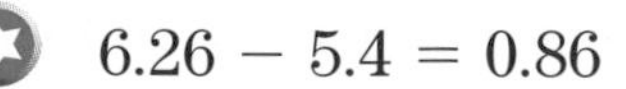

$6.26 - 5.4 = 0.86$

136

Multiplication

Multiplication is a shortcut for adding same-size groups. Imagine that today you are finally going to recycle the cans you have been collecting. There are 869 cans waiting to be turned in for 5¢ each. You certainly don't want to add 5¢ 869 times, so you're really glad you know how to multiply.

Two terms you may hear used in multiplication are **factor** and **product.**

$$\begin{array}{r l} 9 & \longleftarrow \text{factor} \\ \times\ 5 & \longleftarrow \text{factor} \\ \hline 45 & \longleftarrow \text{product} \end{array}$$

137

Multiplying with Whole Numbers

If you know how to multiply 1-digit numbers such as 6×7, you can also multiply larger numbers such as 6×777. That's because you only need to multiply one digit at a time. Each product is called a **partial product**. Just find all the partial products and add them.

138

Multiplying by a 1-Digit Number

There are two ways to multiply a factor greater than 10 by a 1-digit factor. You can list all the partial products and then add. Or, you can use what you know about regrouping.

EXAMPLE: A gross equals 144. How many pencils are in 6 gross?

To solve the problem, multiply 144 by 6.

See 086

ONE WAY You can multiply by listing all the partial products, then adding.

```
   H T O
   1 4 4
×      6
---------
     2 4  ← Multiply the ones. 6 × 4 ones = 24
   2 4 0  ← Multiply the tens. 6 × 4 tens = 240
   6 0 0  ← Multiply the hundreds. 6 × 1 hundred = 600
---------
   8 6 4  ← Add the partial products.
```

ANOTHER WAY You can also multiply without listing the partial products.

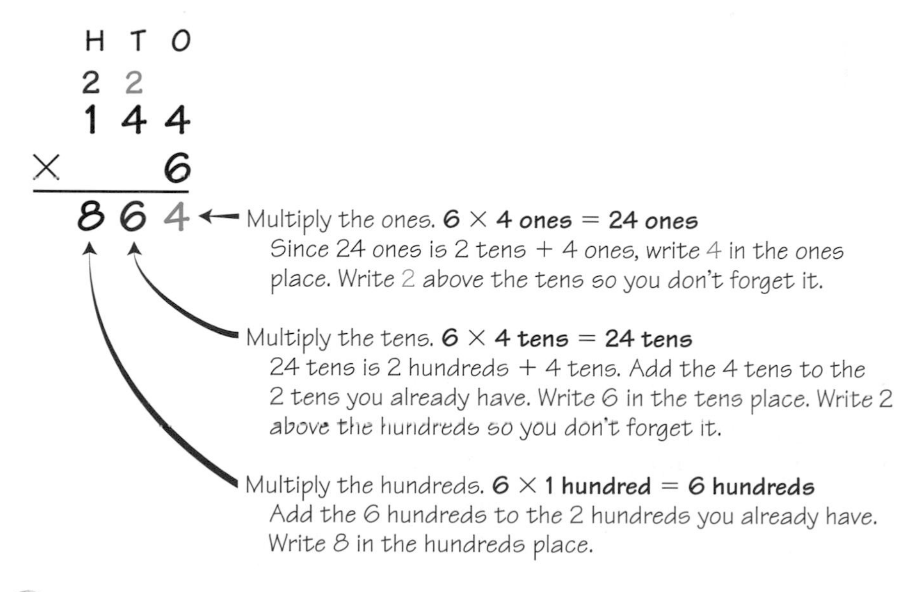

Either way, there are 864 pencils in 6 gross.

139 Two Ways to Multiply Whole Numbers

See 086, 138

EXAMPLE: The school auditorium has 14 rows. Each row has 26 seats. How many seats are in the auditorium?

To solve the problem, multiply 14 by 26.

You can multiply the value of each digit in one factor by the value of each digit in the other factor. List all the partial products and add.

```
   T O
   1 4
 × 2 6     Multiply by the ones.
   2 4       6 × 4 = 24
   6 0       6 × 10 = 60
   8 0     Multiply by the tens.
 2 0 0       20 × 4 = 80
             20 × 10 = 200
 3 6 4     Add the partial products.
```

You can also multiply without listing every partial product.

```
  T O
  2
  1 4
× 2 6
-----
  8 4
```

Multiply by the ones: 6 × 14 = ■
6 × 4 = 24 → 4 ones with 2 tens to regroup
6 × 10 = 60 → 6 tens + 2 tens = 8 tens
So, 6 × 14 = 84

MORE HELP

See 086

```
  2
  1 4
× 2 6
-----
  8 4
2 8 0
-----
3 6 4
```

Multiply by the tens: 20 × 14 = ■
20 × 4 = 80 → 8 tens + 0 ones
20 × 10 = 200 → 2 hundreds
So, 20 × 14 = 280

Add the partial products.

Did you notice that multiplying by 20 is just like multiplying by 2 and tacking on a zero?
2 × 14 = 28
20 × 14 = 280

You can also break apart one of the factors before multiplying.

MORE HELP

See 224–225

1 Break apart one factor into numbers that are easy to multiply.	14 × 26 = (10 + 4) × 26
2 Multiply.	10 × 26 = 260 4 × 26 = 104
3 Add the two products.	260 +104 364

Any way you look at it, there are 364 seats in the auditorium.

140 MATH ALERT Zeros in Factors and Products

Zeros may seem like nothing, but they are important.

EXAMPLE: The web site www.greatsource.com receives an average of 405 visits per week. At this rate, about how many visits would the web site receive in 4 weeks? *(Source: Great Source Education Group)*

To find the answer, you can multiply 405 by 4.

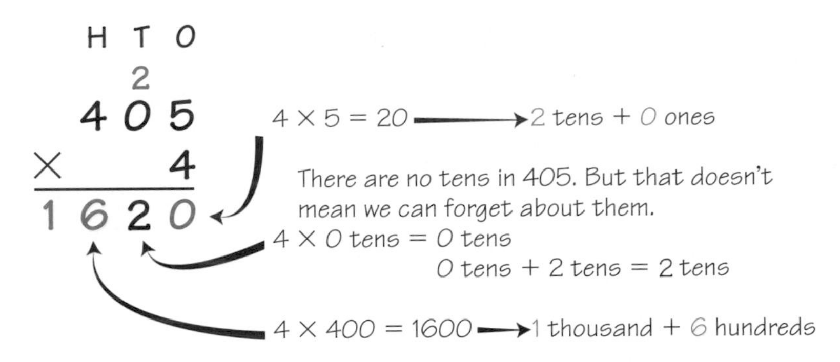

At this rate, the web site would receive about 1620 visits in a month.

141 Checking Multiplication

It's always a good idea to check your work.

EXAMPLE: Does $23 \times 48 = 1104$?

ONE WAY You can check multiplication by reversing the factors.

$$\begin{array}{r} 23 \\ \times 48 \\ \hline 1104 \end{array} \qquad \begin{array}{r} 48 \\ \times 23 \\ \hline 1104 \end{array}$$

If reversing the factors gives the same product, the multiplication is correct.

If reversing the factors does not give the same product, one of the products is not correct.

MORE HELP

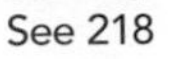

See 218

ANOTHER WAY You can use the lattice method.

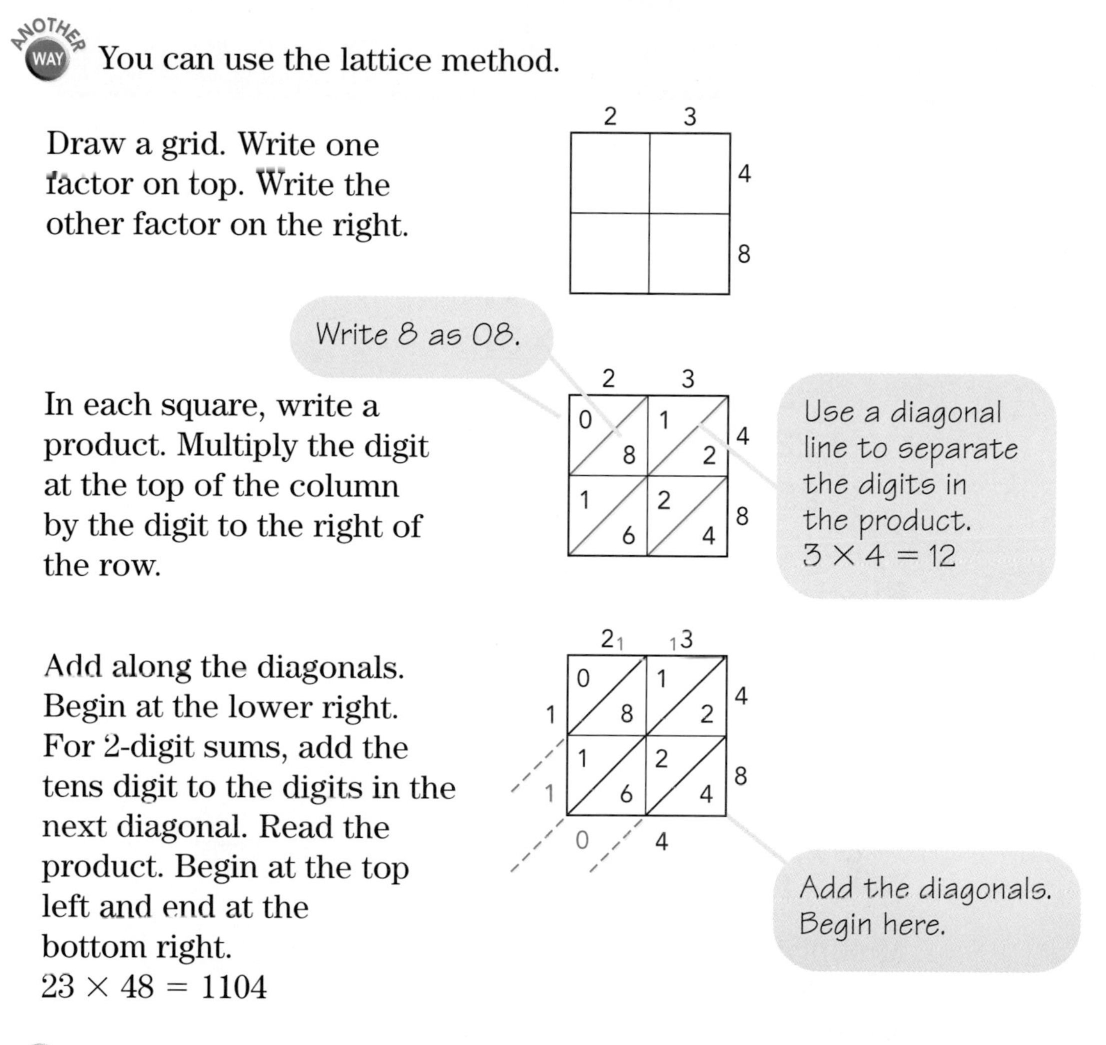

Draw a grid. Write one factor on top. Write the other factor on the right.

In each square, write a product. Multiply the digit at the top of the column by the digit to the right of the row.

Add along the diagonals. Begin at the lower right. For 2-digit sums, add the tens digit to the digits in the next diagonal. Read the product. Begin at the top left and end at the bottom right.
$23 \times 48 = 1104$

Either way, $23 \times 48 = 1104$. The product is correct.

142 Multiplying with Decimals

You multiply decimals the same way you multiply whole numbers. The only difference is that you must correctly place a decimal point in the product.

$$
\begin{array}{r}
74 \\
\times 35 \\
\hline
370 \\
+2220 \\
\hline
2590.
\end{array}
\qquad
\begin{array}{r}
7.4 \\
\times 3.5 \\
\hline
370 \\
+2220 \\
\hline
25.90
\end{array}
$$

The digits in the problems are the same. The steps are the same. But the place values are different because of the decimal points.

143 Placing the Decimal Point in the Product

ONE WAY Decide where to place the decimal point in the product by using an estimate.

MORE HELP
See 139

EXAMPLE 1: Trail mix costs \$2.49 per pound. How much will 1.3 pounds cost?

Multiplying 2.49 by 1.3 will help solve the problem.

❶ Begin by making an estimate.	❷ Multiply as if the factors were whole numbers.	❸ Use estimation to place the decimal point.
1.3 × 2.49 ↓ 1 × 2 = 2 So, the product will be close to 2.	249 × 13 747 2490 3237	3.237 The estimate was about 2, so the decimal point should go after the first 3.

Since the answer needs to be in dollars and cents, round 3.237 to the nearest cent. The cost of the trail mix will be \$3.24.

Think of the decimals as fractions. Multiply the fractions and write the product as a decimal.

See 019, 169

EXAMPLE 2: Multiply 0.04 by 0.7.

❶ Rename the decimals as fractions.	❷ Multiply the fractions.	❸ Write the product as a decimal.
$0.04 = \frac{4}{100}$ $0.7 = \frac{7}{10}$	$\frac{4}{100} \times \frac{7}{10} = \frac{28}{1000}$	$\frac{28}{1000} = 0.028$

$0.04 \times 0.7 = 0.028$

SHORTCUT

Count the places to the right of the decimal point in *both* factors. This will be the number of places to the right of the decimal point in the product.

EXAMPLE 3: Multiply 0.16 by 0.04.

❶ Count the decimal places in both factors.	❷ Multiply as if the factors were whole numbers.	❸ Count from the right of your product to place the decimal point.
0.16 — 2 decimal places 0.04 — + 2 decimal places 4 decimal places	$\begin{array}{r} 16 \\ \times\ 4 \\ \hline 64 \end{array}$	0.0064 4 3 2 1 Tack on zeros between the decimal point and your product if you need to.

$0.16 \times 0.04 = 0.0064$

144

Division

Division involves equal groups. Terms you may hear used in division are **dividend**, **divisor**, and **quotient**.

$$\text{divisor} \rightarrow 2\overline{)16} \leftarrow \text{dividend}, \quad 8 \leftarrow \text{quotient}$$

There are two reasons to divide.

CASE 1 When you know the original amount and the number of shares, you divide to find the **size of each share**.

EXAMPLE 1: A pizza is cut into 8 pieces. Four people want equal shares. How many pieces will each person get?

You know
- original amount (8)
- number of shares (4)

You need to know
- size of one share ($8 \div 4 = 2$)

Each person will get 2 pieces of pizza.

CASE 2 When you know the original amount and the size of one share, you divide to find the **number of shares**.

EXAMPLE 2: A sandwich uses 2 slices of bread. How many sandwiches can you make with 16 slices of bread?

You know
- original amount (16)
- size of one share (2)

You need to know
- how many shares ($16 \div 2 = 8$)

You can make 8 sandwiches.

Relating Multiplication and Division 145

Multiplication is the opposite, or inverse, of division. That means division is the inverse of multiplication.

To multiply with the table, choose one number from the beginning of a row and one number from the top of a column. The product will be where the row and column meet. $3 \times 2 = 6$

$\times$	1	2	3	4
1	1	2	3	4
2	2	4	6	8
3	3	6	9	12
4	4	8	12	16

To divide, choose a number in the table. The quotient of that number and the row number is the column number. $6 \div 3 = 2$

Or, choose a number in the table. The quotient of that number and the column number is the row number. $6 \div 2 = 3$

Dividing with Whole Numbers 146

You use multiplication and subtraction to help you with division examples like these.

$$\begin{array}{r} 4 \\ 3\overline{)12} \\ \underline{-12} \\ 0 \end{array} \qquad 4 \times 3 = 12$$

$$\begin{array}{r} 5 \text{ R2} \\ 6\overline{)32} \\ \underline{-30} \\ 2 \end{array} \qquad 5 \times 6 = 30$$

You can also use multiplication and subtraction to help with *long division* examples like these.

$$\begin{array}{r} 15 \\ 5\overline{)75} \\ \underline{-50} \\ 25 \\ \underline{-25} \\ 0 \end{array}$$

$$\begin{array}{r} 124 \text{ R2} \\ 6\overline{)746} \\ \underline{-600} \\ 146 \\ \underline{-120} \\ 26 \\ \underline{-24} \\ 2 \end{array}$$

Now that's what I call *long division*!

147 Division Without Remainders

Here are some ways to divide by a 1-digit number.

EXAMPLE: You have 75 Beanie Babies®. You want to divide them equally among 5 boxes. How many will be in each box?

You want to put 75 toys into 5 equal groups. To solve this problem, you divide 75 by 5.

ONE WAY You can use models to show the division.

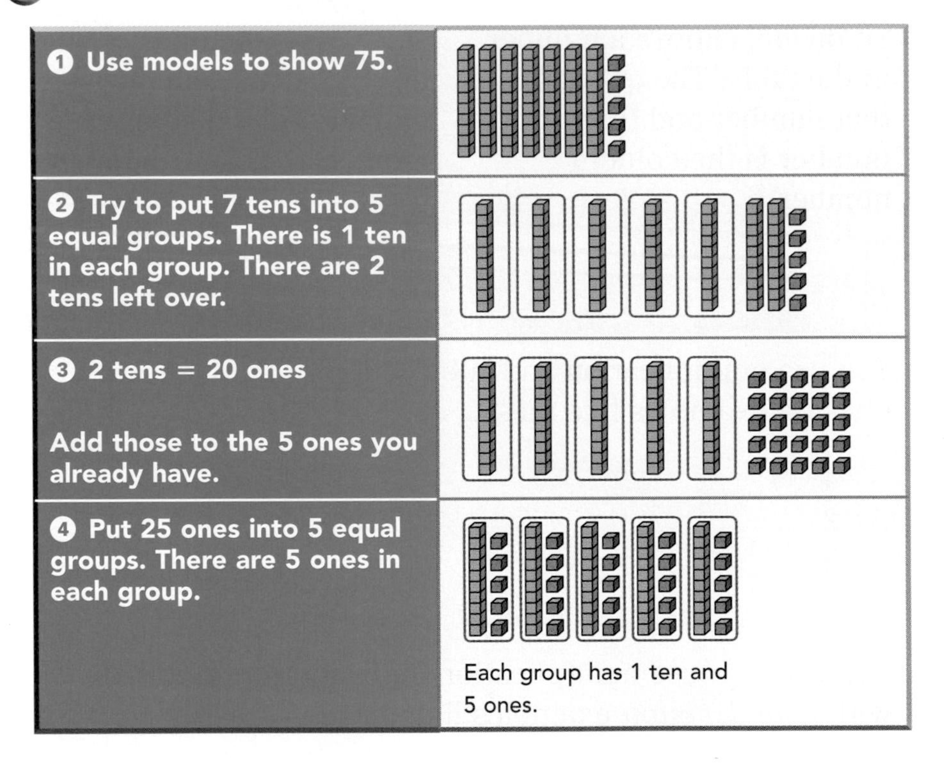

You can divide 75 by 5 without using models.
Think about multiplication.

$75 = 5 \times \blacksquare$ is the same as $5\overline{)75}$ with $\blacksquare$ as the quotient.

1 Divide the tens. $70 \div 5 = \blacksquare$ **Multiply to estimate.** 5×1 ten $= 50$ $5 \times \blacksquare$ tens $= 70$ 5×2 tens $= 100$ 2 tens is too much. Use 5×1 ten. Write 1 in the tens place. Write 50 below 75. **Subtract and compare.** $75 - 50 = 25$ There are 25 ones remaining. $25 > 5$ You have enough to keep dividing.	1 $5\overline{)75}$ -50 ← 5×1 ten 25 ← $75 - 50$
2 Divide the ones. $25 \div 5 = \blacksquare$ **Multiply to estimate.** $5 \times \blacksquare$ ones $= 25$ 5×5 ones $= 25$ Write 5 in the ones place. Write 25 under 25. **Subtract and compare.** $25 - 25 = 0$ There is no remainder. You are finished dividing.	15 $5\overline{)75}$ -50 25 -25 ← 5×5 ones 0 ← $25 - 25$

Either way, there will be 15 Beanie Babies® in each box.

148

Division with Remainders

Sometimes when you try to make equal groups, you can't do it without having some extra or left over. The part left over is called a **remainder**.

EXAMPLE: 74 Girl Scouts are planning a camping trip. Each tent has enough room for 6 people. How many tents will they need?

You want to put 74 people into equal groups of 6. To solve this problem, you can divide 74 by 6.

ONE WAY You can use models to show the division.

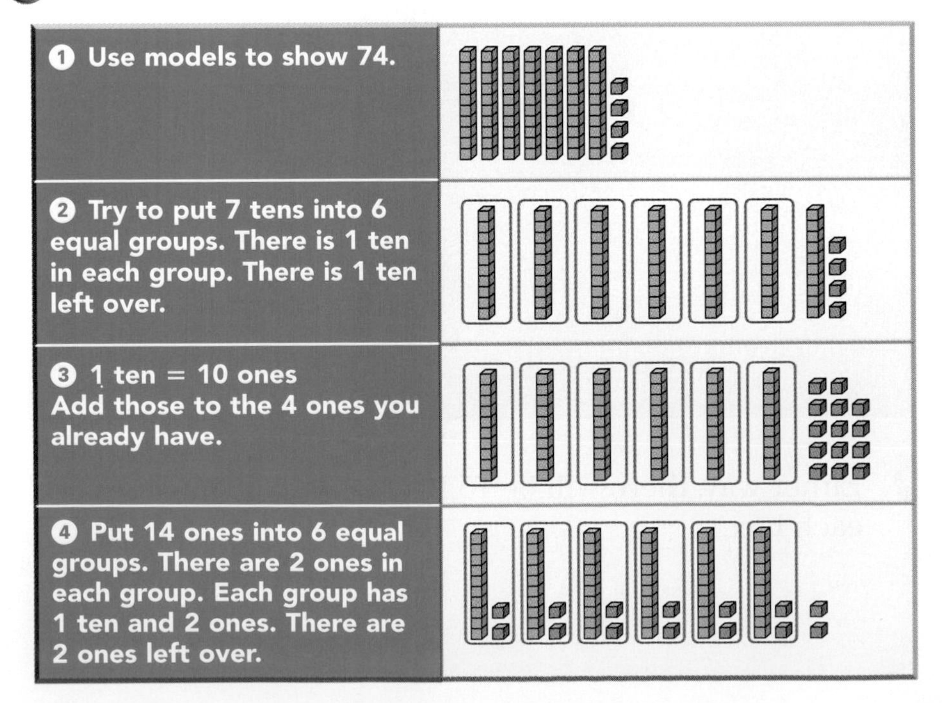

① Use models to show 74.	
② Try to put 7 tens into 6 equal groups. There is 1 ten in each group. There is 1 ten left over.	
③ 1 ten = 10 ones Add those to the 4 ones you already have.	
④ Put 14 ones into 6 equal groups. There are 2 ones in each group. Each group has 1 ten and 2 ones. There are 2 ones left over.	

You can divide 74 by 6 without using models.
Think about multiplication.

$74 = 6 \times \blacksquare$ is the same as $6\overline{)74}$.

1 Divide the tens. $70 \div 6 = \blacksquare$ **Multiply to estimate.** 6×1 ten $= 60$ $6 \times \blacksquare$ tens $= 70$ 6×2 tens $= 120$ 2 tens is too much. Use 6×1 ten. Write 1 in the tens place. Write 60 below 74. **Subtract and compare.** $74 - 60 = 14$ There are 14 ones remaining. $14 > 6$ Keep dividing.	$\begin{array}{r} 1 \\ 6\overline{)74} \\ -60 \\ \hline 14 \end{array}$ -60 ← 6×1 ten 14 ← $74 - 60$
2 Divide the ones. $14 \div 6 = \blacksquare$ **Multiply to estimate.** 6×2 ones $= 12$ $6 \times \blacksquare$ ones $= 14$ 6×3 ones $= 18$ 3 ones is too much. Use 6×2 ones. Write 2 in the ones place. Write 12 below 14. **Subtract and compare.** $14 - 12 = 2$ $2 < 6$ You are finished dividing. The remainder is 2.	$\begin{array}{r} 12 \text{ R2} \\ 6\overline{)74} \\ -60 \\ \hline 14 \\ -12 \\ \hline 2 \end{array}$ -12 ← 6×2 ones 2 ← $14 - 12$

Either way, the Girl Scouts can fill 12 tents and will need another tent for the 2 remaining scouts. They will need 13 tents.

149 Interpreting Quotients and Remainders

When you divide, sometimes you get a remainder. What does the remainder mean? Here are four ways of thinking about remainders.

CASE 1 Ignore the remainder.

EXAMPLE 1: You have 52 ounces of jelly beans. How many 8-ounce bags can you fill?

$$\begin{array}{r} 6\text{ R}4 \\ 8\overline{)52} \\ -48 \\ \hline 4 \end{array}$$

The answer shows that you can fill 6 bags. The remainder, 4, tells you about the weight of the jelly beans in the bag that is not full. You do not need the remainder to answer the question, you only need to know how many 8-ounce bags you can fill.

★ You can fill six 8-ounce bags.

CASE 2 The answer is the next greater whole number.

EXAMPLE 2: On the Shoot-the-Rapids flume ride, each boat holds 6 people. Thirty-four people are in line. How many boats will they need?

$$\begin{array}{r} 5\text{ R}4 \\ 6\overline{)34} \\ -30 \\ \hline 4 \end{array}$$

The answer, 5 R4, shows that you cannot fit all the people on 5 boats. So, the answer is the next whole number, 6.

★ They will need 6 boats.

CASE 3 Use the remainder as the answer.

EXAMPLE 3: You have 217 action hero trading cards. You share the cards equally among 5 friends. You give them as many cards as you can. How many cards do you have left?

$$\begin{array}{r} 43\text{ R}2 \\ 5\overline{)217} \\ -\underline{200} \\ 17 \\ -\underline{15} \\ 2 \end{array}$$

The remainder, 2, shows the number of cards you have left.

You have 2 cards left.

CASE 4 Write the remainder as a fraction.

EXAMPLE 4: Computer keyboarders are needed to input a large book for a publisher. The entire job should take 65 hours. If 4 keyboarders share the job equally, how long will each one work on the book?

$$\begin{array}{r} 16\text{ R}1 \\ 4\overline{)65} \\ -\underline{40} \\ 25 \\ -\underline{24} \\ 1 \end{array}$$

The answer, 16 R1, tells you that each keyboarder will work 16 hours and there will be an extra hour to share.To write the remainder as a fraction, write the remainder over the divisor.

$$\frac{\text{remainder} \longrightarrow 1}{\text{divisor} \longrightarrow 4}$$

They will each work $16\frac{1}{4}$ hours.

MATH ALERT Zeros in the Quotient

MORE HELP
See 004, 006–008, 331

Sometimes you need to write a 0 in the quotient to show that there is nothing in that place. Pay attention to place value and your estimates, and you won't accidentally forget a zero.

EXAMPLE: The area of a kindergarten classroom is 972 square feet. How many square yards is that? (There are 9 square feet in a square yard.)

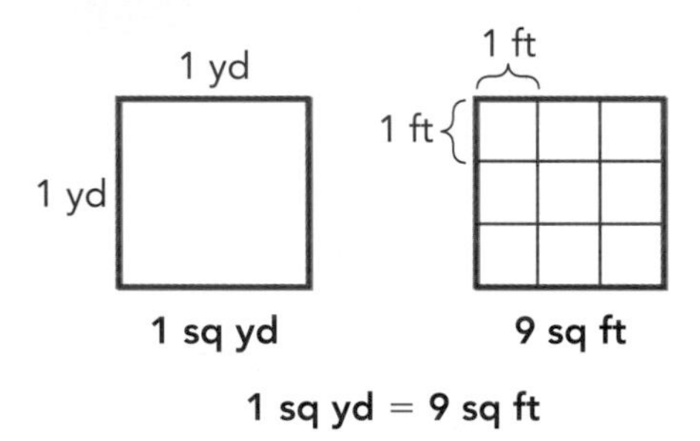

1 sq yd = 9 sq ft

To solve the problem, divide 972 by 9.

Estimate first.
$972 \div 9 \longrightarrow 900 \div 9 = 100$
The answer should be close to 100 square yards.

1 Divide the hundreds. 900 ÷ 9 = ■ **Multiply to estimate.** 9 × ■ hundreds = 900 9 × 1 hundred = 900 Use 9 × 1 hundred. Write 1 in the hundreds place. Write 900 under 972. **Subtract and compare.** 972 − 900 = 72 72 < 9 Keep dividing.	1 9)972 ← 9 × 1 hundred − 900 72 ← 972 − 900
2 Divide the tens. 70 ÷ 9 = ■ **Multiply to estimate.** 9 × 0 tens = 0 9 × ■ tens = 70 9 × 1 ten = 90 1 ten is too much. Use 9 × 0 tens. Write 0 in the tens place.	10 9)972 − 900 72
3 Divide the ones. 72 ÷ 9 = ■ **Multiply to estimate.** 9 × ■ ones = 72 9 × 8 ones = 72 Write 8 in the ones place. Write 72 under 72. **Subtract and compare.** 72 − 72 = 0 There is no remainder. You are finished dividing.	108 9)972 − 900 72 − 72 0

When 7 tens are divided into 9 equal groups, there are 0 tens in each group. You must write the zero.

When you are dividing ones, think of this as 72 ones.

The area of the classroom is 108 square yards. The answer is close to the estimate, 100 square yards.

Another Way to Divide

Here is another way to find a quotient. With this method, you can make different estimates, but you will still get the same answer! Choose estimates that are easy for you to multiply.

MORE HELP

See 112, 226

EXAMPLE: Divide 720 by 30.

❶ Estimate the quotient.	30)720	20
❷ Multiply your estimate by the divisor. If the product is less than the dividend, subtract. If the product is greater than the dividend, try a lower estimate.	30)720	20 −600 120
❸ Keep estimating, multiplying, and subtracting until the difference is less than the divisor. Add your estimates to get the whole number part of the quotient. The remainder is the final difference.	30)720	20 −600 120	2 −60 60	2 −60 0

The quotient is 20 + 2 + 2 = 24. There is no remainder.

152

Checking Division by Multiplying

Checking helps you make sure that you correctly recorded each step in the division, or that you hit the right calculator keys. One way to check your division is to use multiplication. You can do this because multiplication is the opposite, or inverse, of division.

EXAMPLE: Is 28 R3 a correct answer for 143 ÷ 5?

Multiply the whole number part of the quotient by the divisor and add the remainder. The result should be the dividend.

See 137–140

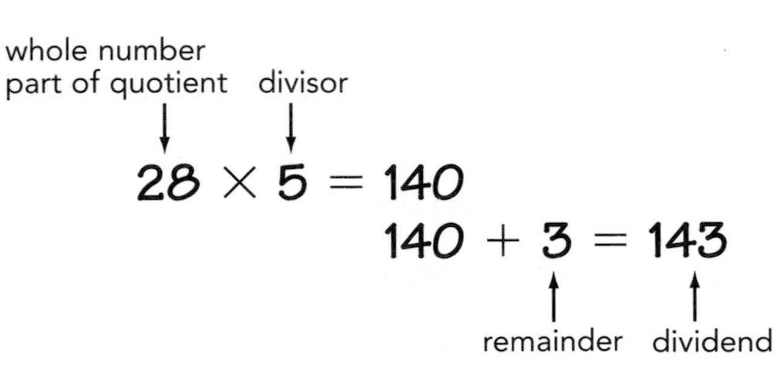

143 = 143, so the division is correct.

Dividing with Decimals 153

You divide decimals the same way you divide whole numbers. The only difference is that you place a decimal point in the quotient. Compare these two division problems.

$$\begin{array}{r} 26 \\ 3\overline{)78} \\ -60 \\ \hline 18 \\ -18 \\ \hline 0 \end{array} \qquad \begin{array}{r} 2.6 \\ 3\overline{)7.8} \\ -6\,0 \\ \hline 1\,8 \\ -1\,8 \\ \hline 0 \end{array}$$

The digits in the problems are the same. The steps are the same. But the place values are different because of the decimal points.

Dividing Decimals by Whole Numbers

To divide a decimal by a whole number, place the decimal point for the quotient above the decimal point in the dividend. Then divide the way you divide whole numbers.

MORE HELP
See 043, 146, 483

EXAMPLE 1: You and three friends buy ice cream. The bill is $6.88. How much does each person pay?

To solve the problem, divide 6.88 by 4.

❶ Estimate the quotient.	❷ Write the decimal point in the quotient.	❸ Divide the same way you divide whole numbers.
$6.88 \div 4$ ↓ $7 \div 4 = 1\frac{3}{4}$ ↓ 1.75	$\require{enclose}\begin{array}{r} . \\ 4 \enclose{longdiv}{6.88} \end{array}$	$\begin{array}{r} 1.72 \\ 4 \enclose{longdiv}{6.88} \\ \underline{-400} \\ 288 \\ \underline{-280} \\ 8 \\ \underline{-8} \\ 0 \end{array}$

You each pay $1.72.

EXAMPLE 2: Divide 18 by 40.

❶ Estimate the quotient.	❷ Write the decimal point in the quotient.	❸ Divide the same way you divide whole numbers.
$18 \div 40$ ↓ $20 \div 40 = \frac{20}{40}$ ↓ 0.5	$\begin{array}{r} . \\ 40 \enclose{longdiv}{18.} \end{array}$ The decimal point in a whole number is after the ones place.	$\begin{array}{r} .45 \\ 40 \enclose{longdiv}{18.00} \\ \underline{-16\ 00} \\ 2\ 00 \\ \underline{-2\ 00} \\ 0 \end{array}$ Tack on zeros and keep dividing until there is no remainder or you can answer the question.

$18 \div 40 = 0.45$. This is close to your estimate.

Dividing by Decimals

See 146, 483

EXAMPLE: A guinea pig weighs 24 ounces. A pygmy shrew weighs 0.16 ounce. How many times heavier is a guinea pig than a pygmy shrew? *(Source: The Sizeaurus)*

To solve the problem, you can divide 24 by 0.16.

❶ Divide as if you had whole numbers.	❷ Estimate the quotient.	❸ Use the estimate to place the decimal point.
$16\overline{)2400}$ with quotient 150	$24 \div 0.16$ ↓ $25 \div 0.2 = 25 \div \frac{2}{10}$ ⟶ $25 \div \frac{1}{5}$ ⟶ $25 \times 5 = 125$	$24 \div 0.16 = 150.$ The estimate, 125, is close to 150, so place the decimal point after the 0.

You can count the number of places in the dividend. This will tell you what power of ten you can multiply by to move the decimal point out of the way.

See 085

❶ Count decimal places in the divisor.	❷ Multiply the divisor and dividend by a power of ten.	❸ Place the decimal in the quotient. Divide.
$0.16\overline{)24}$ (places counted: 2 1) 2 places in the divisor means multiply dividend and divisor by 10^2.	$0.16 \times 10^2 = 16$ $24 \times 10^2 = 2400$ Multiplying by 10^2 moves the decimal point 2 places right.	$16\overline{)2400}$ with quotient 150. -1600 800 -800 0

A guinea pig is 150 times heavier than the pygmy shrew.

156

Computing with Fractions

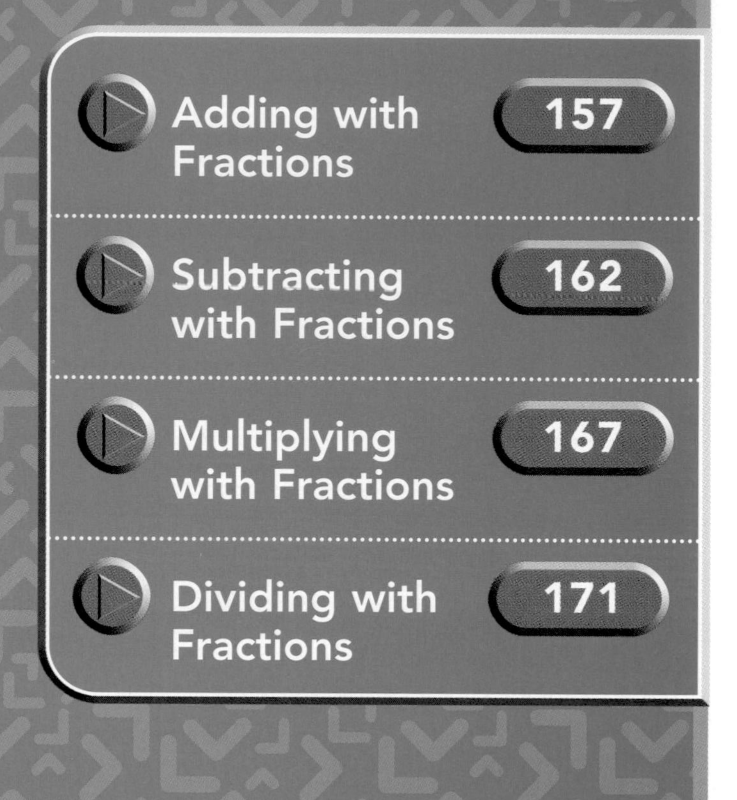

Basketballs are made for throwing, but televisions are not. You treat different things in different ways. Fractions are different from whole numbers, so it makes sense that you treat them differently. If you keep in mind what televisions are, then the rules for what you can and can't do with them make sense. And if you keep in mind what fractions are, then the rules for adding, subtracting, multiplying, and dividing them will make sense.

157

Adding with Fractions

Digits in fractions have different meanings than they do in whole numbers.

MORE HELP See 029

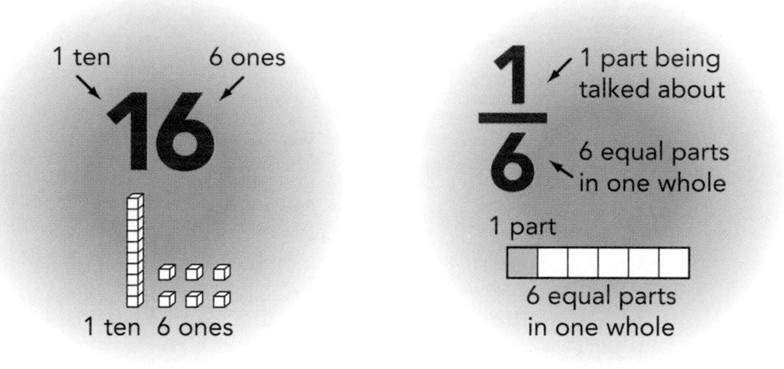

158

MATH ALERT Add ONLY the Numerators

Adding fractions is a lot like adding **bicycles**. When you add 5 **bicycles** to 2 **bicycles** you have 7 **bicycles**—the number of things changes (5 + 2 = 7), but the kind of thing (**bicycles**) doesn't.

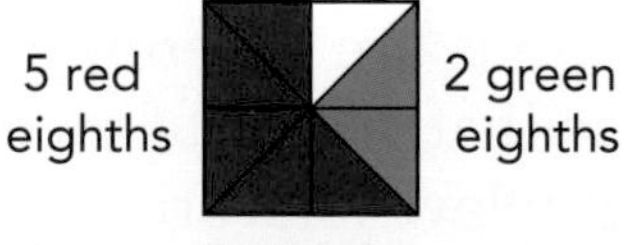

If you add 5 **eighths** and 2 **eighths**, you have 7 **eighths**—the number of things changes (5 + 2 = 7), but the kind of thing (**eighths**) doesn't. The denominators tell the kind of thing you're adding. If the denominators are different, use equivalent fractions to make the denominators the same.

Adding Like Fractions

159

When you add fractions, check the denominators.

If the fractions have like denominators, add the numerators. The denominator stays the same because you are adding the same kind of thing.

See 037, 051, 057–058

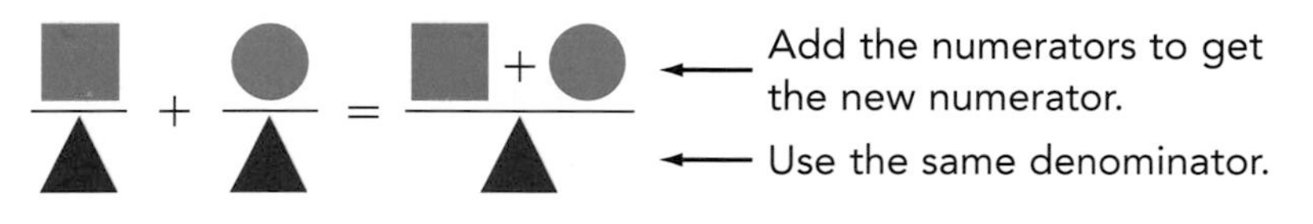

EXAMPLE: A cookie recipe calls for $\frac{3}{8}$ cup of brown sugar and $\frac{1}{8}$ cup of white sugar. If you put all the sugar in a 1-cup measure, how full would it be?

ONE WAY You can draw a picture to show $\frac{3}{8}$ and $\frac{1}{8}$.

To write a fraction in simplest form, divide the numerator and denominator by their greatest common factor.

ANOTHER WAY You can also add $\frac{3}{8}$ and $\frac{1}{8}$ without drawing a picture.

❶ The fractions have like denominators. Write that denominator.	❷ Add the numerators.	❸ Write the sum in simplest form.
$\frac{3}{8} + \frac{1}{8} = \frac{}{8}$	$\frac{3}{8} + \frac{1}{8} = \frac{4}{8}$	$\frac{4}{8} \longrightarrow \frac{4 \div 4}{8 \div 4} = \frac{1}{2}$

Either way, the sugar fills $\frac{4}{8}$ of the cup. This is the same measure as $\frac{1}{2}$ of the cup. $\frac{4}{8} = \frac{1}{2}$

160 Adding Unlike Fractions

Some fractions don't have like denominators. They are sometimes called **unlike fractions**. To add unlike fractions, rewrite them with like denominators.

EXAMPLE: Andrew baked a dozen (12) cookies. He put frosting on $\frac{1}{2}$ of the cookies and sprinkles on $\frac{1}{4}$ of the cookies. He left the rest plain. What fraction of Andrew's cookies are decorated?

You need to add $\frac{1}{2}$ and $\frac{1}{4}$.

ONE WAY You can draw a picture to show $\frac{1}{2} + \frac{1}{4}$.

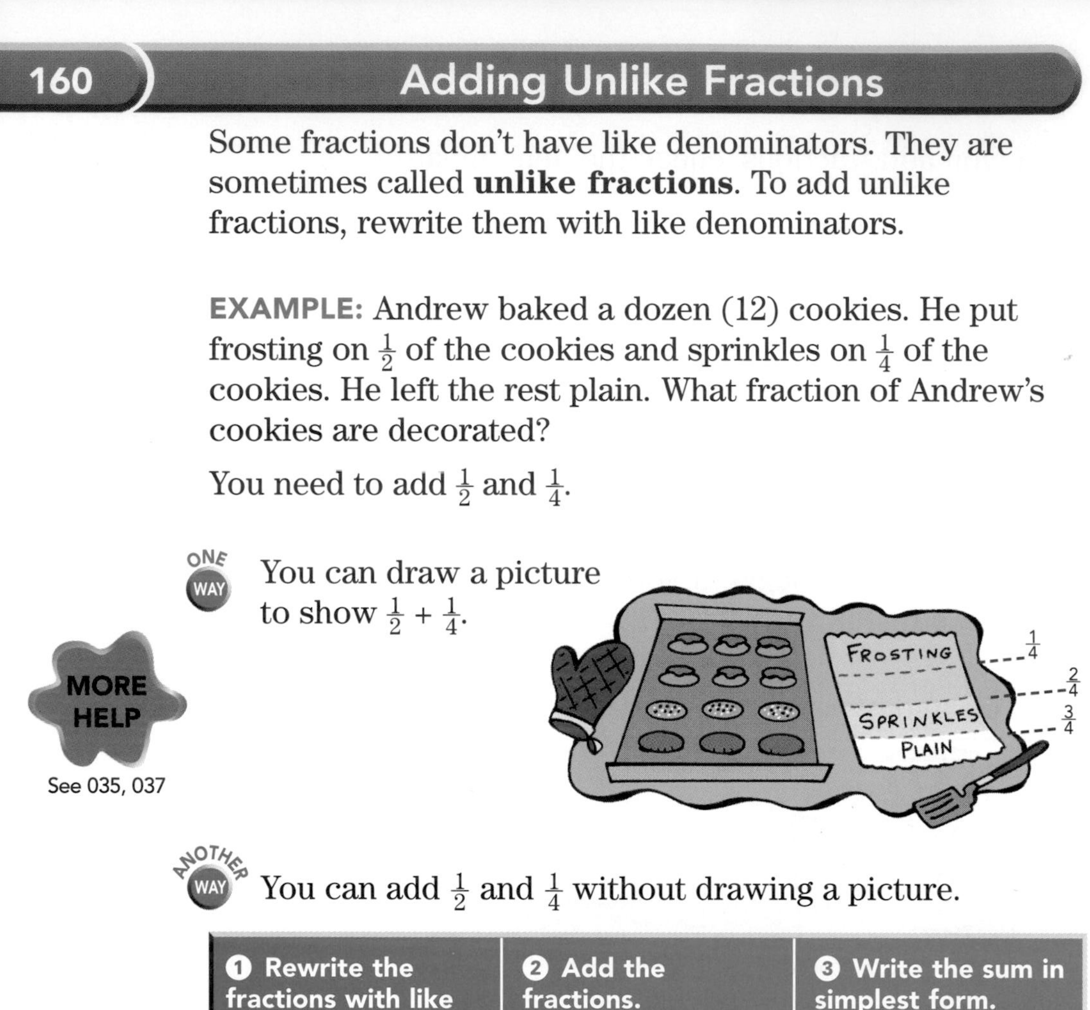

MORE HELP
See 035, 037

ANOTHER WAY You can add $\frac{1}{2}$ and $\frac{1}{4}$ without drawing a picture.

❶ Rewrite the fractions with like denominators.	❷ Add the fractions.	❸ Write the sum in simplest form.
$\frac{1}{2} \longrightarrow \frac{1 \times 2}{2 \times 2} = \frac{2}{4}$ $\frac{1}{4} = \frac{1}{4}$	$\frac{2}{4} + \frac{1}{4} = \frac{3}{4}$	$\frac{3}{4}$ is already in simplest form.

 Either way, $\frac{3}{4}$ of Andrew's cookies are decorated.

161 Adding Mixed Numbers

Mixed numbers have a whole-number part and a fraction part.

EXAMPLE: Lauren is going to Ana's house. She walks the first $1\frac{1}{2}$ miles. Then she jogs the rest of the way, $\frac{3}{4}$ mile. How far does Lauren live from Ana?

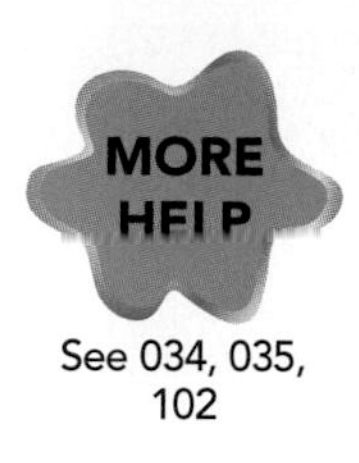

See 034, 035, 102

ONE WAY To solve the problem, add $1\frac{1}{2}$ and $\frac{3}{4}$.

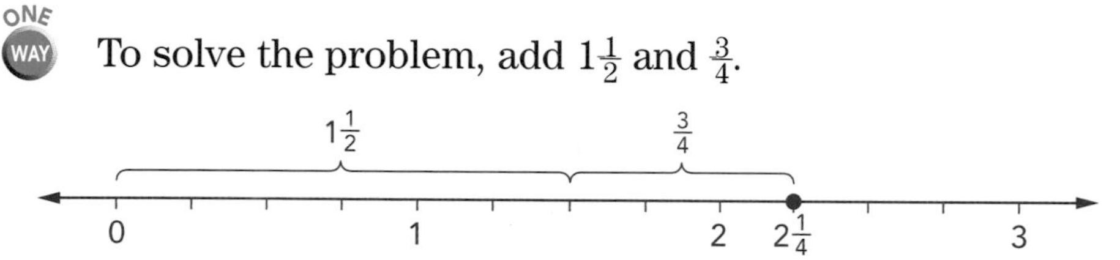

ANOTHER WAY You can rewrite the fractions with like denominators.

1 Rewrite the fractions so that thay have like denominators.	$\frac{1}{2} \rightarrow \frac{1 \times 2}{2 \times 2} = \frac{2}{4}$ $\frac{3}{4} = \frac{3}{4}$
2 Add the fractions. Write in simplest form, as a mixed number if you can.	$1\frac{2}{4} + \frac{3}{4} \rightarrow \frac{5}{4} = 1\frac{1}{4}$
3 Write the fraction part in your sum. Write the whole-number part with the other whole numbers so you don't forget it. Add the whole numbers.	$\overset{1}{1}\frac{2}{4} + \frac{3}{4} = 2\frac{1}{4}$

ANOTHER WAY You can rewrite the mixed numbers as fractions.

1 Rewrite the numbers as fractions with like denominators.	$1\frac{1}{2} \rightarrow \frac{3}{2} \rightarrow \frac{6}{4}$ $+\frac{3}{4} \rightarrow \frac{3}{4} \rightarrow \frac{3}{4}$
2 Add the fractions.	$\frac{6}{4} + \frac{3}{4} = \frac{9}{4}$
3 Write the sum in simplest form.	$\frac{9}{4}$ is in simplest form. It can also be written as $9 \div 4 = 2\frac{1}{4}$.

Any way you look at it, Lauren lives $2\frac{1}{4}$ miles from Ana.

162

Subtracting with Fractions

Digits in fractions have different meanings than they do in whole numbers.

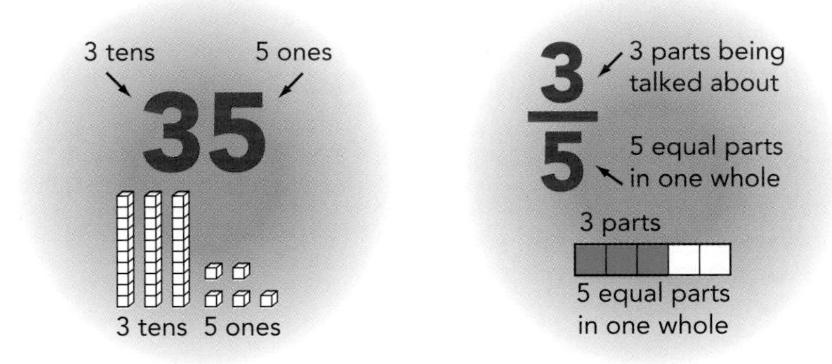

163

Math Alert Subtract ONLY the Numerators

If you subtract 3 **ninths** from 8 **ninths**, you have 5 **ninths**—the number of things changes ($8 - 3 = 5$), but the kind of thing (**ninths**) doesn't. The denominators tell the kind of thing you're subtracting. If the denominators are different, use equivalent fractions to make the denominators the same.

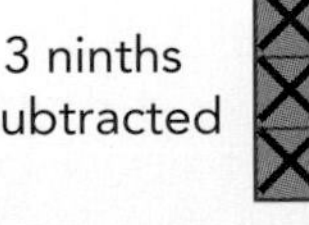

Subtracting Like Fractions 164

When you subtract fractions, check the denominators.

If the fractions have like denominators, subtract the numerators.

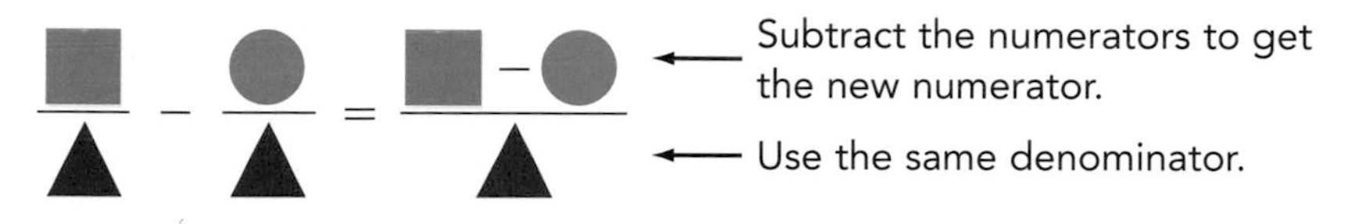

See 037, 057–058

EXAMPLE: A pan of lasagna was cut into 8 equal pieces. After dinner, $\frac{5}{8}$ of the lasagna was left. The next day after school, Marcia and her friends ate 3 more pieces, or $\frac{3}{8}$ of the original lasagna. What fraction of the lasagna was left in the pan?

ONE WAY To solve the problem, subtract $\frac{3}{8}$ from $\frac{5}{8}$.

You can draw a picture to show $\frac{5}{8} - \frac{3}{8}$.

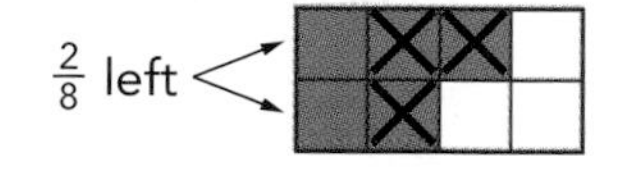

You can subtract $\frac{3}{8}$ from $\frac{5}{8}$ without drawing a picture.

❶ The fractions have like denominators. Write that denominator.	❷ Subtract the numerators.	❸ Write the difference in simplest form.
$\frac{5}{8} - \frac{3}{8} = \frac{}{8}$	$\frac{5}{8} - \frac{3}{8} = \frac{2}{8}$	$\frac{2}{8} \longrightarrow \frac{2 \div 2}{8 \div 2} = \frac{1}{4}$

Either way, $\frac{2}{8}$ of the lasagna was left. You could also say $\frac{1}{4}$ was left.

Subtracting Unlike Fractions

How can you subtract fractions that don't have like denominators? You can rewrite the fractions so they do have like denominators.

See 035, 037

EXAMPLE 1: Tom rides his bike straight home from school every day. He lives $\frac{3}{4}$ of a mile from school. Today after riding for $\frac{1}{2}$ mile, he stops at the candy store. How far is Tom from his house?

To solve the problem, subtract $\frac{1}{2}$ from $\frac{3}{4}$.

You can draw a picture to show $\frac{3}{4} - \frac{1}{2}$.

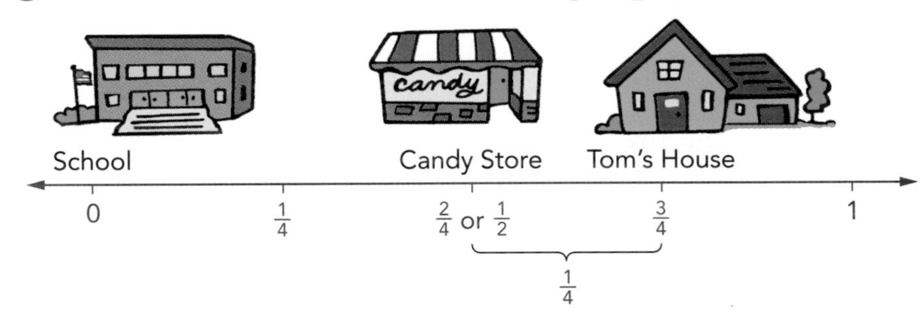

You can subtract $\frac{1}{2}$ from $\frac{3}{4}$ without drawing a picture.

❶ Rewrite the fractions with like denominators.	❷ Subtract the fractions.	❸ Write the difference in simplest form.
$\frac{1}{2} \longrightarrow \frac{1 \times 2}{2 \times 2} = \frac{2}{4}$ $\frac{3}{4} = \frac{3}{4}$	$\frac{3}{4} - \frac{2}{4} = \frac{1}{4}$	$\frac{1}{4}$ is already in simplest form.

Either way, Tom is $\frac{1}{4}$ mile from his house.

EXAMPLE 2: Subtract $\frac{3}{8}$ from $\frac{15}{16}$.

ONE WAY You can draw a picture to show $\frac{15}{16} - \frac{3}{8}$.

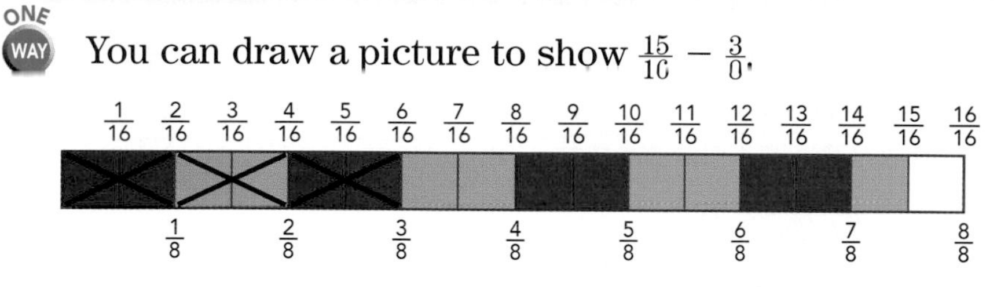

Start with $\frac{15}{16}$. Show $\frac{3}{8}$ crossed out. Count $\frac{9}{16}$ left.

You can subtract $\frac{3}{8}$ from $\frac{5}{16}$ without drawing a picture.

❶ Rewrite the fractions with like denominators.	❷ Subtract the fractions.	❸ Write the difference in simplest form.
$\frac{3}{8} \longrightarrow \frac{3 \times 2}{8 \times 2} = \frac{6}{16}$ $\frac{15}{16} = \frac{15}{16}$	$\frac{15}{16} - \frac{6}{16} = \frac{9}{16}$	$\frac{9}{16}$ is in simplest form.

Either way, $\frac{15}{16} - \frac{3}{8} = \frac{9}{16}$.

EXAMPLE 3: Subtract $\frac{5}{10}$ from $\frac{3}{4}$.

❶ Rewrite the fractions with like denominators.	❷ Subtract the fractions.	❸ Write the difference in simplest form.
$\frac{3}{4} \longrightarrow \frac{3 \times 5}{4 \times 5} = \frac{15}{20}$ $\frac{5}{10} \longrightarrow \frac{5 \times 2}{10 \times 2} = \frac{10}{20}$	$\frac{15}{20} - \frac{10}{20} = \frac{5}{20}$	$\frac{5}{20} \longrightarrow \frac{5 \div 5}{20 \div 5} = \frac{1}{4}$

$$\frac{3}{4} - \frac{5}{10} = \frac{1}{4}$$

Another equivalent fraction for $\frac{5}{10}$ is $\frac{2}{4}$. $\frac{5}{10}$ and $\frac{2}{4}$ are both the same as $\frac{1}{2}$.

166 Subtracting Mixed Numbers

Mixed numbers have a whole-number part and a fraction part. There are several ways you can subtract mixed numbers. Choose the way that's easiest for the numbers you are subtracting.

See 035, 102

EXAMPLE: You have just harvested the honey from your beehives. When it comes from the hive, the honey is in honeycombs. It weighs $4\frac{1}{8}$ pounds. After you separate the honey from the comb, you have $3\frac{1}{4}$ pounds of honey. How much did the comb weigh?

You need to subtract $3\frac{1}{4}$ from $4\frac{1}{8}$.

ONE WAY You can draw a picture to show $4\frac{1}{8} - 3\frac{1}{4}$.

❶ Show $4\frac{1}{8}$. Show $3\frac{1}{4}$.	
❷ Show what's left of the $4\frac{1}{8}$ after you remove $3\frac{1}{4}$.	$\frac{7}{8}$ $4\frac{1}{8}$ is seven eighths more than $3\frac{1}{4}$.

ANOTHER WAY You can rewrite the mixed numbers as fractions.

1 Rewrite the mixed numbers as fractions with like denominators.	$4\frac{1}{8} = 4 + \frac{1}{8}$ ↓ $\frac{32}{8} + \frac{1}{8} = \frac{33}{8}$ $3\frac{1}{4} = 3 + \frac{1}{4}$ ↓ $\frac{12}{4} + \frac{1}{4} = \frac{13 \times 2}{4 \times 2} = \frac{26}{8}$
2 Subtract.	$\frac{33}{8} - \frac{26}{8} = \frac{7}{8}$
3 Write the difference in simplest form.	$\frac{7}{8}$ is already in simplest form.

ANOTHER WAY Subtract the fractions, then subtract the whole numbers. Sometimes you will have to regroup.

1 Rewrite the fractions with like denominators.	$4\frac{1}{8} = 4\frac{1}{8}$ $3\frac{1}{4} \rightarrow 3 + \frac{1 \times 2}{4 \times 2} = 3\frac{2}{8}$
2 Decide whether you need to regroup.	$\frac{1}{8} < \frac{2}{8}$. You need to regroup. $4\frac{1}{8} \rightarrow 3 + \frac{8}{8} + \frac{1}{8} \rightarrow 3\frac{9}{8}$
3 Subtract the fractions. Subtract the whole numbers.	$3\frac{9}{8} - 3\frac{2}{8} = \frac{7}{8}$
4 Write the difference in simplest form.	$\frac{7}{8}$ is already in simplest form.

Any way you look at it, the honeycomb weighed $\frac{7}{8}$ of a pound.

167 Multiplying with Fractions

There are many ways to multiply with fractions. You can multiply fractions by fractions. You can multiply fractions by mixed numbers. Or you can multiply fractions by whole numbers. To understand what this kind of multiplication means, think about the meaning of multiplication and the meaning of fractions.

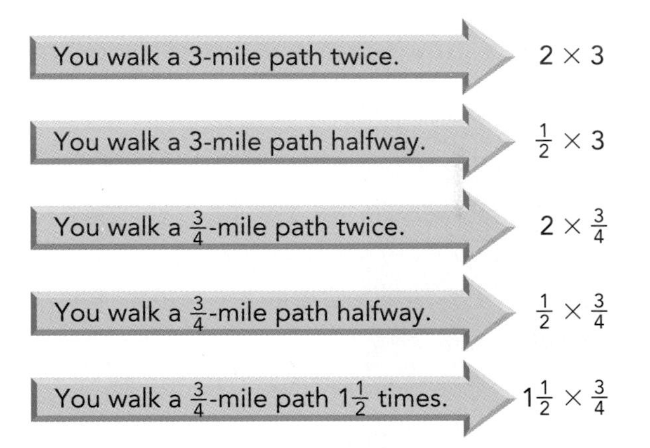

168 Multiplying a Whole Number by a Fraction

Sometimes stores will advertise that items are *a fraction of the original cost.* What they mean is that they started with a price and multiplied it by a fraction to get the new price. When you multiply a whole number by a fraction between zero and one, the product is less than the number you started with.

See 037, 108

$\frac{1}{2} \times 24 = 12$

EXAMPLE: Games are on sale for $\frac{2}{3}$ of the original price. If the original price was $27, what is the sale price?

To solve the problem, you can multiply 27 by $\frac{2}{3}$.

ONE WAY You can draw a picture to show $\frac{2}{3} \times 27$.

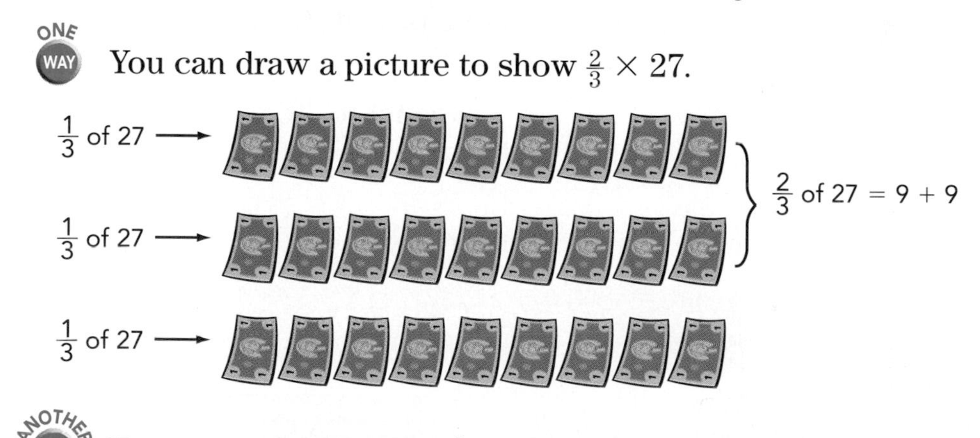

ANOTHER WAY You can multiply 27 by $\frac{2}{3}$ without drawing a picture.

① Write the whole number as a fraction.	② Multiply the numerators. Multiply the denominators.	③ Write the product in simplest form.
$\frac{2}{3} \times 27$ ↓ $\frac{2}{3} \times \frac{27}{1}$	$\frac{2}{3} \times \frac{27}{1} = \frac{54}{3}$	$\frac{54}{3} \rightarrow \frac{54 \div 3}{3 \div 3} = \frac{18}{1} \rightarrow 18$

 Either way, the sale price for the games is $18.

169 Multiplying a Fraction by a Fraction

See 037, 108

Imagine that coins were made of softer metal than they are now. If you needed to pay for something that cost less than your coin was worth, you cut the coin into pieces. If you had half of a $1 coin, you could cut it in half to pay for something that cost a quarter of a dollar! ($\frac{1}{2} \times \frac{1}{2} = \frac{1}{4}$) When you multiply two fractions that are between 0 and 1, the product is smaller than both fractions.

EXAMPLE 1: You have a bag of nuts that is half full. When it was full, the bag weighed $\frac{3}{4}$ of a pound. What does the bag weigh now?

You want to know what $\frac{1}{2}$ of $\frac{3}{4}$ is equal to. To solve the problem, you can multiply $\frac{3}{4}$ by $\frac{1}{2}$.

ONE WAY You can draw a picture to show $\frac{1}{2} \times \frac{3}{4}$.

❶ Start with a picture for $\frac{3}{4}$.	❷ Now show $\frac{1}{2}$ of the $\frac{3}{4}$.
$\frac{1}{4}$ $\frac{1}{4}$ $\frac{1}{4}$ $\frac{1}{4}$ $\frac{3}{4}$ of the whole is shaded blue.	$\frac{1}{2}$ $\frac{1}{2}$ $\frac{6}{8}$ of the whole is shaded blue. $\frac{1}{2}$ of $\frac{6}{8}$ is $\frac{3}{8}$, so $\frac{1}{2}$ of $\frac{3}{4}$ is $\frac{3}{8}$.

ANOTHER WAY You can multiply $\frac{3}{4}$ by $\frac{1}{2}$ without drawing a picture.

❶ Multiply the numerators. Multiply the denominators.	❷ Write the product in simplest form.
$\frac{1}{2} \times \frac{3}{4} = \frac{3}{8}$	$\frac{3}{8}$ is already in simplest form.

Either way, the bag now weighs $\frac{3}{8}$ of a pound.

EXAMPLE 2: Multiply $\frac{3}{8}$ by $\frac{2}{3}$.

ONE WAY You can draw a picture to show $\frac{2}{3} \times \frac{3}{8}$.

❶ Start with a picture for $\frac{3}{8}$.	❷ Now show $\frac{2}{3}$ of the $\frac{3}{8}$.
$\frac{1}{8}$ $\frac{1}{8}$ $\frac{1}{8}$ $\frac{1}{8}$ $\frac{1}{8}$ $\frac{1}{8}$ $\frac{1}{8}$ $\frac{1}{8}$ $\frac{3}{8}$ of the whole is shaded blue.	$\frac{2}{3}$ $\frac{9}{24}$ of the whole is shaded blue. $\frac{2}{3}$ of $\frac{9}{24}$ is $\frac{6}{24}$, so $\frac{2}{3}$ of $\frac{3}{8}$ is $\frac{6}{24}$.

ANOTHER WAY You can multiply $\frac{3}{8}$ by $\frac{2}{3}$ without drawing a picture.

❶ Multiply the numerators. Multiply the denominators.	❷ Write the product in simplest form.
$\frac{2}{3} \times \frac{3}{8} = \frac{6}{24}$	$\frac{6}{24} = \frac{6 \div 6}{24 \div 6} = \frac{1}{4}$

ANOTHER WAY You can simplify the numerators and denominators before multiplying fractions. This is called **canceling**.

❶ Draw a fraction bar between the product of the numerators and the product of the denominators.	$\frac{2}{3} \times \frac{3}{8} \rightarrow \frac{2 \times 3}{3 \times 8}$
❷ Write prime factors of all the factors.	$\frac{2}{3} \times \frac{3}{8} \rightarrow \frac{2 \times 3}{3 \times 2 \times 2 \times 2}$
❸ Cross out ones: any pair of common factors, one from the numerator and one from the denominator.	$\frac{2}{3} \times \frac{3}{8}$ $\frac{\cancel{2}^{1} \times \cancel{3}^{1}}{\cancel{3}_{1} \times \cancel{2}_{1} \times 2 \times 2}$
❹ Multiply the factors that are left. Your product will be in simplest form.	$\frac{1 \times 1}{1 \times 1 \times 2 \times 2} = \frac{1}{4}$

Any way you look at it, $\frac{2}{3} \times \frac{3}{8} = \frac{6}{24}$ or $\frac{1}{4}$.

170 Multiplying Mixed Numbers

See 034, 037, 108

When you multiply a positive mixed number by a fraction between 0 and 1, the product is less than the mixed number.

EXAMPLE: In a cyclocross bicycle race, $\frac{1}{3}$ of the course is road. If a course is $1\frac{1}{4}$ miles long, how much is road?

(Source: The Sports Fan's Ultimate Book of Sports Comparisons)

To solve the problem, multiply $1\frac{1}{4}$ by $\frac{1}{3}$.

You can draw a picture to show $\frac{1}{3} \times 1\frac{1}{4}$.

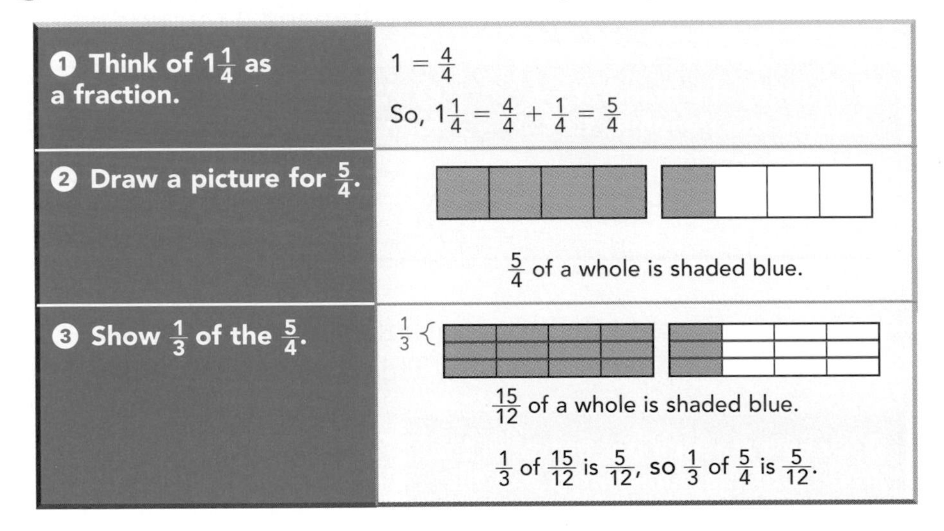

1 Think of $1\frac{1}{4}$ as a fraction.	$1 = \frac{4}{4}$ So, $1\frac{1}{4} = \frac{4}{4} + \frac{1}{4} = \frac{5}{4}$
2 Draw a picture for $\frac{5}{4}$.	$\frac{5}{4}$ of a whole is shaded blue.
3 Show $\frac{1}{3}$ of the $\frac{5}{4}$.	$\frac{1}{3}$ { $\frac{15}{12}$ of a whole is shaded blue. $\frac{1}{3}$ of $\frac{15}{12}$ is $\frac{5}{12}$, so $\frac{1}{3}$ of $\frac{5}{4}$ is $\frac{5}{12}$.

ANOTHER WAY

You can multiply $\frac{1}{3} \times 1\frac{1}{4}$ without drawing a picture.

1 Write the mixed number as a fraction.	2 Multiply the numerators. Multiply the denominators.	3 Write the product in simplest form.
$1\frac{1}{4} \longrightarrow \frac{4}{4} + \frac{1}{4} = \frac{5}{4}$	$\frac{1}{3} \times \frac{5}{4} = \frac{5}{12}$	$\frac{5}{12}$ is already in simplest form.

Either way, the road part of the course is $\frac{5}{12}$ mile.

Dividing with Fractions

To understand what it means to divide with fractions or mixed numbers, think about the meaning of division.

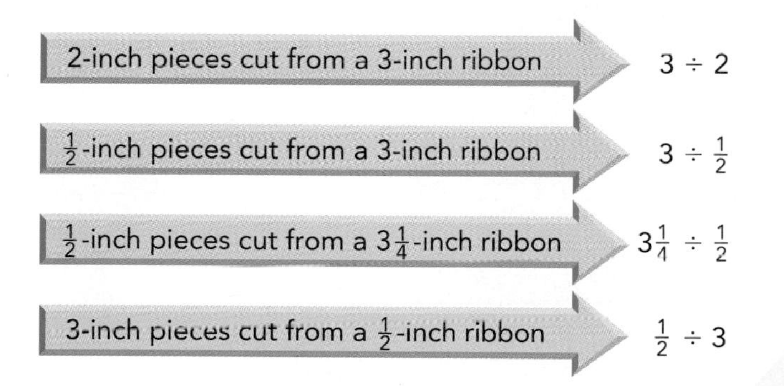

The quotient will be a fraction, but the answer is 0!

Reciprocals

Reciprocals are number pairs that have a product of 1. Here are a few:

See 167–170

$\frac{3}{5}$ and $\frac{5}{3}$

$\frac{3}{5} \times \frac{5}{3} = \frac{15}{15} \rightarrow 1$

5 and $\frac{1}{5}$

$\frac{5}{1} \times \frac{1}{5} = \frac{5}{5} \rightarrow 1$

$2\frac{1}{2}$ and $\frac{2}{5}$

$\frac{5}{2} \times \frac{2}{5} = \frac{10}{10} \rightarrow 1$

173 Dividing a Whole Number by a Fraction

See 037, 144, 172

When you ask, *How many quarters are in a dollar?* you are also asking, *How many fourths are in 1 whole?* This asks the same question as $1 \div \frac{1}{4} = \blacksquare$ and $1 \times 4 = \blacksquare$. But it does NOT ask the same question as $1 \div 4 = \blacksquare$.

EXAMPLE 1: You have 3 cups of pumpkin seeds. You want to put the seeds in bags so that there is $\frac{1}{2}$ cup in each bag. How many bags can you fill?

To solve the problem, divide 3 by $\frac{1}{2}$.

I get it! If you want to know the number of $\frac{1}{2}$s in 3 wholes, you can either divide 3 by $\frac{1}{2}$ or multiply 3 by 2. That's why dividing by a fraction is the same as multiplying by the reciprocal of that fraction.

You can draw a picture to show $3 \div \frac{1}{2}$.

❶ Show 3 wholes.	❷ Mark them in halves.	❸ Color each $\frac{1}{2}$. Count the halves.
		There are 6 halves.

You can divide 3 by $\frac{1}{2}$ without drawing a picture.

① Write the whole number as a fraction.	② Multiply the dividend by the reciprocal of the divisor.	③ Write the product in simplest form.
$3 \div \frac{1}{2} \rightarrow \frac{3}{1} \div \frac{1}{2}$	$\frac{3}{1} \div \frac{1}{2} \rightarrow \frac{3}{1} \times \frac{2}{1} = \frac{6}{1}$	$\frac{6}{1} = 6$

Either way, you can fill six $\frac{1}{2}$-cup bags.

EXAMPLE 2: Divide 6 by $\frac{5}{8}$.

ONE WAY You can draw a picture to show $6 \div \frac{5}{8}$.

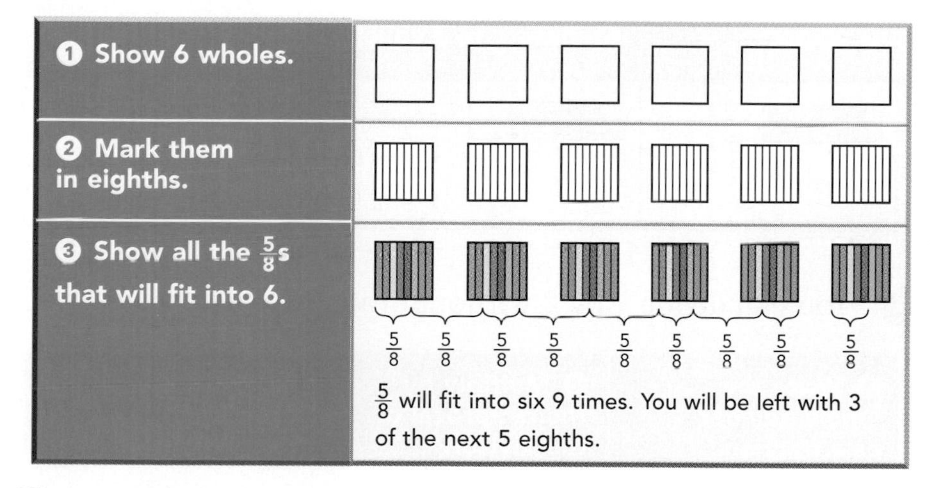

① Show 6 wholes.	
② Mark them in eighths.	
③ Show all the $\frac{5}{8}$s that will fit into 6.	$\frac{5}{8}$ $\frac{5}{8}$ $\frac{5}{8}$ $\frac{5}{8}$ $\frac{5}{8}$ $\frac{5}{8}$ $\frac{5}{8}$ $\frac{5}{8}$ $\frac{5}{8}$ $\frac{5}{8}$ will fit into six 9 times. You will be left with 3 of the next 5 eighths.

ANOTHER WAY You can divide 6 by $\frac{5}{8}$ without drawing a picture.

① Write the whole number as a fraction.	$6 \div \frac{5}{8} \rightarrow \frac{6}{1} \div \frac{5}{8}$
② Multiply the dividend by the reciprocal of the divisor.	$\frac{6}{1} \div \frac{5}{8} \rightarrow \frac{6}{1} \times \frac{8}{5} = \frac{48}{5}$
③ Write the product in simplest form.	$\frac{48}{5} \rightarrow \frac{45}{5} + \frac{3}{5} = 9\frac{3}{5}$

Either way, $6 \div \frac{5}{8} = 9\frac{3}{5}$.

174 Dividing a Fraction by a Whole Number

See 037, 149, 169, 172

Sometimes you want to share something that's smaller than a whole, like $\frac{3}{4}$ of a bag of popcorn. You can divide fractions to help you figure out how to share.

EXAMPLE 1: An avocado grower will ship $\frac{3}{4}$ ton of avocados in two equal shipments. How much will each shipment weigh?

To solve the problem, you can divide $\frac{3}{4}$ by 2.

You can draw a picture to show $\frac{3}{4} \div 2$.

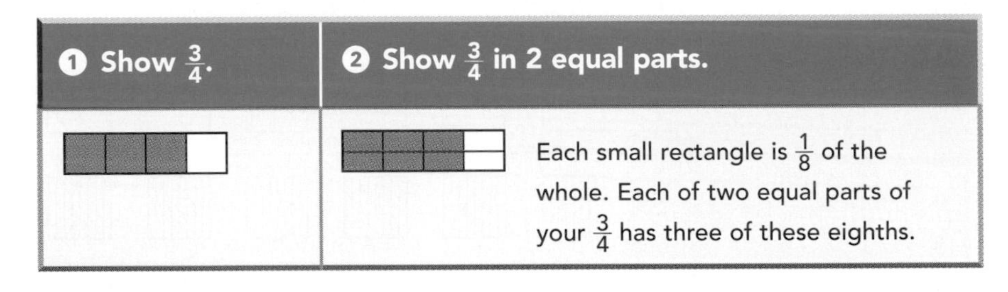

① Show $\frac{3}{4}$.	② Show $\frac{3}{4}$ in 2 equal parts.
	Each small rectangle is $\frac{1}{8}$ of the whole. Each of two equal parts of your $\frac{3}{4}$ has three of these eighths.

ANOTHER WAY You can divide $\frac{3}{4}$ by 2 without drawing a picture.

① Write the whole number as a fraction.	② Multiply by the reciprocal of the divisor.	③ Write the product in simplest form.
$\frac{3}{4} \div 2 \rightarrow \frac{3}{4} \div \frac{2}{1}$	$\frac{3}{4} \div \frac{2}{1} \rightarrow \frac{3}{4} \times \frac{1}{2} = \frac{3}{8}$	$\frac{3}{8}$ is already in simplest form.

Either way, each shipment will weigh $\frac{3}{8}$ ton.

So, dividing by 2 is the same as multiplying by $\frac{1}{2}$.

EXAMPLE 2: Divide $\frac{3}{4}$ by 6.

❶ Write the whole number as a fraction.	❷ Multiply by the reciprocal of the divisor.	❸ Write the product in simplest form.
$\frac{3}{4} \div 6 \rightarrow \frac{3}{4} \div \frac{6}{1}$	$\frac{3}{4} \div \frac{6}{1} \rightarrow \frac{3}{4} \times \frac{1}{6} = \frac{3}{24}$	$\frac{3}{24} \rightarrow \frac{3 \div 3}{24 \div 3} = \frac{1}{8}$

$\frac{3}{4} \div 6 = \frac{1}{8}$

Dividing a Fraction by a Fraction 175

EXAMPLE: You have $\frac{3}{4}$ yard of fabric. Into how many $\frac{3}{8}$-yard pieces can you cut the fabric?

To solve the problem, divide $\frac{3}{4}$ by $\frac{3}{8}$.

See 037

ONE WAY You can draw a picture to show $\frac{3}{4} \div \frac{3}{8}$.

❶ Show $\frac{3}{4}$. Show $\frac{3}{8}$.	❷ Count how many $\frac{3}{8}$s will fit in your $\frac{3}{4}$.
	Two $\frac{3}{8}$s will fit in your $\frac{3}{4}$.

ANOTHER WAY You can divide $\frac{3}{4}$ by $\frac{3}{8}$ without drawing a picture.

❶ Multiply the dividend by the reciprocal of the divisor.	❷ Write the product in simplest form.
$\frac{3}{4} \div \frac{3}{8} \rightarrow \frac{3}{4} \times \frac{8}{3} = \frac{24}{12}$	$\frac{24}{12} \rightarrow \frac{24 \div 12}{12 \div 12} = \frac{2}{1} \rightarrow 2$

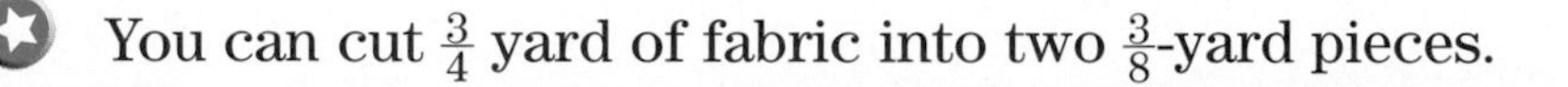
You can cut $\frac{3}{4}$ yard of fabric into two $\frac{3}{8}$-yard pieces.

176 Dividing Mixed Numbers

MORE HELP

See 034, 037, 112–114

You can use what you know about dividing fractions to divide mixed numbers.

EXAMPLE: You have a shelf that is $3\frac{1}{8}$ feet long. How many $\frac{3}{4}$-foot-long boxes can you store on the shelf?

To solve the problem, divide $3\frac{1}{8}$ by $\frac{3}{4}$.

ONE WAY You can draw a picture to show $3\frac{1}{8} \div \frac{3}{4}$.

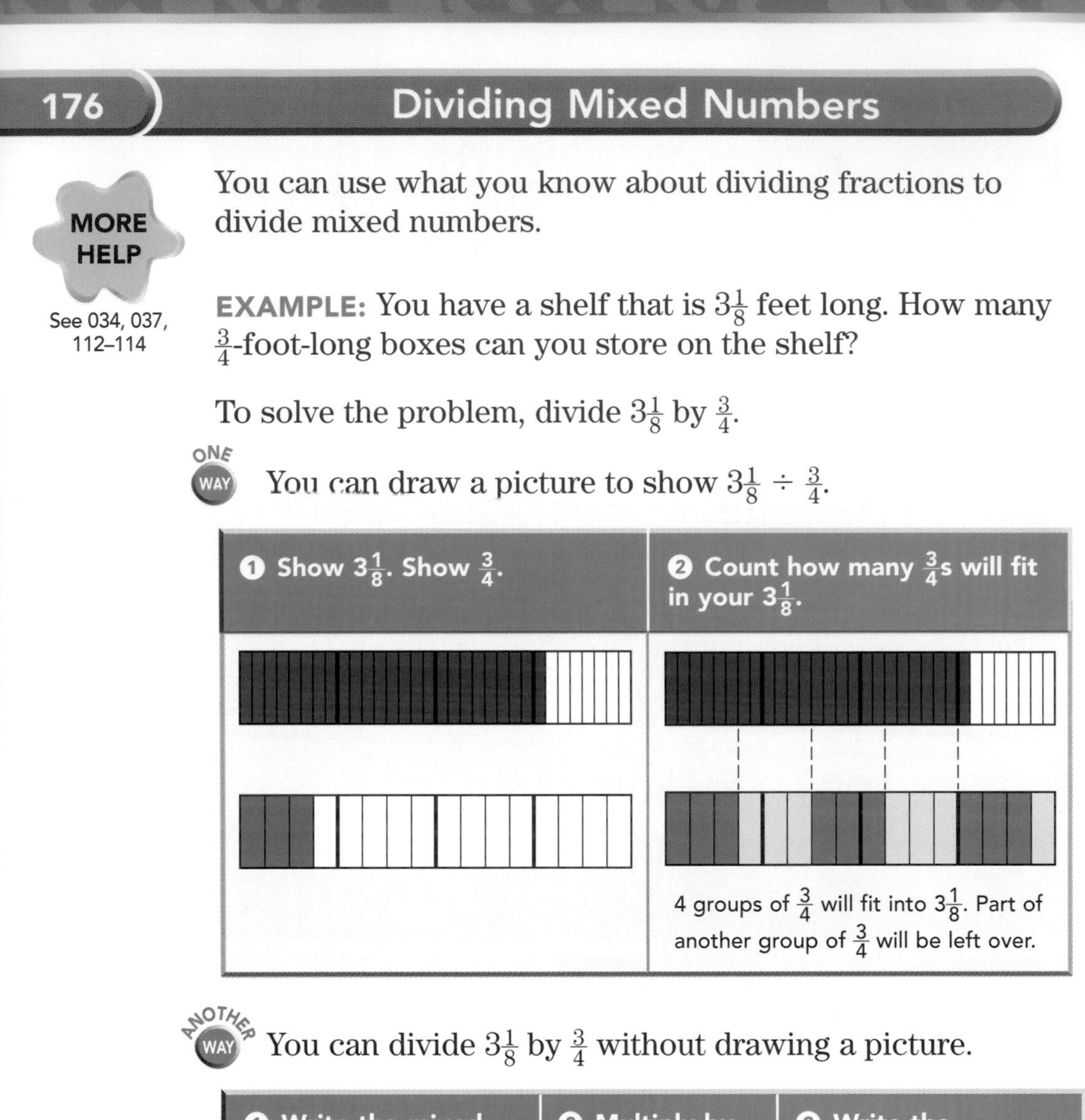

ANOTHER WAY You can divide $3\frac{1}{8}$ by $\frac{3}{4}$ without drawing a picture.

① Write the mixed numbers as fractions.	② Multiply by the reciprocal of the divisor.	③ Write the product in simplest form.
$3\frac{1}{8} \div \frac{3}{4} \rightarrow (\frac{24}{8} + \frac{1}{8}) \div \frac{3}{4}$ ↓ $\frac{25}{8} \div \frac{3}{4}$	$\frac{25}{8} \div \frac{3}{4}$ ↓ $\frac{25}{8} \times \frac{4}{3} = \frac{100}{24}$	$\frac{100}{24} \rightarrow \frac{96}{24} + \frac{4}{24}$ ↓ $4 + \frac{1}{6} = 4\frac{1}{6}$

Either way, you can store four $\frac{3}{4}$-foot-long boxes on a shelf $3\frac{1}{8}$ feet long.

EXAMPLE 2: Divide $4\frac{1}{2}$ by $1\frac{1}{2}$.

ONE WAY You can draw a picture to show $4\frac{1}{2} \div 1\frac{1}{2}$

❶ Show $4\frac{1}{2}$. Show $1\frac{1}{2}$.	❷ Count how many $1\frac{1}{2}$s will fit in your $4\frac{1}{2}$.
	3 groups of $1\frac{1}{2}$ will fit into $4\frac{1}{2}$.

ANOTHER WAY You can divide $4\frac{1}{2}$ by $1\frac{1}{2}$ without drawing a picture.

❶ Write the mixed numbers as fractions.	❷ Multiply the dividend by the reciprocal of the divisor.	❸ Write the product in simplest form.
$4\frac{1}{2} \div 1\frac{1}{2}$ ↓ $(\frac{8}{2} + \frac{1}{2}) \div (\frac{2}{2} + \frac{1}{2})$ ↓ $\frac{9}{2} \div \frac{3}{2}$	$\frac{9}{2} \div \frac{3}{2}$ ↓ $\frac{9}{2} \times \frac{2}{3} = \frac{18}{6}$	$\frac{18}{6} \rightarrow \frac{18 \div 6}{6 \div 6} = \frac{3}{1} \rightarrow 3$

Either way, $4\frac{1}{2} \div 1\frac{1}{2} = 3$.

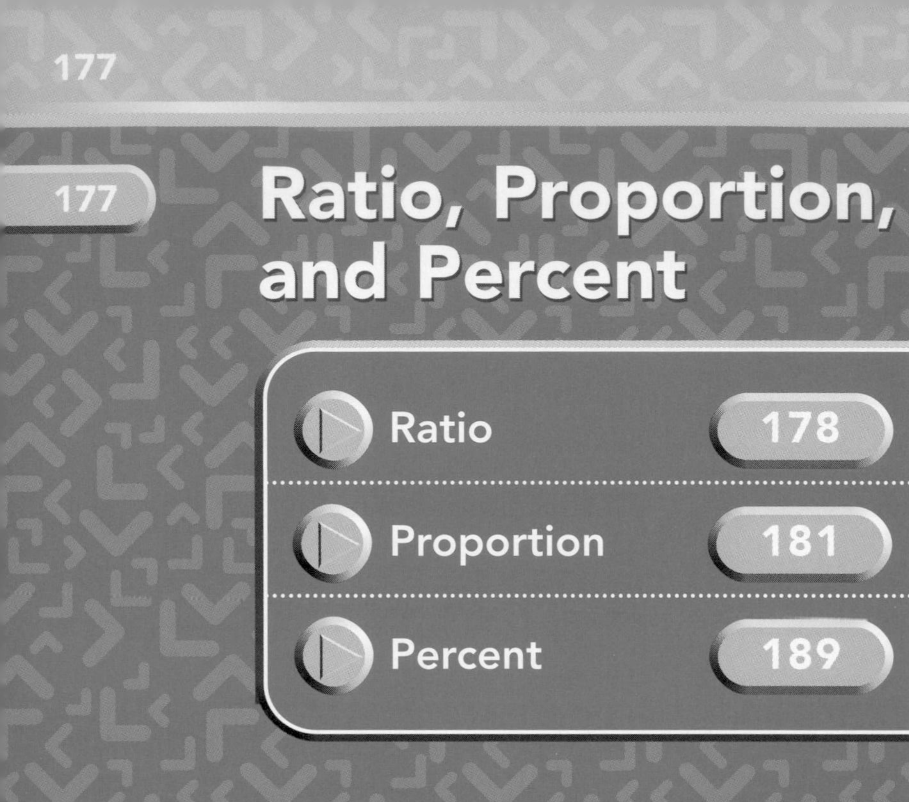

Ratio, Proportion, and Percent

The bottom picture is *out of proportion.* The bottom apple is twice as wide as the first but not twice as long. The ratio of the width of the bottom apple to the width of the top apple is 2 to 1. For the apples to be *in proportion*, their lengths would also have to have a ratio of 2 to 1.

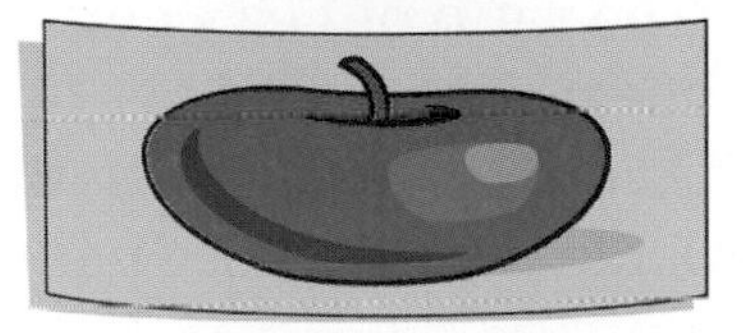

What's wrong with this picture?

Ratio

Ways to Write Ratios

You can compare two quantities two different ways.

EXAMPLE: Lea has 10 pets and Andrew has 6. Compare these two quantities.

ONE WAY You can subtract to find the difference.

Lea	Andrew
dogs	dogs
cats	cats
hamsters	hamsters

MORE HELP
See 028–035

number of pets Lea has − number of pets Andrew has = how many more Lea has
10 − 6 = 4

Lea has 4 more pets than Andrew.

ANOTHER WAY You can also use a ratio to compare two quantities.

$$\frac{\text{number of pets Andrew has}}{\text{number of pets Lea has}} = \frac{6}{10}$$

The ratio of Lea's pets to Andrew's pets is 10 to 6. This means that Lea's pet family is $\frac{10}{6}$, or $1\frac{2}{3}$ times as large as Andrew's pet family. Andrew's pet family is $\frac{6}{10}$ or $\frac{3}{5}$ as large as Lea's pet family.

Ratios compare two numbers.

CASE 1 You can use a ratio to compare part of something to all of it. This ratio is most like a regular fraction.

The ratio of Lea's dogs to all of her pets is 2 to 10. (This can also be written as 2:10 or $\frac{2}{10}$. You can see from the fraction form that the ratio is the same as the ratio 1:5.)

See 037, 180

CASE 2 You can use a ratio to compare one part to another part of the same thing.

The ratio of Lea's dogs to her cats is 2 to 3 (2:3).

CASE 3 You can use a ratio to compare a part of one thing to a part of another thing.

The ratio of Lea's dogs to Andrew's dogs is 2 to 2 (2:2 or 1:1).

CASE 4 You can use a ratio to compare all of something to all of something else.

The ratio of Lea's pet family to Andrew's pet family is 10 to 6 (10:6 or 5:3). The ratio of Andrew's pet family to Lea's is 6 to 10 (6:10 or 3:5).

This means that for every 5 pets Lea has, Andrew has 3 pets. If they want to get more pets but keep the ratios the same, here's how they could do it.

Lea's Pets	Andrew's Pets	Ratio
10	6	10:6 or 5:3
15	9	15:9 or 5:3
20	12	20:12 or 5:3
25	15	25:15 or 5:3

179 Ways to Write Ratios

A car dealer orders sedans and SUVs based on what colors her customers usually buy.

	Sedans	SUVs
Green		
White		
Black		
Red		
Total	32	16

Here are some ways ratios show comparisons of the cars.

Ways to Write	Compare Red Sedans to All Sedans (part to all)	Compare Red Sedans to White Sedans (part to part)	Compare Red Sedans to Red SUVs (part to part)	Compare Sedans to SUVs (all to all)
Use to	8 to 32	8 to 10	8 to 4	32 to 16
Use a colon	8:32	8:10	8:4	32:16
Use fraction form	$\frac{8}{32}$ or $\frac{1}{4}$	$\frac{8}{10}$ or $\frac{4}{5}$	$\frac{8}{4}$ or $\frac{2}{1}$	$\frac{32}{16}$ or $\frac{2}{1}$

MATH ALERT Ratios Differ from Fractions

Ratios can have a numerator and a denominator. They can also be simplified the same way fractions are simplified. A fraction's denominator *always* tells how a whole is divided into equal parts. With ratios, the denominator *may* tell one of three things.

Paul's Class	Girls 17	Boys 13	Total 30
Ana's Class	Girls 16	Boys 15	Total 31

CASE 1 The denominator could tell the number of parts in a whole.

The ratio of girls in Paul's class to students in Paul's class is

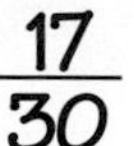

The numerator describes a *part* of the class.
The denominator describes the *whole* class.

CASE 2 The denominator could tell the number of parts in a different whole.

The ratio of girls in Paul's class to girls in Ana's class is

$\frac{17}{16}$ The *whole* for the numerator is girls in Paul's class.
The *whole* for the denominator is girls in Ana's class.

CASE 3 The denominator could describe a different part of the whole than the numerator describes. It could even describe a part of a different whole!

The ratio of girls in Paul's class to boys in Paul's class is

$\frac{17}{13}$ The numerator describes one *part* of the class.
The denominator describes a different *part* of the class.

Ratios don't follow fraction rules when it comes to units of measure. A fraction compares things that have the same units. A ratio *may* compare things with the same units (like cups of lemon juice to cups of water). A ratio *may also* compare things that don't have the same units (like 30 miles traveled to 1 gallon of gas). Whatever you do, don't add or subtract ratios—they're not like fractions when it comes to computing.

181

Proportion

Think about forming teams for a softball tournament. The ratio of teams to players is 1 to 9. If you want to have 4 teams, you will need 36 players. You must keep the ratio of teams to players the same. $\frac{1}{9} = \frac{4}{36}$

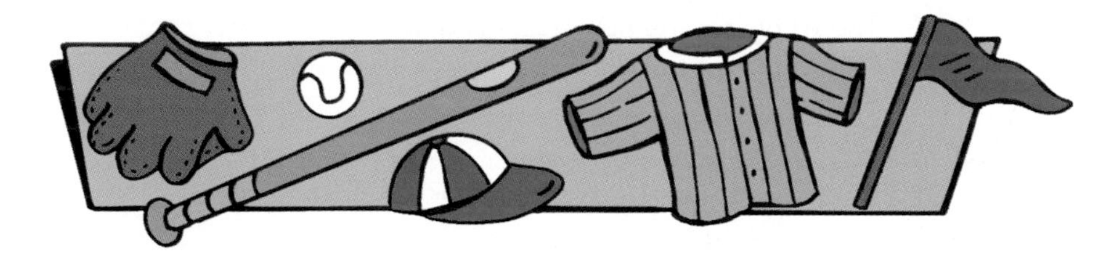

182 Ways to Write Proportions

A **proportion** is an equation showing that two ratios are equal.

EXAMPLE: In an official United States flag, the ratio of the length to the width must be 19 to 10.
(Source: The Flag Book of the United States)

A flag maker wants to make a flag that is two times as long (38 inches). This means the flag must also be two times as wide (20 inches). Use a proportion to show this relationship.

Write proportions the same ways you write ratios.

Ways *to* Write	Compare Ratios of Length to Width
Use *to*	official length to new length = official width to new width 19 to 38 = 10 to 20
Use a colon	official length : new length = official width : new width 19:38 = 10:20
Use fraction form	$\frac{\text{official length}}{\text{new length}} = \frac{\text{official width}}{\text{new width}}$ $\frac{19}{38} = \frac{10}{20}$ $\frac{1}{2} = \frac{1}{2}$

Read proportions the same way you read analogies.

Write: shoe : foot = hat : head
Say: *shoe is to foot as hat is to head*

Write: 19:38 = 10:20
Say: *nineteen is to thirty-eight as ten is to twenty*

Terms of a Proportion 183

Look at this proportion.

new length : official length = new width : official width
38:19 = 20:10

Each element of the proportion is called a **term**. *New length* is a term. So is *official length*. There are four terms in any proportion.

184 Using Proportions to Solve Problems

See 035,183

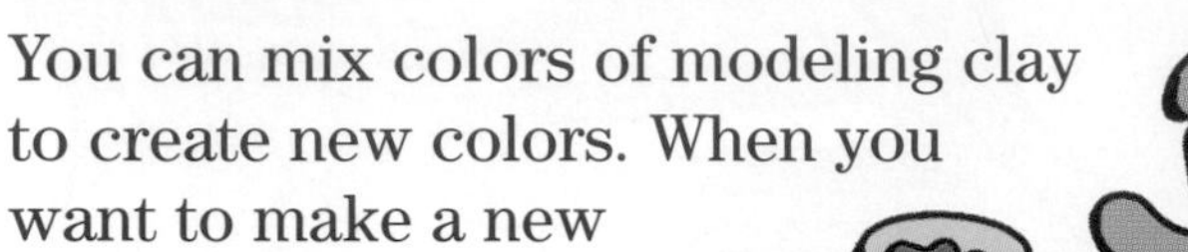

You can mix colors of modeling clay to create new colors. When you want to make a new batch that is *exactly* the same as another batch, you can use proportions.

EXAMPLE: Suppose you made a sample batch of purple clay by mixing 2 sticks of red and 3 sticks of blue. You want to make a larger batch of exactly the same shade. You will use 8 sticks of red. How many sticks of blue should you use?

You can use a proportion to solve this problem.

$$\text{red} \rightarrow \frac{2}{3} = \frac{8}{\blacksquare} \leftarrow \text{red}$$
$$\text{blue} \rightarrow \qquad \leftarrow \text{blue}$$

You now need to **solve this proportion**. This means you need to find a value for the missing term.

ONE WAY You can solve a proportion by finding equivalent fractions.

$$\text{red} \rightarrow \frac{2 \times 4}{3 \times 4} = \frac{8}{12} \leftarrow \text{red / blue}$$

ANOTHER WAY You can solve a proportion by using **cross products**. In any pair of equal ratios, cross products are equal.

$$\frac{2}{3} = \frac{8}{\blacksquare} \qquad 2 \times \blacksquare = 24 \qquad 3 \times 8 = 24$$

Since $2 \times 12 = 24$, $\blacksquare = 12$.

Either way, you should use 12 sticks of blue clay.

Solving Proportions Using Rate

A **rate** is a special ratio. It involves quantities that aren't measured the same way, like miles and hours. In a **unit rate**, the denominator is always 1. A **percent** is a rate where the denominator is always 100. Rates are easy to use in problems with proportions because they are easy to compute with. Here is a list of rates you probably know about.

Rate	Ratio
miles per hour (mph)	$\frac{\text{number of miles traveled}}{\text{1 hour}}$
revolutions per minute (rpm)	$\frac{\text{number of complete turns}}{\text{1 minute}}$
miles per gallon (mpg)	$\frac{\text{number of miles driven}}{\text{1 gallon of gas}}$

EXAMPLE: You're riding your bicycle at a rate of 10 miles per hour. At that speed, how many miles will you go in two hours?

You could set up and solve a proportion.

miles → $\frac{10}{1} = \frac{\blacksquare}{2}$ ← miles
hours → ← hours

$$\frac{10 \times 2}{1 \times 2} = \frac{20}{2}$$

ANOTHER WAY

Since 10 miles per hour is a unit rate, you can multiply the rate by the time to get the distance.
$10 \times 2 = 20$

Either way, you will ride 20 miles in 2 hours.

186 Unit Price

Nt Wt means net weight in pounds—the total weight of the meat without the packaging. The net weight is 1.5 pounds. The unit price is the price for each pound of the meat.

To find the unit price of any item, you can set up and solve a proportion.

$$\frac{\text{total price}}{\text{units in the package}} = \frac{\text{price}}{\text{one unit}}$$

The ratio $\frac{\text{price}}{\text{one unit}}$ is the unit price.

EXAMPLE: What is the unit price for each juice box?

Raspberry unit price ⟶ $\frac{25¢}{8 \text{ oz}} = \frac{3.125¢}{1 \text{ oz}}$

Apple unit price ⟶ $\frac{20¢}{8 \text{ oz}} = \frac{2.5¢}{1 \text{ oz}}$

The raspberry juice costs 3.125 cents per ounce.
The apple juice costs 2.5 cents per ounce.

187  MATH ALERT Better Buy

Unit price may be only part of a buying decision.

EXAMPLE: A 3-ounce can of cat food has a unit price of 25¢ per ounce. A 16-ounce can of a different brand has a unit price of 10¢ per ounce. Which is the better buy?

The 16-ounce container is the better buy only if
- your cat is not a picky eater,
- she eats a lot, and a large container won't go bad,
- both packages are recyclable.

Scale Drawing 188

A **scale** is the ratio of the measurement on a map or model to the measurement of the real object. A **scale drawing** has the same shape but not the same size as the object it represents. The dimensions are increased or decreased proportionally.

See 035, 369

EXAMPLE 1: Your social studies assignment is to make a map of your classroom. The room is 33 feet long. What scale could you use to fit the map onto a piece of paper 11 inches long?

Remember that ratios do not have to use matching units. Set up and solve a proportion.

$$\text{scale} \rightarrow \frac{11 \text{ inches}}{33 \text{ feet}} = \frac{1 \text{ inch}}{\blacksquare \text{ feet}} \leftarrow \text{scale / real}$$

$$\frac{11 \div 11}{33 \div 11} = \frac{1}{3}$$

The scale can be 1 inch to 3 feet.

EXAMPLE 2: You've chosen a scale of 1 inch to 3 feet for your map. The classroom is 20 feet wide. Will you be able to fit the width of the classroom onto the $8\frac{1}{2}$ inch width of your paper?

To find the width of the biggest room that will fit on the page, set up and solve a proportion.

$$\text{scale} \rightarrow \frac{1 \text{ inch}}{3 \text{ feet}} = \frac{8.5 \text{ inches}}{\blacksquare \text{ feet}} \leftarrow \text{scale / real}$$

$$\frac{1 \times 8.5}{3 \times 8.5} = \frac{8.5}{25.5}$$

This tells you that your page is wide enough to show 25.5 feet. Since you need to represent only 20 feet, your map will fit on the paper.

189

Percent

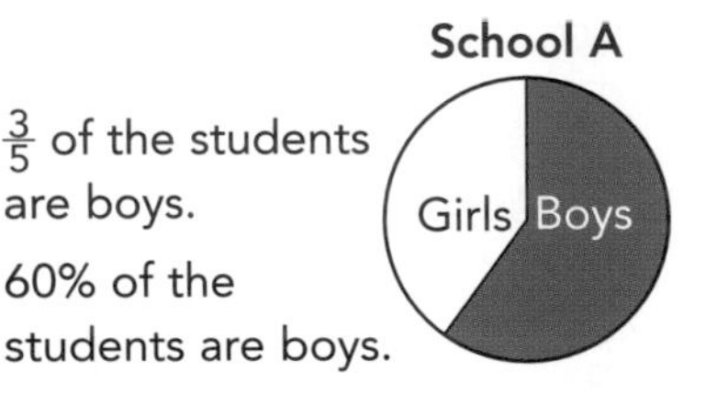

$\frac{3}{5}$ of the students are boys.

60% of the students are boys.

School B

$\frac{5}{8}$ of the students are boys.

62.5% of the students are boys.

Girls Boys

Percents can make it easier to see that School B has a greater ratio of boys to girls than School A does.

190 Reading and Writing Percents

Percent means *per hundred.*

See 019–020, 043–044

- You have 100 marbles and 25 of them are blue. 25% of the marbles are blue.

  Write: 25%

  Say: *twenty-five percent*

- Your class goal was to wash 100 cars in one day. You washed 150 cars on that day. You achieved 150% of your goal!

Today's Goal: 100 Cars

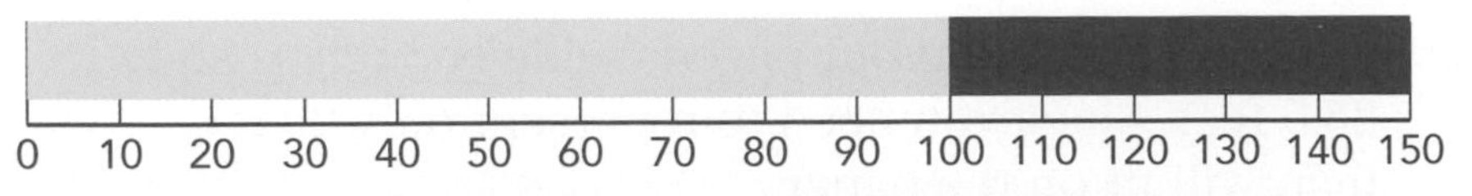

A percent is like a ratio in which the second term is 100. You can compare a number to 100 three ways, as a percent, as a fraction, and as a decimal.

Percent	Fraction	Decimal
25%	$\frac{25}{100}$	0.25
200%	$\frac{200}{100} = 2$	2.00
12.5%	$\frac{12.5}{100} = \frac{125}{1000}$	0.125

Comparing Percents 191

CASE 1 If the percents are describing the same-size things, then comparing them makes sense.

EXAMPLE 1: In an average year in Des Moines, Iowa, 59% of the days have some sunshine. In an average year in Abilene, Texas, 70% of the days have some sunshine. Can you tell which town usually has more sunny days?
(Source: National Weather Service, San Francisco)

These percents both describe days in an average year, so you can compare them. 70% > 59%, so Abilene has more sunny days than Des Moines.

CASE 2 If the percents are describing different-size things, then comparing may not make sense.

EXAMPLE 2: About 5% of the people who live in Des Moines are 10–13 years old. About 5.6% of the people who live in Abilene are 10–13 years old. Can you tell which city has more 10–13 year olds? *(Source: U.S. Census Bureau, 1990 Census data)*

Since you don't know how many people live in each city, you can't tell if more 10–13 year-olds live in Des Moines or in Abilene.

192 Computing with Percents

Percent statements usually involve three numbers.

1. The whole
2. The percent
3. The part (the part *can* be the same size as, or even bigger than, the whole!)

If you know any two of these three numbers, you can find the third.

Percent	of	whole	=	part
■%	of	■	=	■
10%	of	50	=	5

193 Finding a Percent of a Number

You can make a budget to help you plan how you'll use your money. Here's the Jones family's weekly entertainment budget for their daughter, Joanie.

See 035,184

To find how much money Joanie can spend in each category, she needs to find an amount that represents a percent of $20.

EXAMPLE: Joanie Jones likes to see a movie every week. It costs \$5.50 to go to a movie theater. Does her budget allow her to go to the movie theater each week?

ONE WAY You can understand the problem by drawing a picture.

\$0 | \$5 | \$10 | \$15 | \$20

25% of \$20 is \$5.

0% | 25% | 50% | 75% | 100%

ANOTHER WAY You can write the percent as a fraction and then solve a proportion using equivalent fractions.

❶ Write the percent in fraction form.	❷ Write the proportion.	❸ Solve the proportion using equivalent fractions.
$25\% = \frac{25}{100}$	part → whole → $\frac{25}{100} = \frac{■}{20}$ ← part ← whole	$\frac{25 \div 5}{100 \div 5} = \frac{5}{20}$

You can write the percent as a fraction and then solve a proportion using cross products.

❶ Write the percent in fraction form.	❷ Write the proportion.	❸ Solve the proportion using cross products.
$25\% = \frac{25}{100}$	part → whole → $\frac{25}{100} = \frac{■}{20}$ ← part ← whole	$25 \times 20 = 500$ $100 \times ■ = 500$ $■ = 5$

Any way you look at it, Joanie only has \$5.00 to spend on movies. Since the movie costs \$5.50, she should find another theater, rent movies at the video store, or use some of her miscellaneous money.

194 What Percent One Number Is of Another

MORE HELP

See 035, 184

Here's the nutrition information from a can of chicken broth.

This means that 1 cup of this broth has 25 calories. 15 of those calories are from the fat that's in the broth.

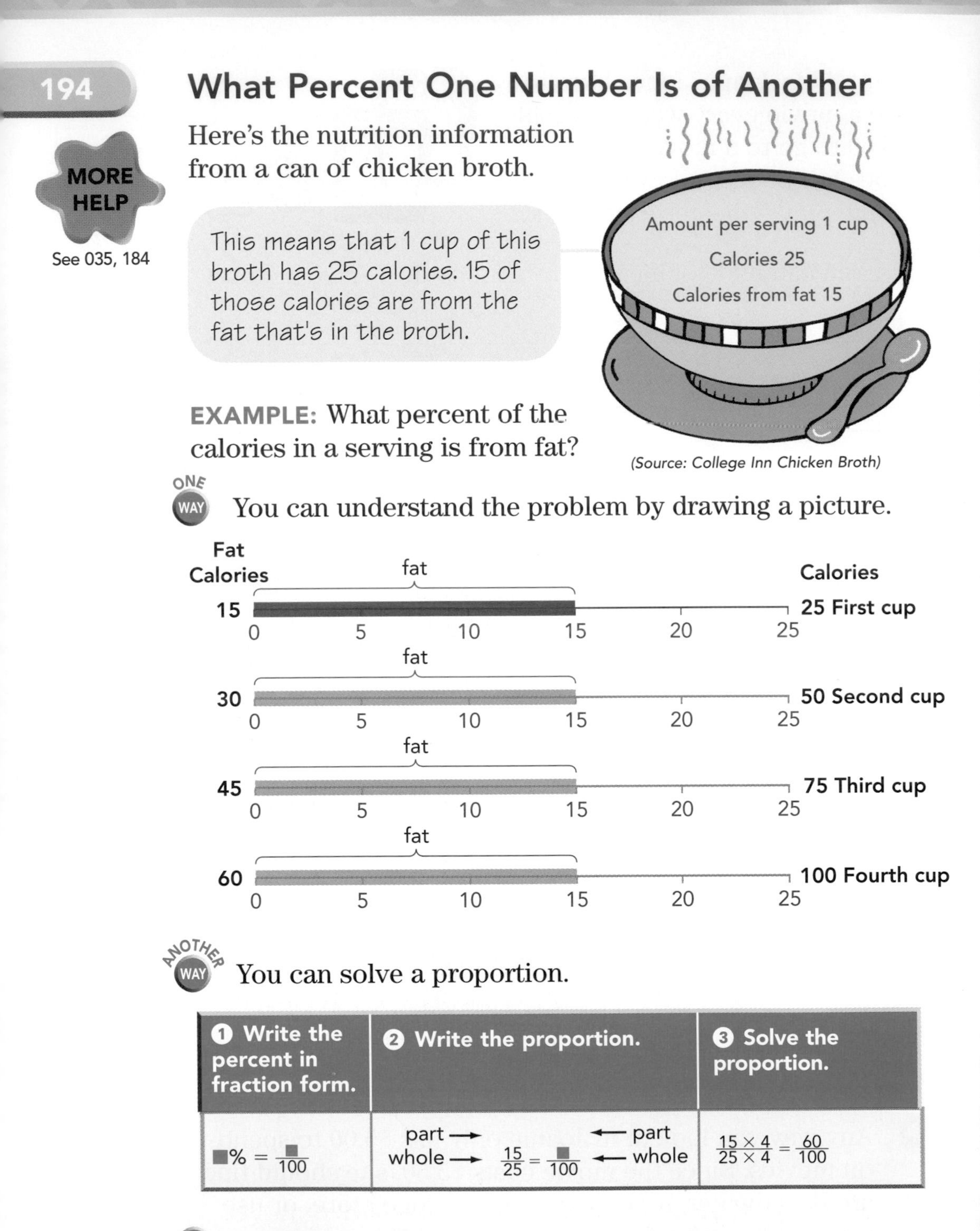

(Source: College Inn Chicken Broth)

EXAMPLE: What percent of the calories in a serving is from fat?

ONE WAY You can understand the problem by drawing a picture.

ANOTHER WAY You can solve a proportion.

❶ Write the percent in fraction form.	❷ Write the proportion.	❸ Solve the proportion.
■% $= \frac{■}{100}$	part → whole → $\frac{15}{25} = \frac{■}{100}$ ← part ← whole	$\frac{15 \times 4}{25 \times 4} = \frac{60}{100}$

Either way, 60% of the calories in this broth come from fat.

Finding the Total When the Percent Is Known

195

You can compute to find the whole amount when all you know is the part and the percent. Here's more nutrition information from the chicken broth label.

See 035, 037, 184

Total Fat 1.5 grams % Daily value 2%

This means that a serving of this chicken broth has 1.5 grams of fat. This is 2% of the amount of fat the government recommends per day.

EXAMPLE: How many grams of fat does the government recommend per day?

ONE WAY You can understand the problem by drawing a picture.

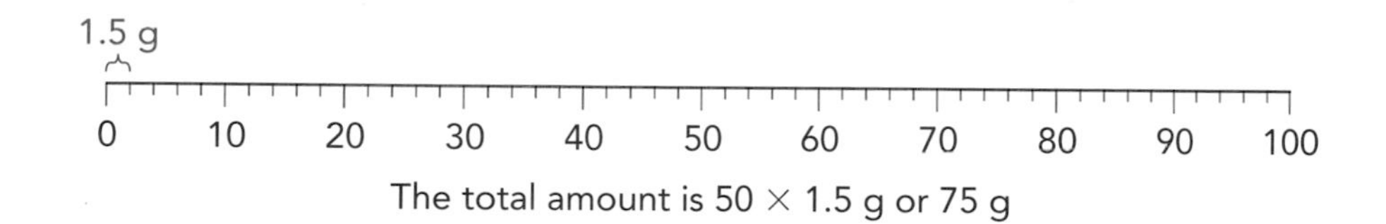

The total amount is 50 × 1.5 g or 75 g

ANOTHER WAY You can solve a proportion.

❶ Write the percent in fraction form.	❷ Write the proportion.	❸ Solve the proportion.
$2\% = \frac{2}{100}$ $\frac{2}{100} = \frac{1}{50}$	part → whole → $\frac{1}{50} = \frac{1.5}{■}$ ← part ← whole	$50 \times 1.5 = 75$ $1 \times ■ = 75$ $■ = 75$

Either way, the government recommends 75 grams of fat per day.

Simple Interest

See 239, 246

Interest is the amount that someone pays to use someone else's money. If you invest money in a savings account, the bank pays interest to you. This is because they are glad to have your money—they'll use it to earn money for themselves. If you borrow money from the bank, you pay interest on that money to the bank. The amount of money borrowed or invested is called the **principal**. The **interest rate** is the percent charged or paid during a given period of time.

EXAMPLE: Suppose you borrow $500 to buy a mountain bike. The simple interest rate is 12% per year. You want to make payments for 2 years. How much interest will you pay? How much will you pay to the bank in all?

To find the amount of interest that you will pay, you can use this formula:

interest (I) = principal (p) × annual rate (r) × time in years (t)

$$I = prt$$
$$= 500 \times 0.12 \times 2$$
$$= 120$$

12% = 0.12

You will pay $120 in interest. Since the principal is $500, you will pay $500 + $120, or $620, to the bank.

Discounts

197

When a store has a sale, they usually discount their regular prices. A **discount** is a percent or a fraction of the original price that is subtracted to make the sale price.

See 246

EXAMPLE: The regular price for a pair of bicycle shorts is \$40. During a sale, the store gives a discount of 25%. How much do you save when you buy a pair of bicycle shorts on sale? How much will you pay for the shorts?

To solve the problem, find 25% of \$40.

original number × percent = percent of number

$$40 \times 25\% = \blacksquare$$

$$40 \times 0.25 = 10.00$$

You will save \$10 when you buy the shorts on sale. You will pay \$40 − \$10, or \$30, for those shorts.

198

Pre-Algebra

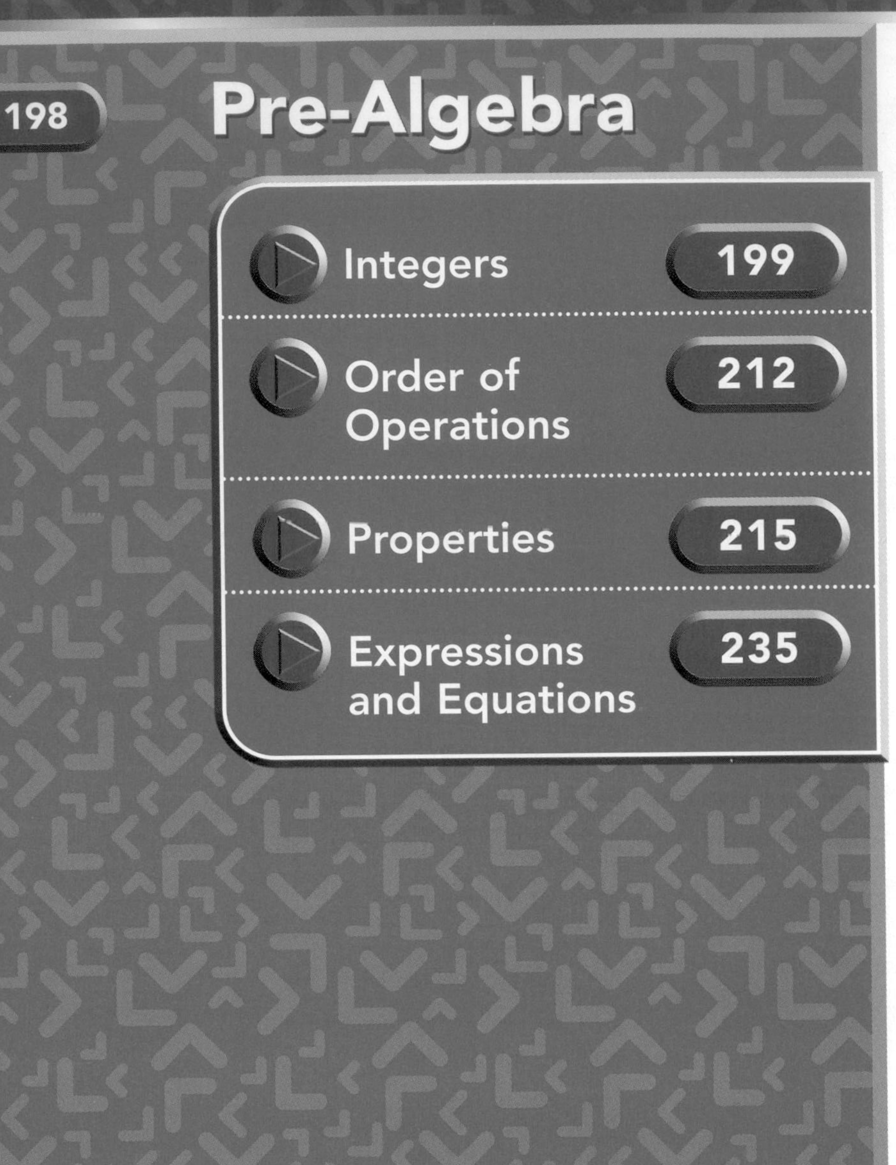

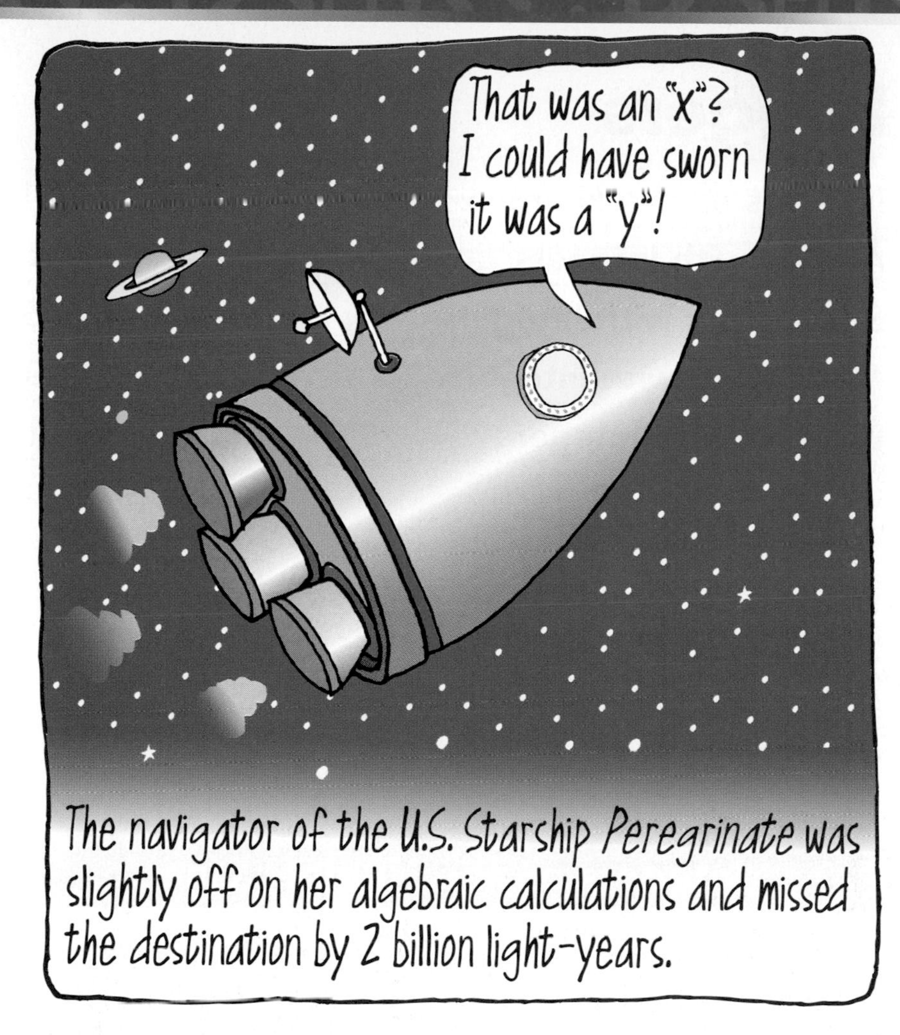

Things change. When you throw a ball, it takes time to go up, level off, then come down. When you wash more cars at a fund-raiser, you raise more money. The speed of the ball and the amount of money vary. But, you can still describe them. To describe things that vary, mathematicians invented **algebra**. Algebra makes it easier to say exactly how two changing things (like dollars earned and hours worked) are related.

Our English word *algebra* is derived from two Arabic words: *al* (the) + *jabara* (to reunite). This makes sense since algebra helps us tie together many mathematical ideas. *(Source: Webster's New World College Dictionary)*

199 Integers

MORE HELP
See 045–048

Integers are all of the whole numbers and all of their opposites. Six and $^{-}6$ are both integers. They are opposites. Zero is an integer. It is its own opposite. When you compute with integers it's like finding your way in a new part of town—you have to watch the signs.

200 Integers on the Number Line

Think of the symbol ($^{+}$ or $^{-}$) on an integer as a direction sign. If the symbol is a direction sign, the number line is an integer highway. The on-ramp is at zero. Left from zero is negative. Right from zero is positive.

NEGATIVE ZERO POSITIVE

$^{-}6$ $^{-}5$ $^{-}4$ $^{-}3$ $^{-}2$ $^{-}1$ 0 $^{+}1$ $^{+}2$ $^{+}3$ $^{+}4$ $^{+}5$

Usually, we don't put the little plus symbol above and to the left of a positive number. You need to remember: if you don't see a direction sign, you're looking at a positive number.

Adding with Integers

Remember, positive 1 is one unit to the right of zero on the number line. Negative 1 is one unit to the left of zero on the number line. If you start at zero and go one unit right, then one unit left ($1 + {}^-1$), you end up at zero. Positive 1 and $^-1$ are sometimes called a **zero pair** because their sum is zero.

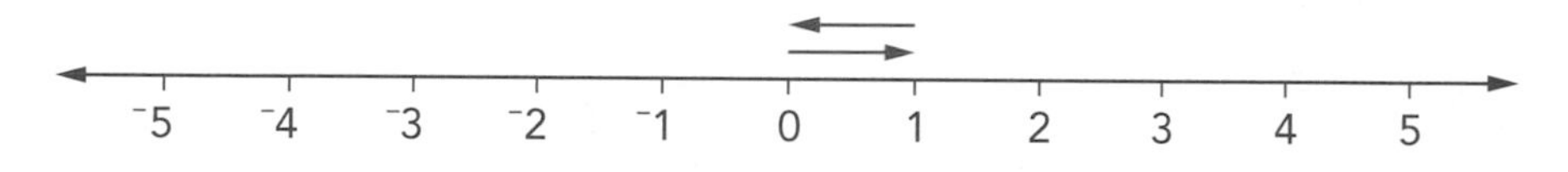

You can use counters to add integers. To model adding integers with counters, use two colors. One color stands for positive and a different color stands for negative.

Understanding Positive Addends

EXAMPLE: Add. $3 + 6 = \blacksquare$

❶ Show counters for the first addend.	❷ Add counters for the second addend.	❸ Look for zero pairs of counters.
First addend / Second addend / Zero pairs	First addend / Second addend / Zero pairs	First addend / Second addend / Zero pairs
3 is a positive number, so use 3 positive counters.	6 is a positive number, so use 6 positive counters.	All the counters are positive. There are no zero pairs.

All the remaining counters will be one color. Count them. If the counters are blue, the sum is positive. If the counters are red, the sum is negative.

$3 + 6 = 9$

203

Understanding Positive and Negative Addends

EXAMPLE 1: Add. $^{-}3 + 6 = ■$

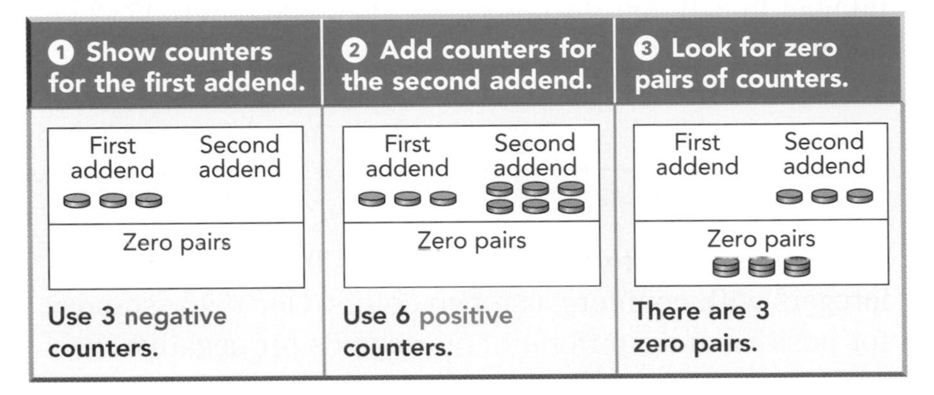

All the remaining counters will be one color. Count them. The counters are blue, so the sum is positive.

$^{-}3 + 6 = 3$

EXAMPLE 2: Add. $3 + {}^{-}6 = ■$

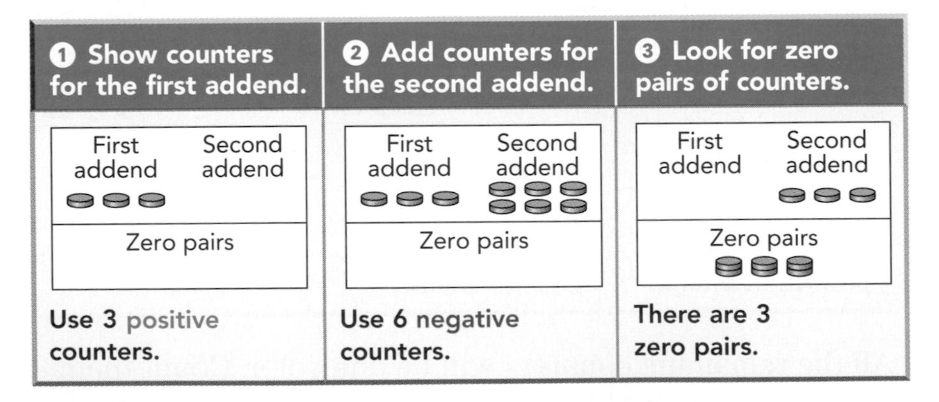

All the remaining counters will be one color. Count them. The counters are red, so the sum is negative.

$3 + {}^{-}6 = {}^{-}3$

Understanding Negative Addends

EXAMPLE 3: Add. $^{-}3 + {}^{-}6 = \blacksquare$

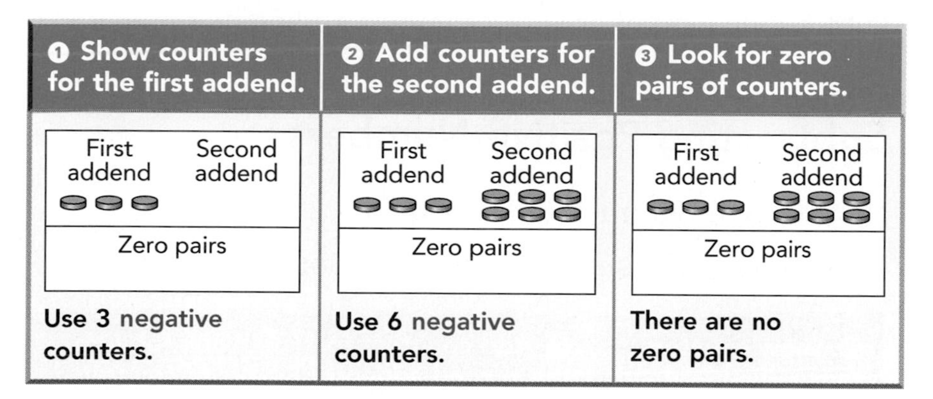

All the remaining counters will be one color. Count them. The counters are red, so the sum is negative.

$^{-}3 + {}^{-}6 = {}^{-}9$

Rules for Adding Integers

You can add integers without using counters.

- If the signs are the same, pretend the signs aren't there. Add the numbers. Then put the addends' sign on your sum.

  $^{-}25 + {}^{-}75 = {}^{-}100$ and $25 + 75 = 100$

- If the signs are different, pretend the signs aren't there. Subtract the smaller number from the larger one. Look at the number you subtracted from. What sign did it have before you pretended it didn't have one? That's the sign that goes on your answer.

  $^{-}25 + 75 = 50$ and $25 + {}^{-}75 = {}^{-}50$

  Same sign Same sign

206 Subtracting with Integers

See 201–205

You can rewrite subtraction as addition because subtracting a number is the same as adding its opposite.

207 Subtracting Positive Numbers

EXAMPLE 1: Subtract. $6 - 4 = \blacksquare$

First, rewrite the subtraction as addition. $6 - 4 = 6 + {}^{-}4$

❶ Show counters for the first addend.	❷ Add counters for the second addend.	❸ Look for zero pairs of counters.
First addend / Second addend / Zero pairs	First addend / Second addend / Zero pairs	First addend / Second addend / Zero pairs
Use 6 positive counters.	Use 4 negative counters.	There are 4 zero pairs.

$6 - 4 = 2$

EXAMPLE 2: Subtract. ${}^{-}6 - 4 = \blacksquare$

First, rewrite the subtraction as addition. ${}^{-}6 - 4 = {}^{-}6 + {}^{-}4$

❶ Show counters for the first addend.	❷ Add counters for the second addend.	❸ Look for zero pairs of counters.
First addend / Second addend / Zero pairs	First addend / Second addend / Zero pairs	First addend / Second addend / Zero pairs
Use 6 negative counters.	Use 4 negative counters.	There are no zero pairs.

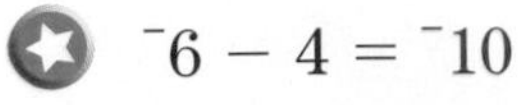

${}^{-}6 - 4 = {}^{-}10$

Subtracting Negative Numbers

EXAMPLE 1: Subtract. $^-6 - {}^-4 = \blacksquare$

First, rewrite the subtraction as addition. $^-6 - {}^-4 = {}^-6 + 4$

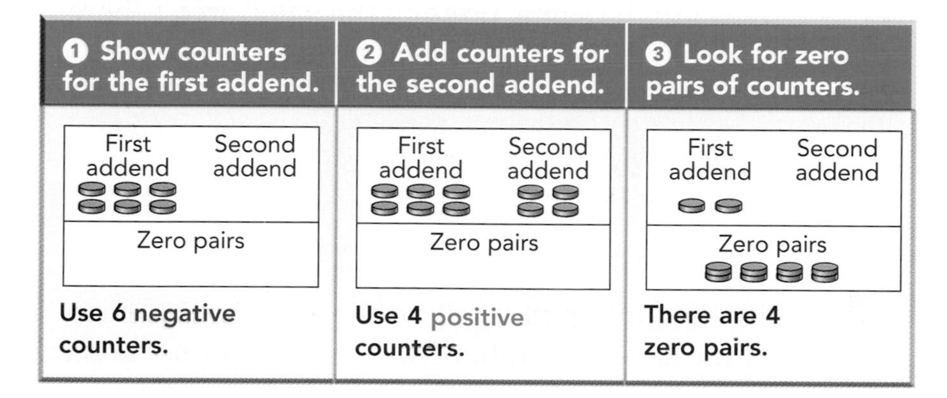

$^-6 - {}^-4 = {}^-2$

Subtracting a negative number is just like using a double negative in English: If you *don't* have *no* homework, you *do* have homework!

EXAMPLE 2: Subtract.
$6 - {}^-4 = \blacksquare$

First, rewrite the subtraction as addition. $6 - {}^-4 = 6 + 4$

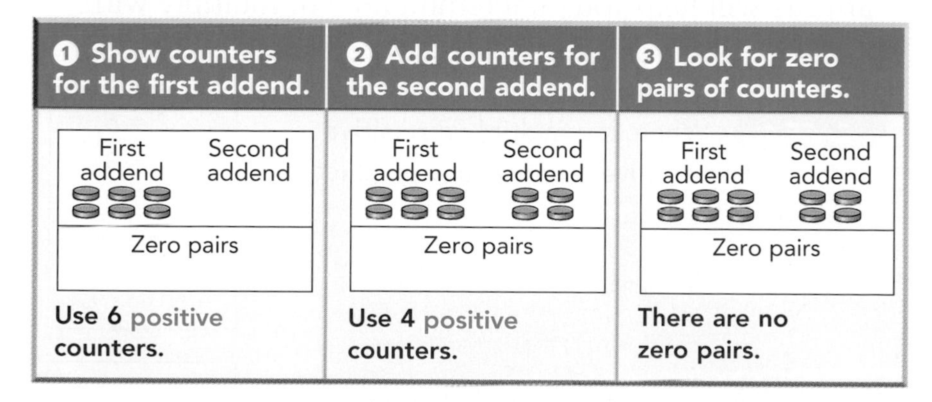

$6 - {}^-4 = 10$

209 Rules for Subtracting Integers

You can subtract integers without using counters.
First, rewrite the subtraction as addition.

- If the signs are the same, pretend the signs aren't there. Add the numbers. Then put the addends' sign on your sum.

$$^{-}25 - 75 \longrightarrow {^{-}25} + {^{-}75} = {^{-}100}$$

Same signs

- If the signs are different, pretend the signs aren't there. Subtract the smaller number from the larger one. Look at the number you subtracted from. What sign did it have before you pretended it didn't have one? That's the sign that goes on your answer.

$$^{-}25 - {^{-}75} \longrightarrow {^{-}25} + 75 = 50$$

210 Multiplying with Integers

Patterns will help you understand how to multiply with integers.

CASE 1 When you multiply a positive number by a negative number, the product is negative.

$3 \times 2 = 6$
$3 \times 1 = 3$
$3 \times 0 = 0$
$3 \times {^{-}1} = {^{-}3}$
$3 \times {^{-}2} = {^{-}6}$

CASE 2 When you multiply two negative numbers, the product is positive.

$2 \times {^{-}3} = {^{-}6}$
$1 \times {^{-}3} = {^{-}3}$
$0 \times {^{-}3} = 0$
${^{-}1} \times {^{-}3} = 3$
${^{-}2} \times {^{-}3} = 6$

SHORTCUT

You can multiply integers without using the pattern.

- When factors have the same signs, pretend the signs aren't there. Multiply. The product is always positive.

$^{-}400 \times {}^{-}50 = 20{,}000$ and $400 \times 50 = 20{,}000$

- When factors have different signs, pretend the signs aren't there. Multiply. The product is always negative.

$^{-}400 \times 50 = {}^{-}20{,}000$ and $400 \times {}^{-}50 = {}^{-}20{,}000$

Dividing with Integers 211

Multiplication and division are related. You can divide with integers by rewriting division equations as multiplication equations.

MORE HELP
See 145

Division Equation	Related Multiplication Equation
$30 \div 5 = 6$	$6 \times 5 = 30$
$^{-}30 \div 5 = {}^{-}6$	$^{-}6 \times 5 = {}^{-}30$
$^{-}30 \div {}^{-}5 = 6$	$6 \times {}^{-}5 = {}^{-}30$
$30 \div {}^{-}5 = {}^{-}6$	$^{-}6 \times {}^{-}5 = 30$

EXAMPLE: Divide. $^{-}28 \div 7 = ■$

$^{-}28 \div 7 = ■$ asks the same question as $■ \times 7 = {}^{-}28$. The product is negative. This means that one factor is positive and one is negative. The factor you know is positive. So, the missing factor must be negative.

If the signs are the same, the quotient is positive. If the signs are different, the quotient is negative.

 Since $^{-}4 \times 7 = {}^{-}28$, $^{-}28 \div 7 = {}^{-}4$.

212

Order of Operations

Order of operations is a set of rules. It tells you the order in which to compute so that you'll get the same answer that anyone else would get.

EXAMPLE: Compute. $6 + 4 \times 3 \div 6 = \blacksquare$

If you worked in order from left to right, you'd do this.	But, if you knew about order of operations, you'd do this.
$\underbrace{6 + 4}_{10} \times 3 \div 6$	$6 + \underbrace{4 \times 3}_{12} \div 6$
$\underbrace{10 \times 3}_{30} \div 6$	$6 + \underbrace{12 \div 6}_{2}$
$30 \div 6 = 5$	$6 + 2 = 8$

MORE HELP

See 214

$6 + 4 \times 3 \div 6$ really does equal 8. The order of operations makes sure that there is only one correct answer for this or any other computation.

Using Parentheses to Show Order of Operations 213

Sometimes, when you write a problem, you don't want to follow the regular order of operations. You can use parentheses to say *do this first.*

EXAMPLE: Each of 5 friends got a full box of snacks and an extra 6 snacks. Write an equation to show how many snacks are in all those boxes and all those extra snacks.

Even if you don't know how many snacks are in a box, you can write an expression to show how many.

See 235–239

The order of operations would tell you to multiply 5 by SNACKS then add 6. But every friend has a *sum* of snacks (SNACKS + 6) and you want to multiply the *sum* by 5.

- Use parentheses to group the *sum*: $5 \times$ (SNACKS $+ 6$).
  So, if SNACKS $= 4$, you compute like this:

  $5 \times (4 + 6)$

  ↓

  $5 \times 10 = 50$

214 Rules for Order of Operations

To make sure that everyone finds the same answer when computing, we have rules called **order of operations**.

MORE HELP

See 064, 067, 442

This silly sentence may help you remember: "Please pay real money down at Store24."

1. Compute inside the parentheses.
2. Do powers or roots.
3. Multiply or divide in order.
4. Add or subtract in order.

EXAMPLE 1: Five friends have collected 300 cans. They take them to the store that pays 6¢ each for aluminum cans. Fifteen of the cans are not aluminum. If they share the money equally, how much does each friend get?

- Set up the computation.

300 cans, 15 of them worthless ⟶ $300 - 15$

Each aluminum can is worth 6¢ ⟶ $6 \times (300 - 15)$

Five friends share equally ⟶ $6 \times (300 - 15) \div 5$

- Follow the order of operations.

1. Compute inside parentheses ⟶ $6 \times (285) \div 5$
2. Do the powers and roots ⟶ no powers or roots
3. Multiply or divide left to right ⟶ $1710 \div 5 = 342$

Each friend will get 342¢ or $3.42 for the cans.

EXAMPLE 2: Compute. $16 \div 4 + (4 - 3) \times 2^2$

1. Compute inside parentheses	$16 \div 4 + 1 \times 2^2$
2. Do powers and roots	$16 \div 4 + 1 \times 4$
3. Multiply or divide	$4 + 4$
4. Add or subtract	8

$16 \div 4 + (4 - 3) \times 2^2 = 8$

Properties

When you're working with numbers, there are some things that are always true about how they behave. These things are called **properties**.

Commutative Properties

Sometimes order doesn't matter. You could put on your shoes before your belt and it wouldn't make any difference in how you looked. The addends in addition can be placed in any order. The factors in multiplication can also be placed in any order. The answer will be the same either way. These operations are **commutative**.

Sometimes order does matter. If you put on your socks after your shoes, you'd look a little strange. For subtraction and division, if you put the numbers in a different order, your answer will be different. $20 \div 4 \neq 4 \div 20$ and $35 - 10 \neq 10 - 35$, so subtraction and division are *not* commutative.

Commutative sounds like commute. That means go back and forth.

217 Commutative Property of Addition

MORE HELP
See 118–126

The **Commutative Property of Addition** is sometimes called the Order Property of Addition. It says that changing the order of addends does not change the sum.

EXAMPLE: Add. 25 + 147 + 75 = ■

ONE WAY You can follow the order of operations rules.

MORE HELP
See 214

① Add the first two addends.	② Add the third addend to the sum of the first two.
25 + 147 + 75; 25 + 147 = 172	25 + 147 + 75; 172 + 75 = 247

ANOTHER WAY Use the Commutative Property to switch the order of 147 + 75. Then you can use mental math.

25 + 147 + 75

25 + 75 + 147

100 + 147 = 247

The addition is friendlier this way because you are using compatible numbers.

Either way, the sum is 247.

218 Commutative Property of Multiplication

The **Commutative Property of Multiplication** is also called the Order Property of Multiplication. It says that changing the order of factors does not change the product.

5 rows of 3 is the same as 3 rows of 5.

EXAMPLE: Multiply. $5 \times 29 \times 2 = ■$

ONE WAY You can follow the order of operations rules.

❶ Multiply the first two factors.	❷ Multiply the product by the third factor.
$5 \times 29 \times 2 = ■$ 5 × 29 45 100 145	$5 \times 29 \times 2 = ■$ 145 × 2 290

You can use the Commutative Property to switch the order of 29 and 2. Then you can use mental math.

$5 \times 29 \times 2$

$5 \times 2 \times 29$

$10 \times 29 = 290$

Either way, the product is 290.

Math Alert Subtraction and Division Are Not Commutative

Is $16 - 4$ equal to $4 - 16$? No. Is $16 \div 4$ equal to $4 \div 16$? No. If you can find even *one* case where a property is not true, then the property doesn't work for that kind of computation. So, subtraction and division are *not* commutative.

MORE HELP

See 206–209

220 Associative Properties

If you have a bunch of chores to do, you might save time by grouping them in a certain way. The idea of grouping tasks to make things easier works in math, too. The **Associative Properties** say you can keep the order of addends or factors but group them for easy computing.

221 Associative Property of Addition

The **Associative Property of Addition** says that changing the grouping of three or more addends does not change the sum. So, $(6 + 2) + 4 = 6 + (2 + 4)$.

The 2 associates with 6, then with 4, but the order of the addends does not change.

EXAMPLE: Add. $57 + 25 + 25 = ■$

ONE WAY You can follow the order of operations rules.

❶ Add the first two addends.	❷ Add the third addend to the sum of the first two.
57 25 + 25 57 + 25 82	82 + 25 107

ANOTHER WAY You can use the Associative Property to change the grouping so you can use mental math.

$(57 + 25) + 25$

↓

$57 + (25 + 25)$

↓

$57 + 50 = 107$

Either way, the sum is 107.

Associative Property of Multiplication

222

The **Associative Property of Multiplication** says that changing the grouping of factors does not change the product. So, $(6 \times 4) \times 2 = 6 \times (4 \times 2)$.

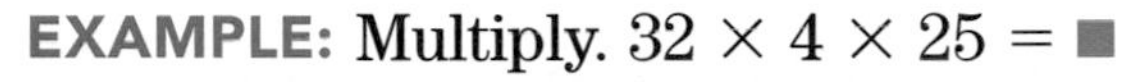

EXAMPLE: Multiply. $32 \times 4 \times 25 = ■$

ONE WAY You can follow the order of operations rules.

❶ Multiply the first two factors.	❷ Multiply the product by the third factor.
$32 \times 4 \times 25 = ■$ 32 × 4 128	$32 \times 4 \times 25 = ■$ 128 × 25 640 2560 3200

ANOTHER WAY You can use the Associative Property. Change the grouping so you can use mental math.

$(32 \times 4) \times 25$

↓

$32 \times (4 \times 25)$

↓

$32 \times 100 = 3200$

Either way, the product is 3200.

MATH ALERT Subtraction and Division Are Not Associative

Is $(16 - 4) - 2$ equal to $16 - (4 - 2)$? No. Is $(16 \div 4) \div 2$ equal to $16 \div (4 \div 2)$? No. If you can find even one case where a property is not true, then the property doesn't work for that kind of computation. So, subtraction and division are *not* associative.

224 Distributive Property

When you distribute things, you spread them out. The Distributive Property lets you spread out numbers so they're easier to work with.

225 Distributive Property of Multiplication

Any number can be written as a sum or difference of other numbers. The **Distributive Property of Multiplication** says that you can multiply a sum by multiplying each addend separately, then adding their products.
So, $8 \times (20 + 3) = (8 \times 20) + (8 \times 3)$.

This property is also true when you multiply a difference. Multiply each number in the difference, then subtract the products. So, $7 \times (20 - 1) = (7 \times 20) - (7 \times 1)$.

EXAMPLE 1: Find the number of eggs in 11 dozen.

You can think of 11 as 10 + 1 and multiply mentally.

$$12 \times 11 = 12 \times (10 + 1)$$
$$\downarrow$$
$$(12 \times 10) + (12 \times 1)$$
$$\downarrow \qquad \downarrow$$
$$120 + 12 = 132$$

There are 132 eggs in 11 dozen.

EXAMPLE 2: Find 6 × 98 mentally.

$$6 \times 98 = 6 \times (100 - 2)$$
$$\downarrow$$
$$(6 \times 100) - (6 \times 2)$$
$$\downarrow \qquad \downarrow$$
$$600 - 12 = 588$$

$6 \times 98 = 588$

Using the Distributive Property with Division

226

Remember, dividing by a number is the same as multiplying by its reciprocal. So $4 \div 2$ has the same result as $4 \times \frac{1}{2}$. This means that you can rewrite any division problem as multiplication. Then, if you want to use the distributive property, you can.

See 171–176

EXAMPLE 1: Compute. $(17 + 3) \div 5 = ■$

❶ Rewrite the division as multiplication.	❷ Does the distributive property make computing easier?	❸ Compute.
$(17 + 3) \div 5$ ↓ $(17 + 3) \times \frac{1}{5}$	No, $3 \times \frac{1}{5}$ is a fraction. But $20 \times \frac{1}{5}$ is easy.	$(17 + 3) \times \frac{1}{5}$ ↓ $20 \times \frac{1}{5} = \frac{20}{5} = 4$

$(17 + 3) \div 5 = 4$

EXAMPLE 2: You bring 36 decorated eggs to the egg-rolling contest. Your friend brings 18 more. Show how to share the eggs among 6 chidren.

Show the equation: $(36 + 18) \div 6 = ■$

❶ Rewrite the division as multiplication.	❷ Does the distributive property make computing easier?	❸ Compute.
$(36 + 18) \div 6$ ↓ $(36 + 18) \times \frac{1}{6}$	Sure, you can multiply 36 and 18 by $\frac{1}{6}$ in your head.	$(36 + 18) \times \frac{1}{6}$ ↓ $(36 \times \frac{1}{6}) + (18 \times \frac{1}{6})$ ↓ ↓ $6 + 3 = 9$

Each child gets 9 eggs.

227 Identity Elements

Identity elements are numbers that combine with other numbers, in any order, without changing the original number.

CASE 1 The Identity Element for Addition is 0 because when you add 0 to any number you end up with that number.
$5 + 0 = 5$ and $0 + 5 = 5$

CASE 2 The Identity Element for Multiplication is 1 because when you multiply any number by 1 you end up with that number.
$5 \times 1 = 5$ and $1 \times 5 = 5$

228 MATH ALERT No Identity Elements for Division and Subtraction

See 219

CASE 1 Zero is not an identity element for subtraction.

Is any number minus 0 always that number? Yes.
Is 0 minus any number always that number? No. One good example is $0 - 1 = {}^{-}1$ (not 1). So, you can't say that zero is an identity element for subtraction.

CASE 2 One is not an identity element for division.

Is any number divided by 1 always that number? Yes.
Is 1 divided by any number always that number? No.
One good example is $1 \div 4 = \frac{1}{4}$ (not 4). So, you can't say that 1 is an identity element for division.

229 Inverse Elements

See 046

Inverse elements are numbers that combine, in any order, with other numbers and result in identity elements—1 or zero.

CASE 1 Additive Inverse

Additive inverses are also called **opposites**. A number added to its opposite always equals 0: 8 and $^-8$ are opposites.
$8 + {}^-8 = 0$ and ${}^-8 + 8 = 0$

CASE 2 Multiplicative Inverse

Multiplicative inverses are also called **reciprocals**. A number multiplied by its reciprocal always equals 1: 3 and $\frac{1}{3}$ are reciprocals.
$3 \times \frac{1}{3} = 1$ and $\frac{1}{3} \times 3 = 1$

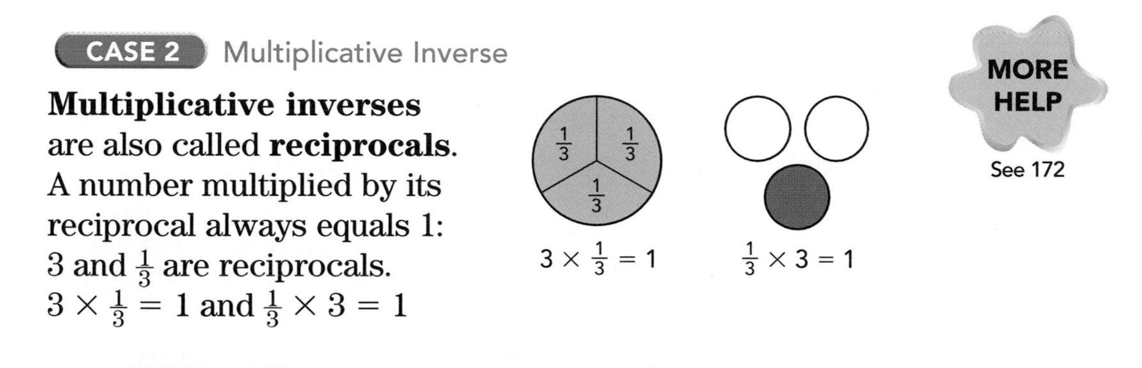

MORE HELP See 172

Zero Property 230

The **zero property of multiplication** says that the product of any number and 0 is 0. So, $5 \times 0 = 0$ and $0 \times 5 = 0$. Also $7642.063 \times 0 = 0$ and $0 \times 7642.063 = 0$.

MATH ALERT Don't Try to Divide by Zero

Think about division as undoing multiplication. This will help you see why we don't divide by zero.

EXAMPLE: Try to divide 245 by 0.
Think of the related multiplication equation.
$245 \div 0 = \blacksquare$ asks the same question as $\blacksquare \times 0 = 245$.

When you think about it this way, you can see that you're really stuck. *Any* number times zero is zero! So, *no* number times zero is 245. Mathematicians say that division by zero is undefined.

232 Equality Properties

See 237–240

The **equality properties** say that you can add to or multiply the expression on one side of the equals sign. But, to keep the equation true, you must do the same thing to the amount on the other side.

233 Addition Property of Equality

Think of an equation as a scale that is balanced.

The amount in one pan balances the amount in the other pan.
$2 = 2$

If you add something to one pan, the scale becomes unbalanced.
$2 + 1 \neq 2$

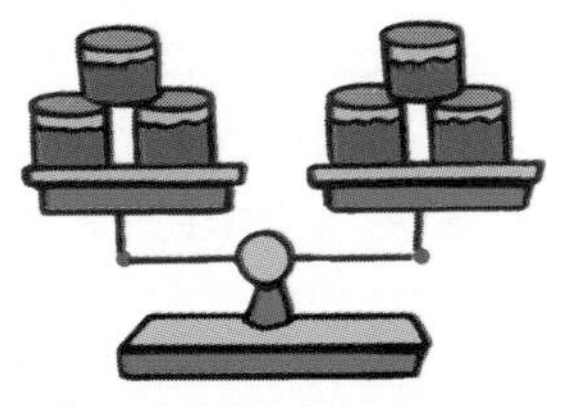

If you add the same thing to the other pan, the scale balances again!
$2 + 1 = 2 + 1$

You can also subtract the same number from both sides of an equation without putting it out of balance.

234 Multiplication Property of Equality

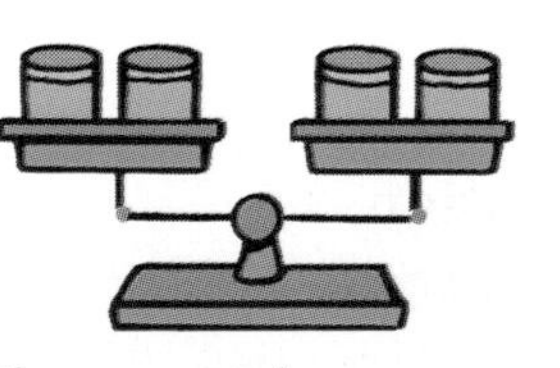

The amount in one pan balances the amount in the other pan.
$2 = 2$

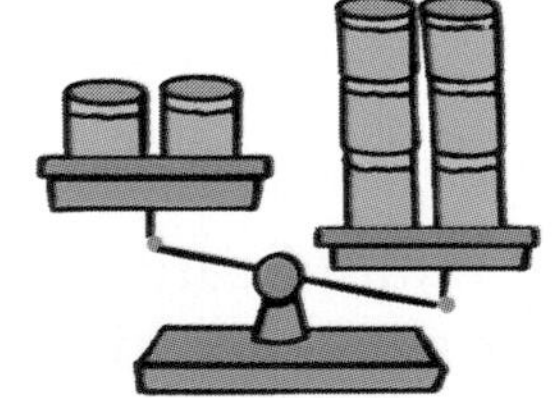

If you multiply the amount in one pan by 3, the scale becomes unbalanced.
$2 \neq 2 \times 3$

If you multiply the amount in the other pan by 3, the scale balances again!
$2 \times 3 = 2 \times 3$

See 172

You can also divide both sides of an equation by the same number without putting it out of balance.

Expressions and Equations

An **expression** names a number. An **equation** describes a relationship between two expressions. When you don't know all the numbers, an equation names these unknown numbers with letters, called **variables**.

Variables and Constants

A **constant** is a quantity that stays the same. A **variable** is a quantity that can change.

Constants stay the same.

- **25** number of cents in a quarter
- **12** number of inches in a foot
- **12** number of months in a year

Variables can change.

- number of inches tall you are
- amount of time you spend on homework
- number of cents in your pocket

You can choose any letter you like to stand for a variable. Sometimes the first letters of the important words make the meaning of the equation easier to remember.

237 Expressions

An expression names a number. Sometimes an expression is a number, like 6. Sometimes an expression is a variable, like n. Sometimes an expression shows an arithmetic operation, like $6n$ or $6 + n$.

238 Writing Expressions

To write an expression that describes what's going on in a word problem think about a word expression. Use numbers when you know what they are. Use variables when you don't know the numbers.

MORE HELP

See 043, 206–209

EXAMPLES:

Problem	Word Expression	Algebraic Expression
Show a full box of pencils and 3 extra pencils.	full box + 3 pencils	$b + 3$
Show a box of pencils with 3 pencils missing.	full box − 3 pencils	$b - 3$
Show 3 full boxes of pencils.	full box × 3	$3b$ When your expression has variables, write multiplication without the ×.
Show a full box of pencils shared equally among 3 people.	full box ÷ 3	$\frac{b}{3}$ When your expression has variables, write division in fraction form.

Evaluating Expressions

When you **evaluate** an expression, you substitute a number for every variable in the expression. Then, you can compute.

Expression	Evaluate if $b = 12$
$b + 3$	$12 + 3 = 15$
$b - 3$	$12 - 3 = 9$
$3b$ A number next to a variable means multiplication.	$3 \times 12 = 36$
$\frac{b}{3}$ The fraction bar means division.	$12 \div 3 = 4$

Sometimes the "X" for multiplication gets confused with a variable. So, to show that A is the product of l and w you can write: $A = lw$, $A = l(w)$, or $A = l \cdot w$.

Equations

An **equation** is a mathematical sentence. It always says that two expressions are equal. Some equations have only one variable. Then, there may be only one number that will make the equation true. But in an equation with more than one variable, there is often more than one way to make it true. Then, the variables really do vary!

When you **solve an equation** you find values for the variables that make the equation true. Sometimes there is only one solution, and sometimes there is more than one solution.

241 Writing Equations

To write an equation, think about which two quantities are equal to each other. Then write an expression for each quantity.

EXAMPLES:

Problem	Word Equation	Algebraic Equation
How many pieces of paper do I have if I have a ream and 6 loose sheets?	ream + 6 pieces = ■ pieces	$r + 6 = p$
How many pieces of paper are left after you use 6 pieces out of a ream?	ream − 6 pieces = ■ pieces	$r - 6 = p$
How many pieces of paper are in 6 reams of paper?	6 × ream = ■ pieces	$6r = p$
How many pieces of paper will each of us get, if all 6 of us equally share a ream?	ream ÷ 6 = ■ pieces	$r \div 6 = p$

Solving Addition and Subtraction Equations

To solve addition and subtraction equations, think about missing addends and sums. If the solution isn't obvious with mental math, you can use the guess and check problem-solving strategy. Whenever you solve an equation, you find values for the variables. Check the values you find by substituting them back into the equation to be sure they work.

EXAMPLE 1:
Solve $3 + r = 5$.

Think about missing addends. *3 + what number will give me 5?*

Since $5 - 3 = 2$,
then $3 + 2 = 5$

$r = 2$
Check: $3 + 2 = 5$

EXAMPLE 2:
Solve $m - 5 = 8$.

Think: *What number can I subtract 5 from and get 8?*

Since $8 + 5 = 13$,
then $13 - 5 = 8$

$m = 13$
Check: $13 - 5 = 8$

EXAMPLE 3: Carol is growing tomatoes. She counted 257 tomatoes after she threw away 17 that the squirrels had chewed. Write and solve an equation to find how many tomatoes she had before the squirrels got to them.

Think: *What number* − 17 = 257?
$t - 17 = 257$

Since $257 + 17 = 274$, then $274 - 17 = 257$.
$t = 274$

Check: $274 - 17 = 257$

There were 274 tomatoes before the squirrels got to them.

243

Solving Multiplication and Division Equations

To solve multiplication and division equations, you can think about missing factors and products. If the solution isn't obvious with mental math, you can use the guess and check problem-solving strategy.

EXAMPLE 1: Solve $10t = 250$

Think: *How many tens in 250?*
$250 \div 10 = 25$

$t = 25$
Check: $10 \times 25 = 250$

EXAMPLE 2: Solve $\frac{m}{2} = 450$

Think: *What number is twice as big as 450?*
$2 \times 450 = 900$

$m = 900$
Check: $900 \div 2 = 450$

EXAMPLE 3: 168 children showed up at the park to join the soccer league. The coaches grouped them equally into 12 teams. Write and solve an equation to show how many children they put on each team.

ONE WAY

You can think about this as a multiplication problem.
12 teams times how many children on a team will equal 168 children?

$12k = 168$
$k = 14$
Check: $12 \times 14 = 168$

ANOTHER WAY You can think about this as a division problem.
168 kids divided into what-size groups will equal 12 groups?

$$\frac{168}{k} = 12$$
$$k = 14$$

Check: $168 \div 14 = 12$

Either way, 14 children are on each team.

Making a Table of Values

244

You can think of many pairs of numbers with a sum of 80. That means there is more than one solution to the equation $x + y = 80$. When an equation has two variables, there can be many pairs of numbers that make the equation true. A table of values will show some of these pairs

EXAMPLE: The force of gravity on Pluto is $\frac{1}{25}$ the force of gravity on Earth. An equation that describes this relationship is $p = \frac{e}{25}$. A table of values will show weight on Pluto for several Earth weights.

Weight on Earth in pounds (*e*)	25	50	75	100
Weight on Pluto in pounds (*p*)	1	2	3	4

(Source: The World Almanac and Book of Facts)

245 Making a Graph from a Table of Values

You can make a graph to show how the values for the variables in an equation are related.

See 266

EXAMPLE: Here's a math puzzle about counting animals by counting their legs. There are chickens and horses in a farm yard and there are 24 legs altogether. Show the different possible combinations of chickens and horses.

First, write an equation.

Words	Expression or Equation
Multiply 2 legs by the number of chickens.	$2c$
Multiply 4 legs by the number of horses.	$4h$
The sum of chicken legs and horse legs is 24.	$2c + 4h = 24$

Next, make a table of values. Remember, $2c + 4h$ must equal 24.

c	h	$2c + 4h$
12	0	$24 + 0 = 24$
10	1	$20 + 4 = 24$
8	2	$16 + 8 = 24$

Then plot the points on the graph.

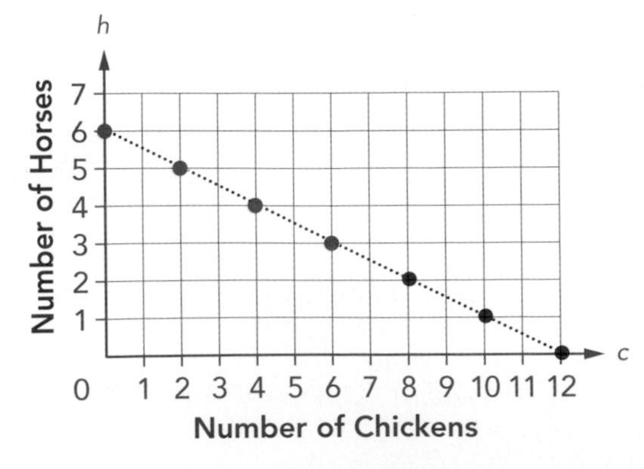

There are 7 possible combinations of chickens and horses: 12 and 0, 10 and 1, 8 and 2, 6 and 3, 4 and 4, 2 and 5, and 0 and 6. The problem says there are chickens *and* horses, so the best answers are 8 chickens, 2 horses; 6 chickens, 3 horses; 4 of each; and 2 chickens, 5 horses.

Using Formulas

Look at the formula for the area of a rectangle, $A = lw$. This is an equation. It has two expressions that name equal values.

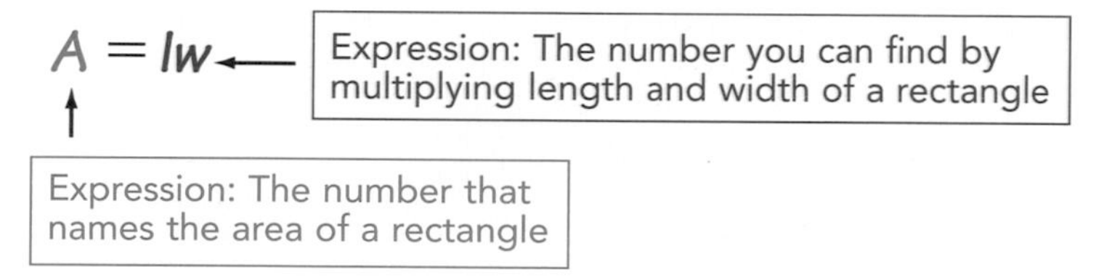

Because a formula is an equation, you can solve it just as you can an equation.

EXAMPLE: When you buy grass seed, you need to know the area of the lawn you want to plant. Here is a diagram of Ricardo's rectangular back yard. Is one sack of seed enough to plant the whole yard?

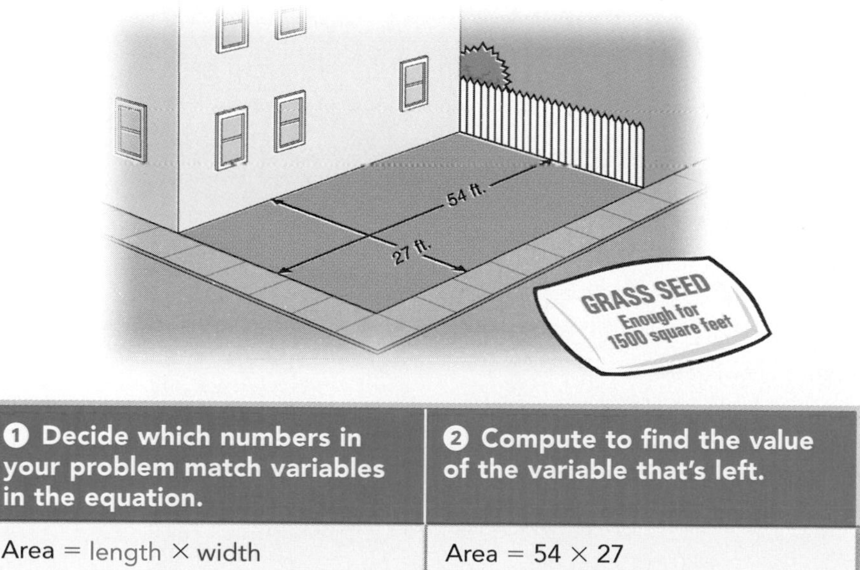

❶ Decide which numbers in your problem match variables in the equation.	❷ Compute to find the value of the variable that's left.
Area = length × width = 54 × 27	Area = 54 × 27 = 1458

The area of the yard is 1458 square feet. $1458 < 1500$, so, one sack of seed is enough.

247

Graphing, Statistics, and Probability

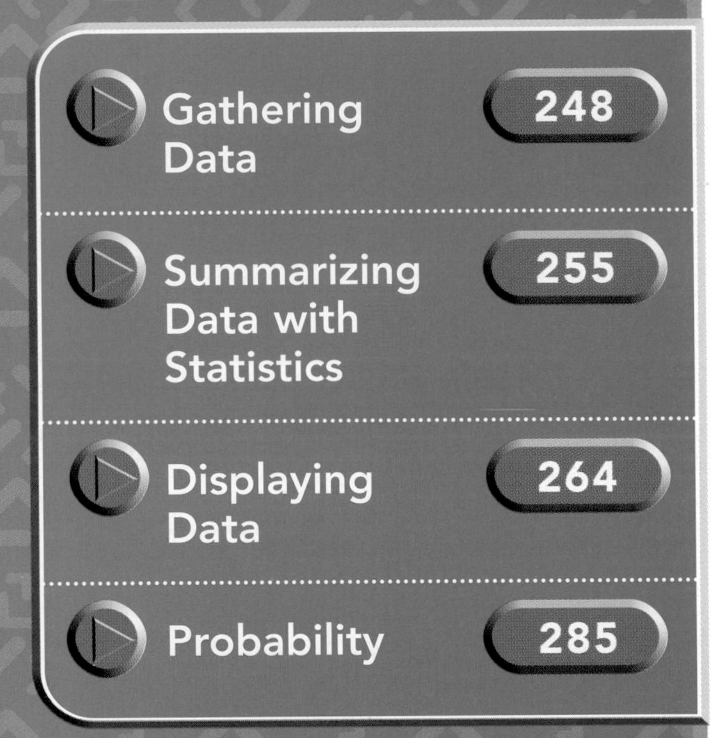

Will it rain tomorrow? Which batteries really last longest? How healthy or unhealthy is my lunch? Many questions like these don't have simple answers. Brand X batteries may usually last longer than Brand Y, but you might be unlucky and buy a package of Brand X batteries that weren't made so well. Still, you have a better chance with Brand X than with Brand Y. When you collect information about a group of people or things and try to tell what's *typical* about them, you are using **statistics**. For example, 65% of households in Columbus, Ohio, have cable TV. That's a statistic. It means that if you look at lots of households in that city, about 65 out of every 100 have cable. This does *not* mean that exactly 65 of the first 100 houses you see as you drive through Columbus have cable.

(Source: Nielsen Media Research)

248

Gathering Data

When you measure your height, you are collecting information. When you ask friends about their favorite TV shows, you are also collecting information. These collected pieces of information are called **data**.

Did You Know...

Data is a funny word. It means *more than one bit of information*. So, you write and say *data are* instead of *data is*. To talk about one bit of information, you can say *piece of data, bit of data, data point,* or *datum*.

249

Taking Samples

There are over 98 million households in the U.S. You can't ask everyone in every household every week what they're watching on TV. But the Nielsen company *can* ask everyone in about 5000 households what they're watching. The Nielson company is taking a **sample**. Nielsen tries to make their sample be a picture of all the different kinds of U.S. households. *(Source: Nielsen Media Research)*

250

Populations

When you are gathering information about a group, that group is called the **population**. The population can be small, such as fifth-graders in a school. The population can also be large, such as everyone living in the U.S. Populations can also be animals, or even toys.

Sample Size

251

You take a sample in order to get an idea about the whole population. You need to get enough data to be sure your conclusions are reliable.

EXAMPLE: What information will help you judge this statement?

One thing you should know is how many dentists were surveyed. A sample size of 300 or 3000 dentists should make you trust the results much more than a sample size of only 3 dentists.

Random Samples

252

A **random sample** is a group that represents the whole population.

EXAMPLE: You want to know what foods most students in your town like. You want a random sample.

Whom would you ask? The members of the boys' baseball teams? But that wouldn't include any girls. The students at one elementary school? But that wouldn't include older students. It also might not include students from all the neighborhoods in your town. So, if you're being honest, you'll randomly select people who represent the entire population of students, including boys and girls from different cultures and neighborhoods.

One way to do this is to have a person stand outside each school some morning and survey every tenth student walking by. The random sample would be all the students surveyed.

253 Recording Data

When you are gathering data, you need a way to keep track of that data. It helps if you collect information in an organized way. If you don't, you may miss something important about the data. Using a table or a chart helps you record your data.

EXAMPLE: You are a weather watcher for your local TV station. You report the temperature and the amount of new precipitation (rain or snow) each morning at 6 A.M.

Here's one table that will help you keep track of the data you collect.

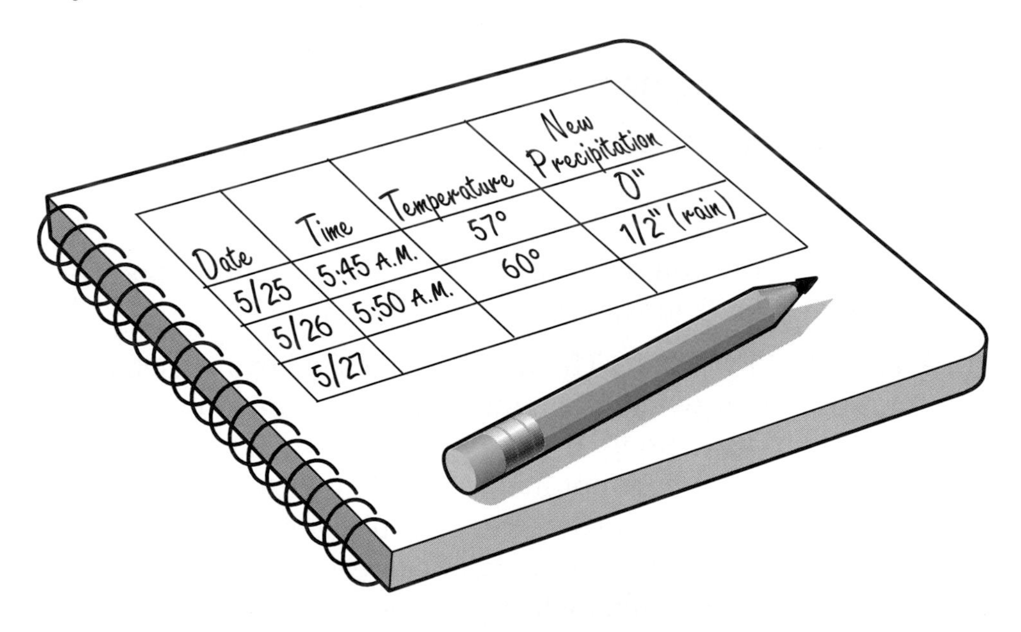

Tallies

254

You've probably been doing tallies since kindergarten. They're an easy way to keep track of things as you are counting them. Tallies are really helpful when you're counting events as they happen, and you can't keep track of the data in your head. Tallies are not so helpful when the data are already collected and written down where you can count them.

EXAMPLE: Every year, people help the Audubon Society by counting birds in December. You can help by counting all the birds that come to your backyard feeder between certain times. A good way to be sure you count every bird you see is to use a tally.

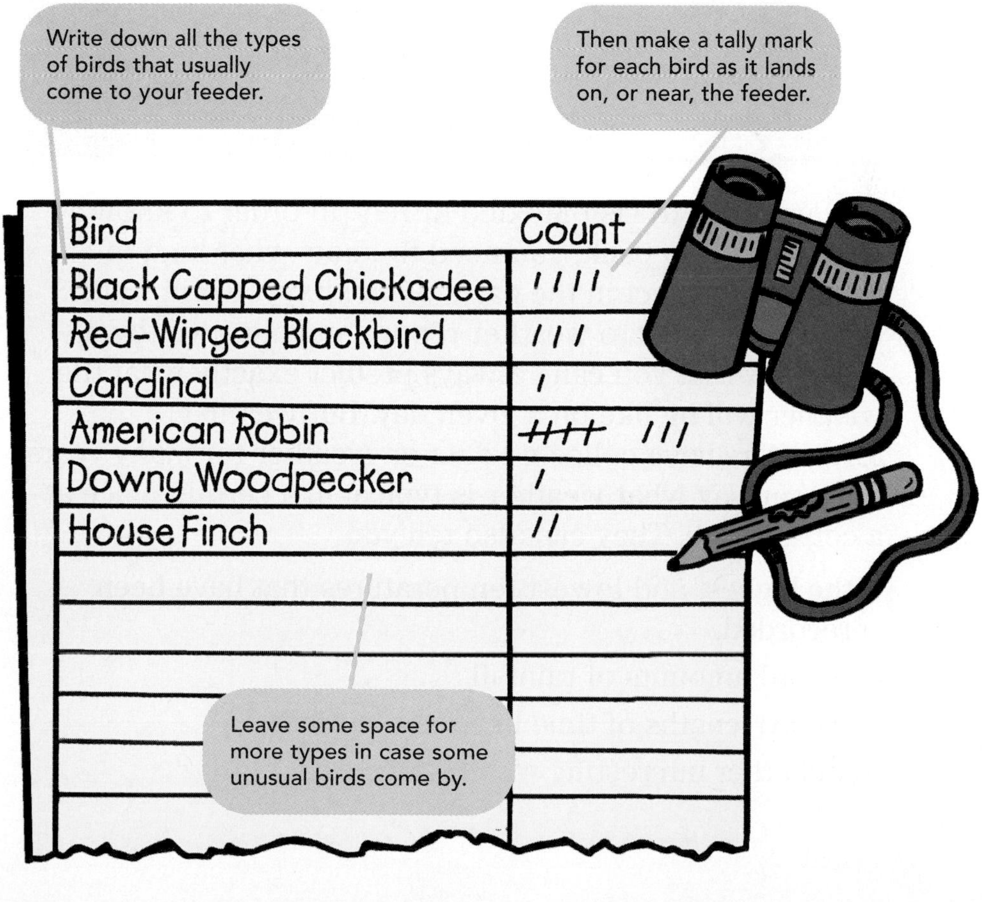

Write down all the types of birds that usually come to your feeder.

Then make a tally mark for each bird as it lands on, or near, the feeder.

Bird	Count
Black Capped Chickadee	////
Red-Winged Blackbird	///
Cardinal	/
American Robin	卌 ///
Downy Woodpecker	/
House Finch	//

Leave some space for more types in case some unusual birds come by.

255

Summarizing Data with Statistics

You are going to visit Mexico in July. In order to know what clothes to pack, you need to know what kind of weather to expect in the parts of Mexico you will visit. If you've listened to weather reports in your own town, you know that you can't always predict exactly what the weather will be like on a given day. But weather forecasters have collected data on weather for many years. They can say what weather is typical in a certain place at a certain time. They can also tell you

- the highest and lowest temperatures that have been recorded,
- record amounts of rainfall,
- record lengths of time between rainfalls,
- and other interesting weather records.

Variability of Data

256

Not everyone in your class is the same height, but the heights are all between 1 foot and 10 feet. How much the data in a group are spread out, or vary, is called **variability**. Range and outliers are measures of the variability of data. These measures can help you understand data. They can also help you make a graph the right size so all your data fit.

Range

The **range** of a set of data is the difference between the greatest and least numbers in the set. If the range is a small number, the data are close together.

EXAMPLE: Here are some fishing statistics from the San Diego area for one week in June of 1998. What is the range of the data?

Lake	People Fishing
Barrett	194
Hodges	677
Jennings	767
Miramar	205
Murray	182
Lower Otay	696

(Source: www.sdfish.com)

To find the range, put the numbers in order from least to greatest. 182, 194, 205, 677, 696, 767

The difference between the least and greatest numbers is $767 - 182 = 585$.

 The range is 585 people. You can say that the number of people fishing is very different among the lakes.

258 Outliers

Did you ever have a video-game score that was incredibly better or worse than all your other scores? In statistics, a piece of data that seems to float too far out at one end of the range is called an **outlier**. Outliers can affect how you interpret your data.

EXAMPLE: The people fishing at Lower Otay Lake one week in June 1998 caught (and kept) the fish shown in the table. Does the set of data contain any outliers?

Type of Fish	Number Caught
Bass	7
Bluegill	3203
Channel Catfish	17
Sunfish	10
Crappie	2

(Source: www.sdfish.com)

To decide, in an informal way, whether there are any outliers, place the numbers on a number line. Then check to see how the data are grouped. If there is one number much less or much greater than the others, you have found an outlier.

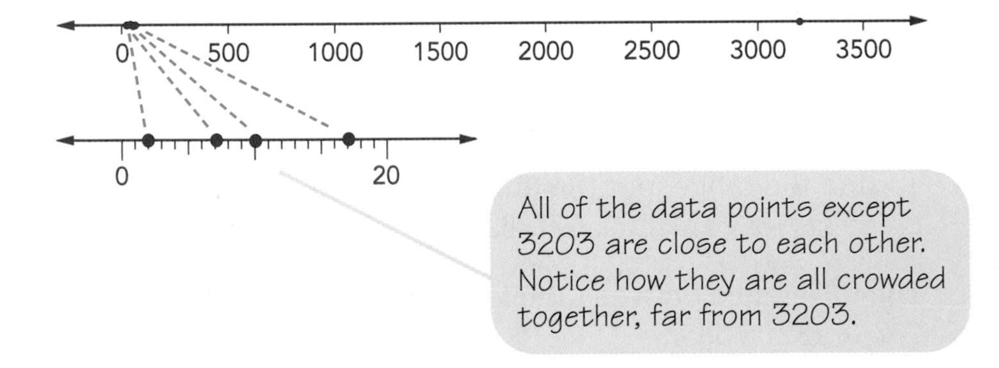

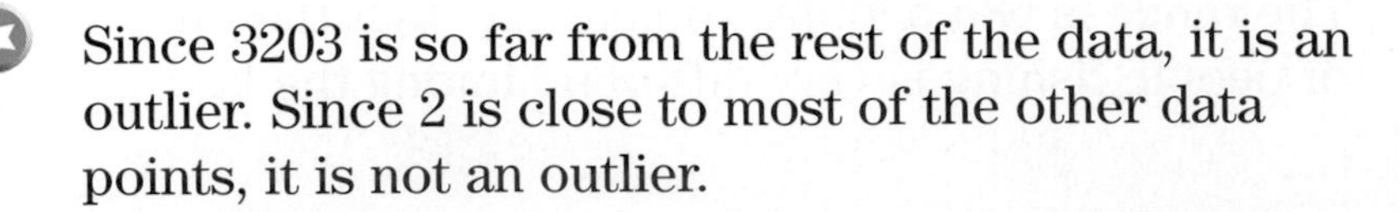

Since 3203 is so far from the rest of the data, it is an outlier. Since 2 is close to most of the other data points, it is not an outlier.

Types of Average

If you were to pick a number that best describes all the heights of students in your class, what number would you pick? To find a number like that, people often put the data in order from least to greatest. Then they choose a number somewhere in the middle of the data or a number with a lot of data clustered around it. This number is an average. There are three types of **average**: *mean*, *median*, and *mode.*

The three main ways to describe average are useful in different situations.

1. For sets of data with no very high or low numbers, the mean usually works well.
2. For data sets with a couple of points much higher or lower than most of the others, the median may be a good choice.
3. For sets of data with many identical data points, the mode may be the best description.

Look at the question you are trying to answer to decide which type of average to use.

What's the average height of the players on this team?

260 Mean

In math, the **mean** is an average. If the numbers in a set of data were *evened out* so that all the numbers were the same, that evened-out number would be the mean.

CASE 1 When all the data are close together, the mean is close to all the data.

EXAMPLE 1: The table shows the number of students absent from Ms. Dori's class each day for one week. What was the mean number of students absent per day that week?

Day	Number of Students Absent
Monday	2
Tuesday	5
Wednesday	2
Thursday	1
Friday	5

ONE WAY Find the mean by evening out the numbers.

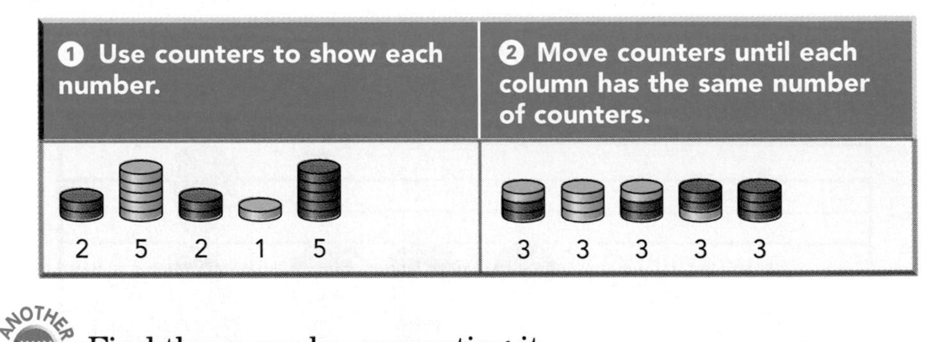

ANOTHER WAY Find the mean by computing it.

① Add to find the sum of the numbers.	② Divide the sum by the number of addends.
$2 + 5 + 2 + 1 + 5 = 15$	$15 \div 5 = 3$

Either way, the average number of students absent was 3 per day.

CASE 2 When one piece of data is much bigger or much smaller than the rest, it can move the mean away from the main group of data.

EXAMPLE 2: In order to convince your parents that they should increase your allowance, you collect data from your friends. What is their mean weekly allowance?

See 258

Notice that one of your friends has a much greater weekly allowance than any of the others. First, look at what the mean would be without that outlier.

❶ Add to find the sum of the amounts.	❷ Divide the sum by the number of addends.
\$4.50 + \$6.50 + \$6.00 + \$5.00 = \$22.00	\$22.00 ÷ 4 = \$5.50

Now look at what the outlier does to the mean.

❶ Add to find the sum of the amounts.	❷ Divide the sum by the number of addends.
\$4.50 + \$6.50 + \$6.00 + \$50.00+ \$5.00 = \$72.00	\$72.00 ÷ 5 = \$14.40

The mean weekly allowance is \$14.40.

When you found the mean of the amounts that were close together, it was near all the amounts. Then you found the mean with an outlier. The outlier pulled the mean up so far that it was larger than all of the amounts in the group except the outlier itself. But, none of your friends has an allowance anywhere near \$14.40.

261 Median

Arrange a set of data in order from least to greatest. The **median** is the number that falls exactly in the middle. If there aren't any big gaps in the middle of the data, the median may be the best number to describe all the data.

CASE 1 When there is an odd number of data points, the median is the middle number.

EXAMPLE 1: What is the median time of songs on this CD? Why is median a good average for these data?

Song	Time (min:s)
New San Antonio Rose	2:45
Texas Swing	2:45
The Yellow Rose of Texas	2:58
Remember the Alamo	2:48
No Place But Texas	2:59
El Paso	4:58
Panhandle Rag	2:46
Ballad of the Alamo	2:52
Lone Star Rag	2:59
Deep in the Heart of Texas	2:50
Cross the Brazos at Waco	2:47

(Source: Great Songs About the Great State of Texas)

See 010, 260

❶ Arrange the times in order from least to greatest.	2:45, 2:45, 2:46, 2:47, 2:48, 2:50, 2:52, 2:58, 2:59, 2:59, 4:58,
❷ Find the middle time.	2:45, 2:45, 2:46, 2:47, 2:48, 2:50, 2:52, 2:58, 2:59, 2:59, 4:58,

The median for this set of data is 2:50.

The mean is about 3:03. Most of the times are less than 3 minutes. So, *median* is a better average time.

CASE 2 An even number of numbers in a set of data has two middle numbers. In this case you need to find the mean of the two middle numbers. This creates a *fake* middle number for the set of data.

EXAMPLE 2: A music store is open six days a week, Monday through Saturday. The table below shows the number of CDs the store sold in one particular week in the summer. What is the median number sold per day?

	Mon.	Tues.	Wed.	Thurs.	Fri.	Sat.
Number of CDs Sold	313	2301	395	412	221	686

❶ Arrange the numbers in order from least to greatest. Find the two middle numbers.	❷ Find the mean of the two middle numbers. This is the *fake* middle number.
221, 313, 395, 412, 686, 2301	395 + 412 = 807 807 ÷ 2 = 403.5

The median is 403.5 CDs.

Many more CDs were sold on Tuesday than on any of the other days. That is because the store had its biggest sale of the year on that day! If you used the mean to describe the average number of CDs sold per day, the sale day would make the average much higher. The median is not affected by the one high number. So you can say that on a typical summer day, the store sells about 404 CDs.

262 Mode

Sometimes, the best way to describe what is typical about a set of data is to use the value that occurs most often. This value is called the **mode**. For example, in the set of data 2, 3, 5, 5, 6, the mode is 5 because it appears more often than any other number.

CASE 1 Sometimes there is one value that occurs more often than any other.

EXAMPLE 1: The table shows the result of a survcy taken in Mr. Dombrowski's class. What is the mode of the data?

Number of Children in Family	Number of Students
1	𝍸
2	𝍸 𝍸 ||
3	𝍸 |
4	|||
5	|

If you look at the data, more students had 2 children in their family than had any other number.

The mode is 2.

To find the median, list all the data points in order.

1 1 1 1 1 2 2 2 2 2 2 2 2 2 2 2 2 3 3 3 3 3 3 4 4 4 5

↑ median

The median for these data is the 14th number, which is also 2. So, it makes sense, when talking about the families of the students in Mr. Dombrowski's class, to say that the typical family has 2 children.

CASE 2 Sometimes there is more than one value that occurs most often. In this case, all these highest values are modes for the set of data.

EXAMPLE 2: The table shows the ages of the scouts in one Boy Scout troop. What is the mode?

Age	Number of Scouts
9	1
10	4
11	4
12	2

You could say that the ages are **bimodal** (two modes: 10 and 11). There are 4 scouts that are 10 and 4 scouts that are 11.

CASE 3 Sometimes there is no value that occurs more often than the others. In this case, there is no mode. When there is no mode, you can use the mean or median to describe the set of data.

MATH ALERT Statistics Can Be Misleading

Before you jump to conclusions based on statistics, think carefully. Statistics can mislead you.

EXAMPLE: The public swimming pool announces that the mean age of swimmers on Monday mornings is 15. Can you be sure that mostly teenagers swim then?

There are many groups of ages that could give a mean of 15. Here are two:

- 12 13 13 14 16 17 17 18
- 4 5 6 6 24 24 25 26

No, you can't be sure that mostly teenagers swim on Monday mornings.

264

Displaying Data

Displaying data clearly can help you prove a point. It can also help you learn things from your data. Clear displays can help you see trends, make predictions, and compare ideas. And if you know how to read graphs carefully, you can even tell whether someone's trying to fool you!

A big part of showing data clearly is choosing which kind of display to use. If you want to show how your friends' family pet data are related to each other, you might use a Venn diagram. If you want to show how the price of a pack of gum has changed over the years, you might choose a line graph. To compare the numbers of games your favorite teams have won, you might select a bar graph. Whatever kind of data you have, there is probably a diagram or graph that will help you see the big picture hidden in your data.

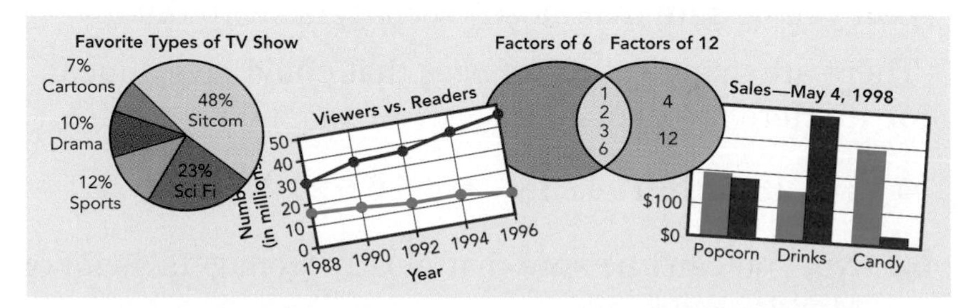

Coordinate Geometry 265

A grid is very useful. In some cities, streets are laid out in a grid and it is very easy to tell someone how to get from one place to another.

A coordinate grid is a way to locate points in a plane. To draw a coordinate grid, draw a horizontal number line and a vertical number line.

You can name any point on this plane with two numbers. These two numbers are called **coordinates**. The pair is *always* named in order (first x, then y), so it's called an **ordered pair**.

Write: (3, 4)
Say: *point three four*

The vertical number line is called the **y-axis** or the **vertical axis**.

The second number is the ***y*-coordinate**. It tells the distance along the *y*-axis.

y-axis

5
4
3
2
1
0

(3, 4)

0 1 2 3 4 5

x-axis

The horizontal number line is called the **x-axis** or the **horizontal axis**.

The point where the axes meet (0, 0) is called the **origin**.

The first number is the ***x*-coordinate**. It tells the distance from the origin along the *x*-axis.

266 Points on the Coordinate Grid

CASE 1 Sometimes you have the coordinates of a point and you need to place the point (**plot** it) on the grid. When you're plotting a point, start where the axes cross. On this grid, this is point (12,30). Let the x-coordinate tell you how far to move horizontally (→). Then, let the y-coordinate tell you how far to move vertically (↑).

MORE HELP

See 269–271

EXAMPLE 1: You're making a graph to show how the temperature changes during the afternoon. At 1 P.M., it is 40°. Plot the point (1, 40).

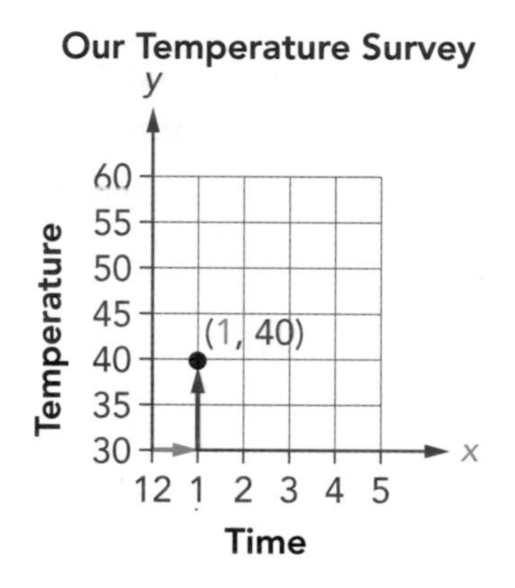

CASE 2 Sometimes you have the point and need to name it.

EXAMPLE 2: The grid shows the locations of different spots in a town. What ordered pair describes the location of the movie theater?

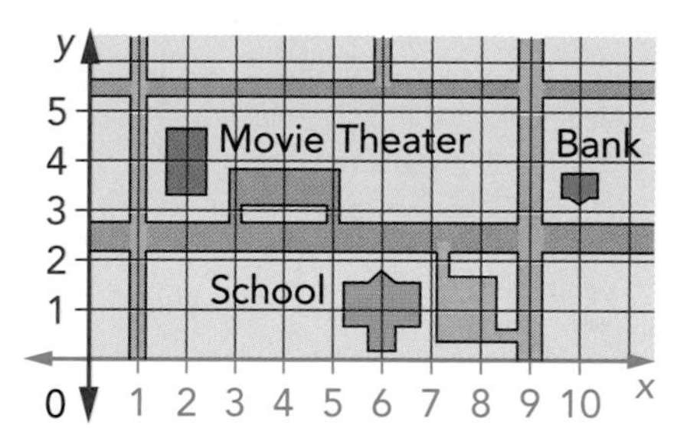

1. Find the distance along the x-axis. Put your finger on the movie theater. Move straight down to the x-axis. Read the number, 2.
2. Find the distance along the y-axis. Put your finger back onto the movie theater. Move straight over to the y-axis. Read the number, 4.

The movie theater is at (2, 4).

Data in Tables

267

Suppose you have lots of data about a topic. You can make a table that makes it easy to find and compare pieces of data.

EXAMPLE: Find the number of bronze medals won by Germany in the 1998 Winter Olympics.

Country	Gold	Silver	Bronze	Total
Germany	12	9	8	29
Norway	10	10	5	25
Russia	9	6	3	18

(Source: Facts on File)

Move along the row for Germany until you find the number in the column for bronze. The number is 8.

Germany won 8 bronze medals.

Frequency Tables

268

A **frequency table** is a way to show how often an item, a number, or a range of numbers occurs.

EXAMPLE: Some words and phrases get in the way of a good oral report if you say them too often. So, as you practice, your best friend counts those words and phrases. How can you record the data?

Make a frequency table. List each item in the data. Then count and record the number of times each item occurs.

"Like I said, um, my plants grew faster when I, um, watered them. Y'know, like, I mean, every, um, day . . ."

269 Making Graphs

When you make a bar graph or a line graph, you start with a grid, a set of crossed lines. To make that grid a graph, you need to give meaning to the lines.

270 Labeling the Axes

See 265–266

The **axes** on a graph are the reference lines. They are a horizontal line and a vertical line that cross. The lines in the grid all have names or numbers. If they have numbers, the lowest number is usually at the place where the axes cross. How you label the axes depends on your data.

EXAMPLE: Suppose you want to make a graph that shows the average daily temperature in your town for each month for a year. How would you label the two axes?

Since you want to show the temperatures for each month, one axis would list the months. The other axis would list the temperatures.

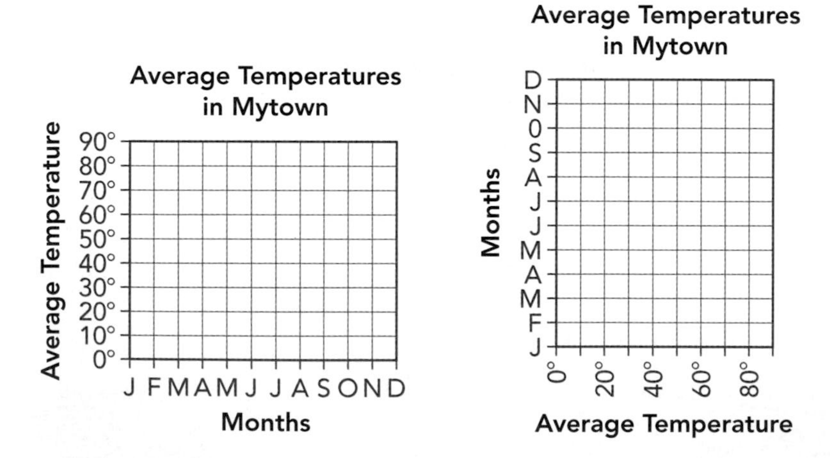

A bar graph might use either of these setups. The first grid would give you a graph with vertical bars. The other grid would give you a graph with horizontal bars. On line graphs, time is usually shown on the horizontal axis.

Choosing the Scales

271

After you label the axes, choose the **scale**—the numbers running along a side of the graph. The difference between numbers from one grid line to another is the **interval**. The interval will depend on the range of your data and the number of lines on your graph paper. When you can, you should choose simple scales, with the numbers starting at 0 and increasing by 1 or some other easy number.

See 257

EXAMPLE: Choose the scales for a graph of these data.

Temperatures in Sometown												
Month	J	F	M	A	M	J	J	A	S	O	N	D
Average High Temperature (°C)	0	1	9	17	23	28	30	29	25	19	9	1

You can list the months across the horizontal axis and the temperatures along the vertical axis. The range of temperatures is 30°. You can choose the interval between grid lines.

- Choosing an interval of 1 will make a tall, skinny graph.
- Choosing an interval of 5 will make a more squarish graph.
- Choosing an interval of 10 will make a short, wide graph.

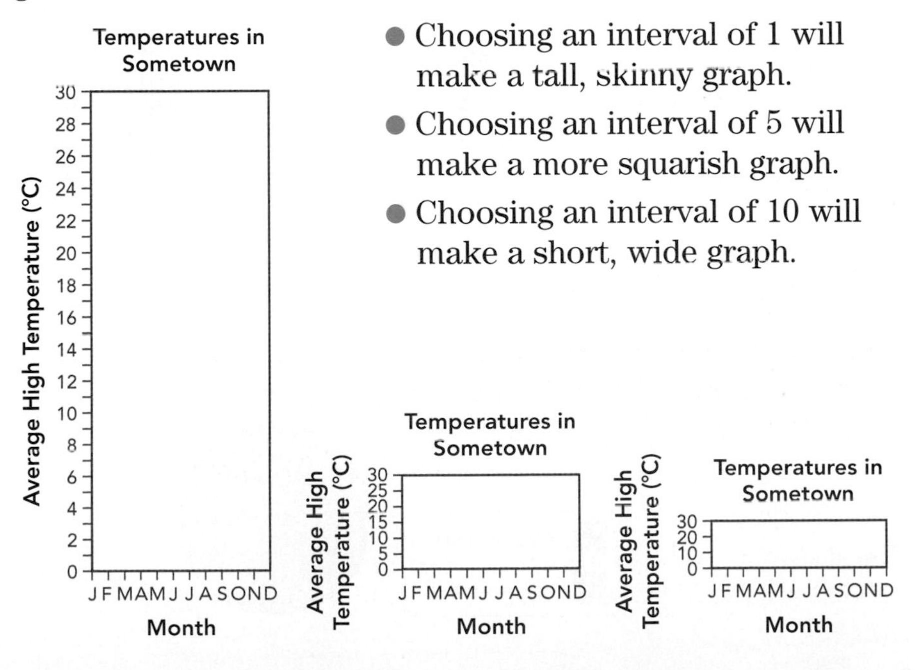

272 Graphs That Compare

Some graphs illustrate how data compare.

- Graphs can show the same kind of data at different times or places.

- Graphs can show different sets of data at the same time or place.

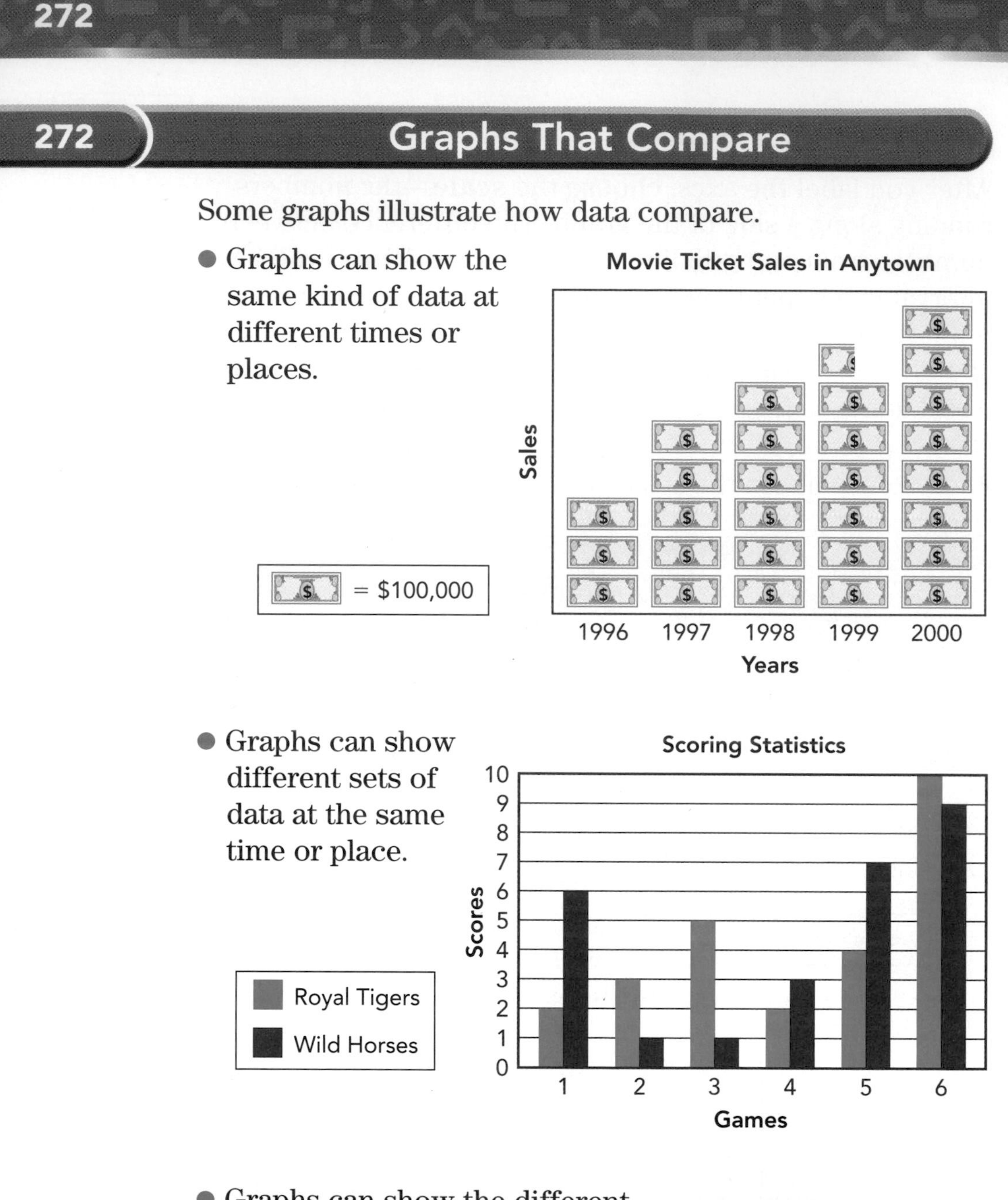

- Graphs can show the different kinds of data that make up 100% of one group of data.

Ages of Kids in Fourth Grade

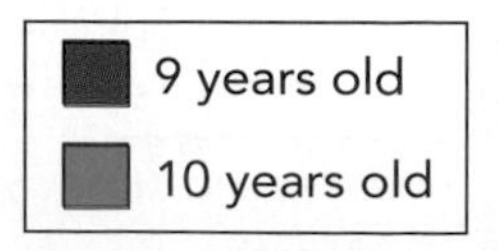

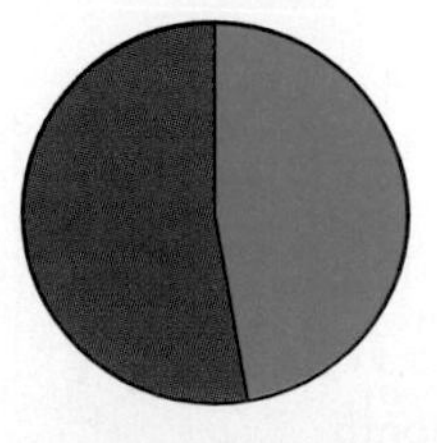

Single-Bar Graphs

In a bar graph the length of bars represents numbers.

Titanic made more money in its opening weekend than any of the other three movies.

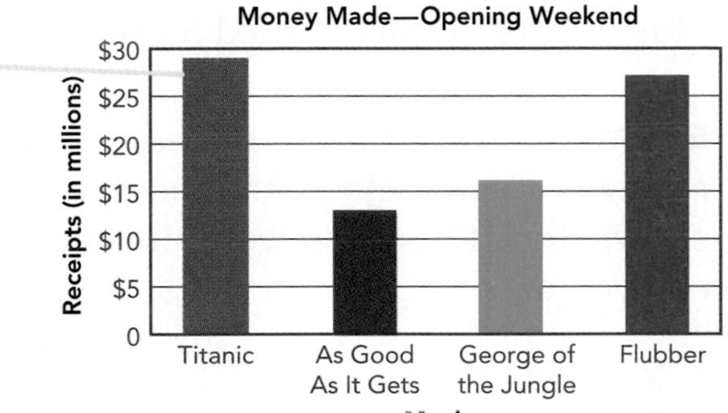

EXAMPLE: Here is a list of the top four opening three-day weekend moneymakers as of May, 1999. Make a bar graph to compare these amounts.

Movie	Opening Weekend Sales (to nearest million dollars)
The Lost World	$93,000,000
Star Wars Episode I	$65,000,000
Mission: Impossible	$57,000,000
Godzilla	$56,000,000

(Source: www.boxofficeguru.com)

1. **Title your graph. Draw and label your axes.**
2. **Choose increments for the scale. The data range from 56 million to 93 million.**
3. **Write the movie titles on the horizontal axis. Estimate where each amount would fall on the vertical axis, then draw a bar to that height.**

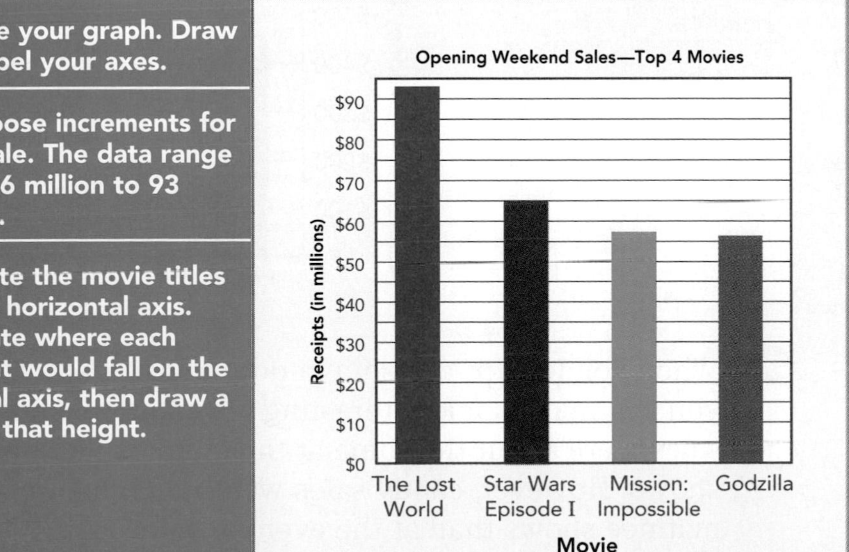

274

Double-Bar Graphs

Sometimes, to compare sets of data, a bar graph with pairs of bars works better than a bar graph with individual bars. This kind of graph is called a **double-bar graph**.

EXAMPLE: The manager of *Anytown Movies* keeps records of daily refreshment sales in a table. The graph shows the information from the table. What observations can you make?

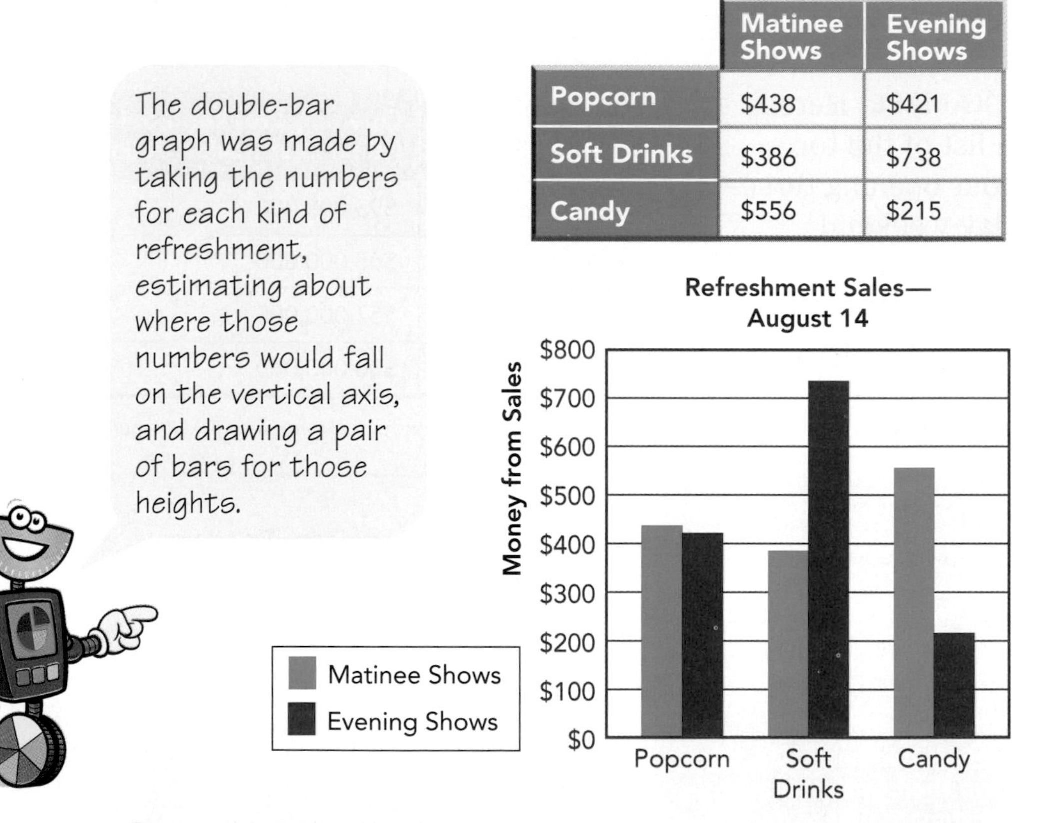

	Matinee Shows	Evening Shows
Popcorn	$438	$421
Soft Drinks	$386	$738
Candy	$556	$215

When you look at the data in double-bar graph form, you can make some interesting observations. Popcorn sales were about the same at the matinee and evening shows. However, candy sales were much higher at the matinee shows than at the evening shows.

Pictographs

Do you want to compare data, but do it in a way that's more eye-catching than a bar graph? A **pictograph** uses *pictures* or symbols to compare data.

EXAMPLE: You take a random survey to find out what kinds of movies students in your school like best. Make a pictograph that shows the results.

Kind of Movie	Number of Votes
Drama	15
Comedy	25
Musical	10
Adventure	45
Science Fiction	20
Animation	15

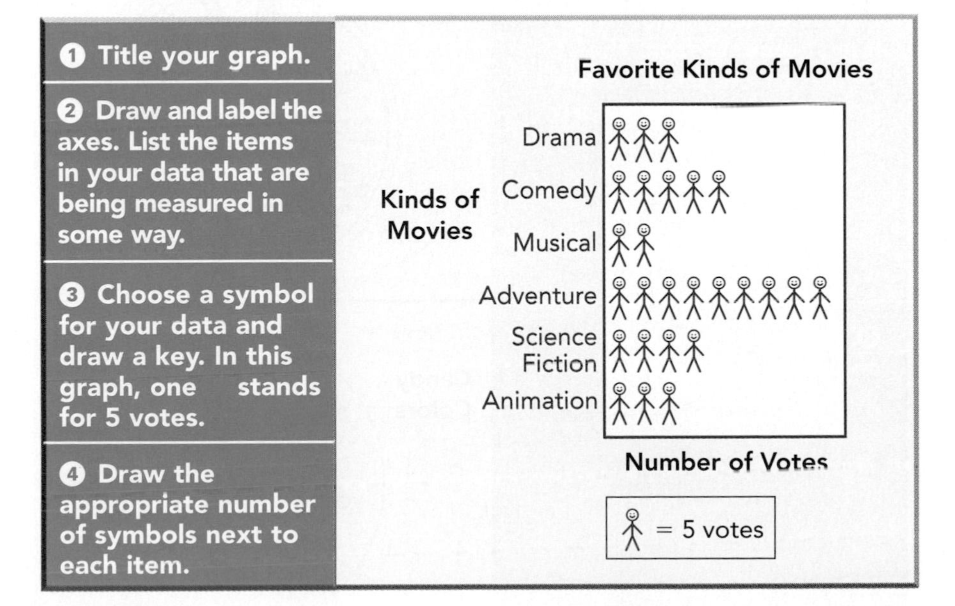

276 Circle Graphs

A **circle graph** is a helpful way to organize data. Circle graphs are sometimes called pie charts.

EXAMPLE: A bag of hard candies contains 15 candies of the following colors: 7 red, 2 yellow, 1 orange, 3 green, and 2 purple. Make a circle graph to show these data.

ONE WAY

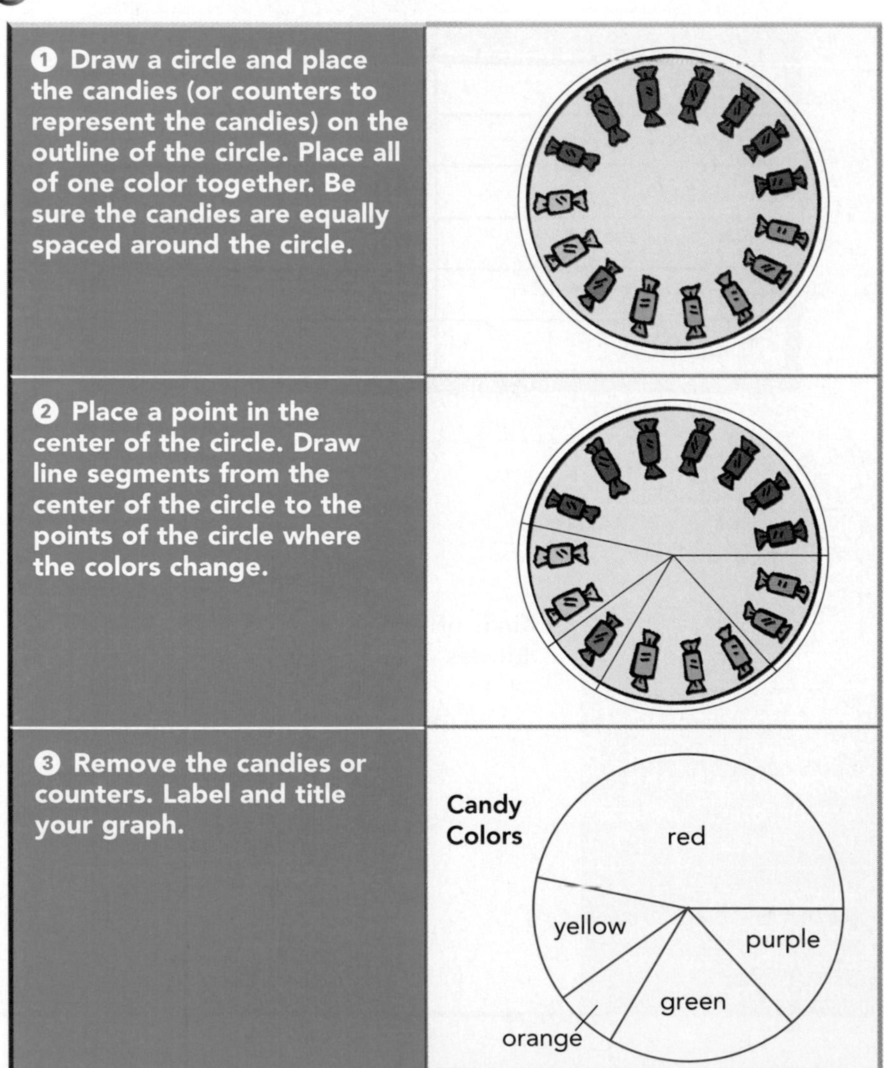

ANOTHER WAY

MORE HELP

See 028, 033, 167–169, 200, 466

❶ Draw a circle. Mark the center.	
❷ For each color, write a fraction that shows what part of all the candies are that color.	red $\frac{7}{15}$ yellow $\frac{2}{15}$ orange $\frac{1}{15}$ green $\frac{3}{15}$ purple $\frac{2}{15}$
❸ Multiply each fraction by 360° to find out how many degrees of the circle you'll need for each color.	$\frac{7}{15} \times 360° = 168°$ $\frac{2}{15} \times 360° = 48°$ $\frac{1}{15} \times 360° = 24°$ $\frac{3}{15} \times 360° = 72°$ $\frac{2}{15} \times 360° = 48°$
❹ Use a protractor to draw a central angle for the first color. Make sure it is the size you calculated in Step 3.	
❺ Draw angles for the rest of the colors. Label and title your graph.	Candy Colors: red, yellow, purple, green, orange

The circle graph shows that almost half of the candies are red and that there aren't many orange candies. On a bar graph, it would still be easy to see that there aren't many orange candies. But it wouldn't be as easy to see that almost half of the candies are red.

277 Graphs That Show Change Over Time

Some graphs are best at showing how things change over time. The most widely used graphs for showing change over time are line graphs.

278 Single-Line Graphs

Most line graphs have some sort of time measure like minutes, days, or years on the horizontal axis. The vertical axis will usually have some other measure. When you look at the line in a **line graph**, you can tell whether something has increased, decreased, or stayed the same as time has passed.

EXAMPLE 1: From 1992 through 1996, in-line skating was one of the country's fastest-growing sports. During which year did the number of in-line skaters grow the most?

Year	Number of Skaters (to the nearest 100,000)
1992	9,400,000
1993	12,600,000
1994	18,800,000
1995	22,500,000
1996	27,500,000

(Source: Sporting Goods Manufacturers Association)

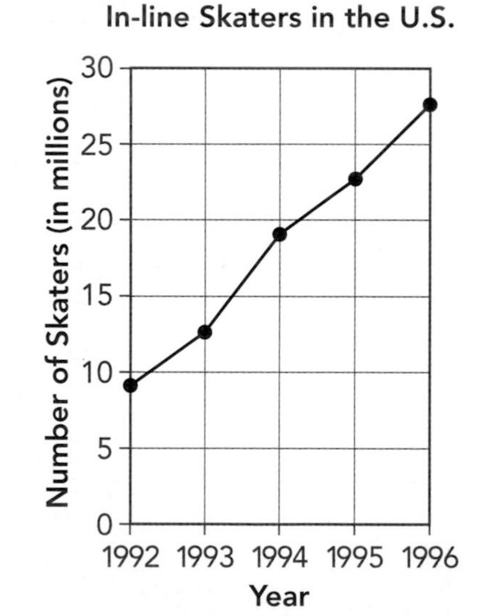

To find the year in which in-line skating increased the most, find the steepest line segment. The steepest line segment runs from 1993–1994.

 The number of in-line skaters increased the most from 1993 to 1994.

EXAMPLE 2: This table shows the total number of skating injuries from 1992 through 1996. Make a line graph with these data.

Year	Number of Injuries (to the nearest thousand)
1992	29,000
1993	37,000
1994	75,000
1995	100,000
1996	103,000

(Source: National Electronic Injury Surveillance System)

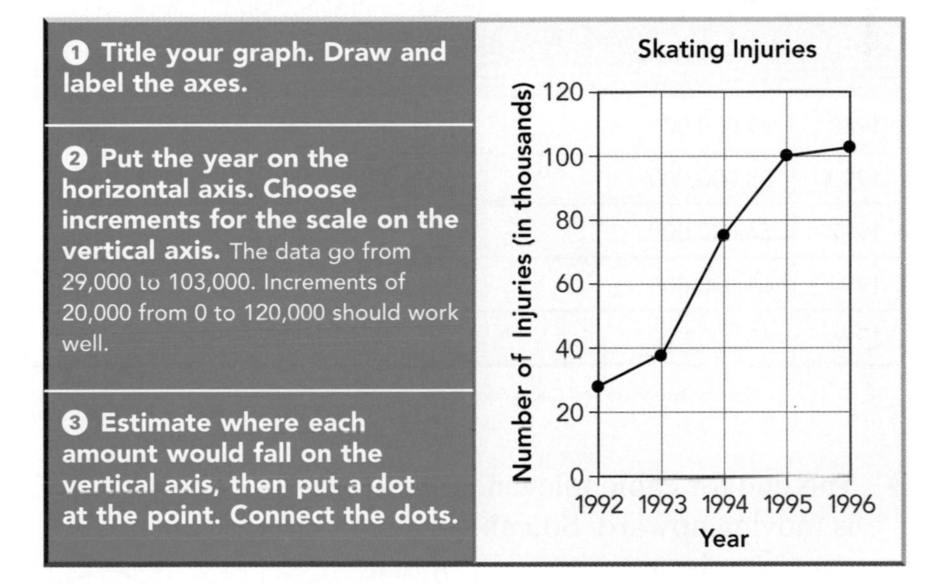

The graph shows that the number of skating injuries increased each year. This is not that surprising since the number of in-line skaters also increased during the same time period.

279 Multiple-Line Graphs

Suppose you want to compare two or more quantities that are increasing or decreasing over time. You can use a **multiple-line graph**. Each line shows one set of data.

EXAMPLE: People's entertainment and news habits are changing. The table and graph show the numbers of cable television subscribers and the daily newspaper circulation from 1988–1996. What does the graph tell you?

Year	Number of Cable TV Subscribers (to the nearest million)	Daily Newspaper Circulation (to the nearest million)
1988	49,000,000	63,000,000
1990	55,000,000	62,000,000
1992	56,000,000	60,000,000
1994	61,000,000	60,000,000
1996	65,000,000	57,000,000

(Source: Television and Cable Fact Book; Editor and Publisher International Yearbook; The World Almanac and Book of Facts)

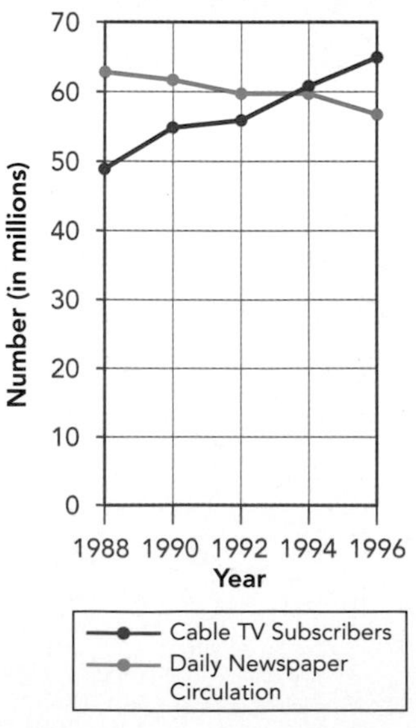

The line for cable television is moving upward. So cable television is getting more popular. The line for newspapers is dropping or staying the same. So newspapers are not getting more popular. If both trends continue, the number of cable TV subscribers will exceed the daily newspaper circulation more and more each year. These data don't tell us *why* the trend is this way.

Time Lines

A time line is a graph. It is really a number line with numbers that are years or dates or times of day. Events that happened in the past can be graphed on a time line. Events that are planned for the future can also be graphed on a time line. You can even graph your daily schedule on a time line.

EXAMPLE: Here's Joe's plan for the day. Make a time line with the data.

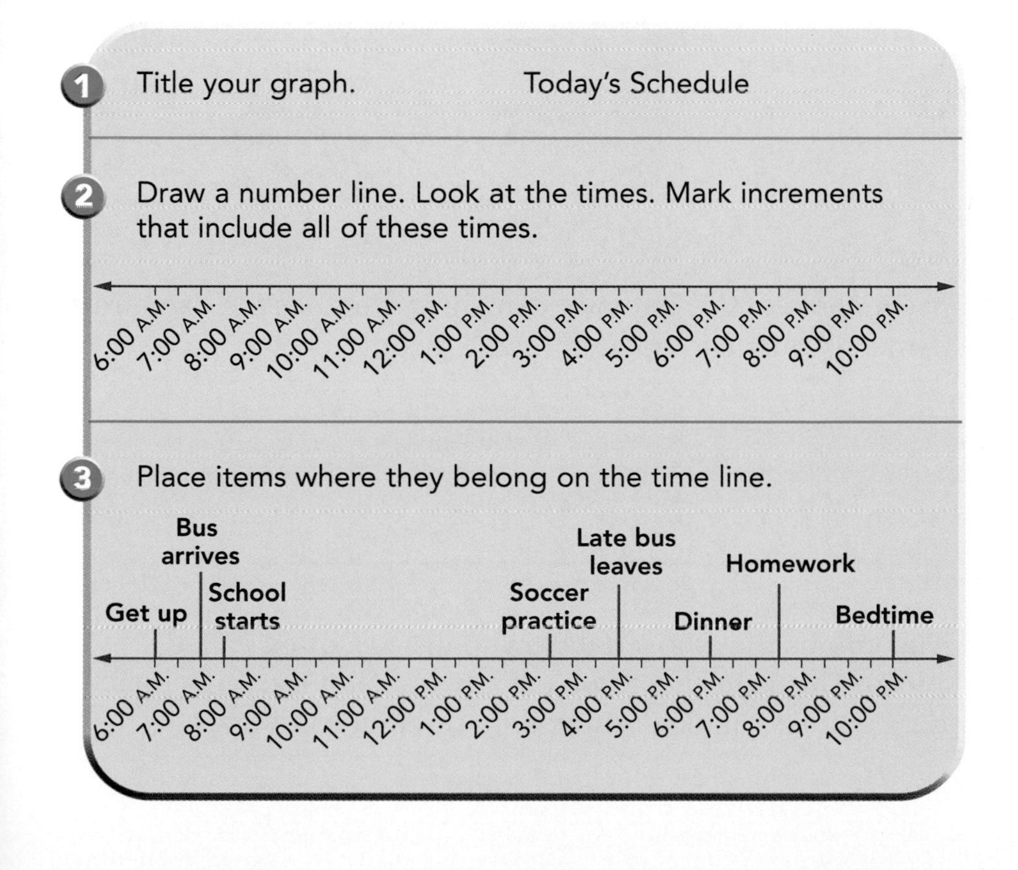

281 Graphs That Show How Data Are Grouped

Some graphs do a very good job of letting you see how data are grouped.

- Graphs like this Venn diagram can show you which of your data belong together.

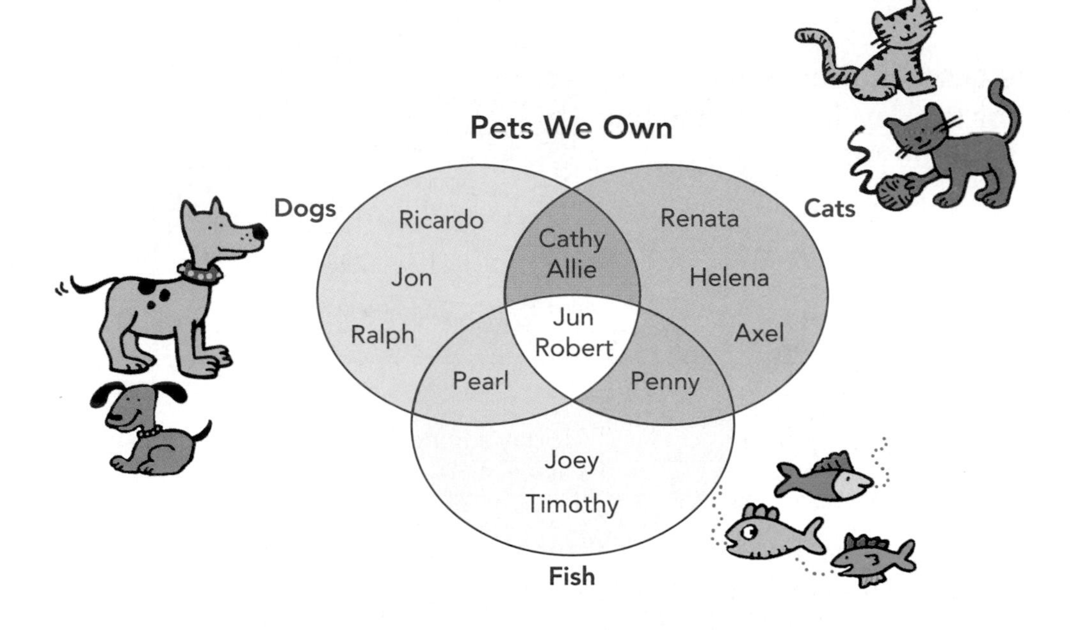

- Graphs like this line plot can show whether the data are mostly bunched up or spread way out.

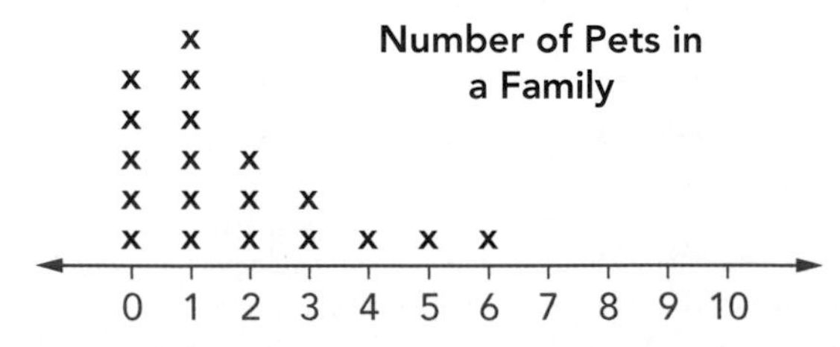

The graphs that show data groupings are often called **plots**. This is because you usually just plot the individual data points; you don't draw bars or connected lines.

Line Plots

Sometimes, instead of comparing data or showing trends, you want to show the spread of the data. You can do this with a **line plot**. On a line plot, you can quickly identify the range, mode, and any outliers.

Students	Hours
Anna	14
Barbara	16
Carlos	12
Debra	14
Eric	14
Franny	11
Glenna	20
Han	12
Jules	10
Kami	16
Lee	15
Mark	17
Nancy	5
Oren	15
Patty	10

See 257–258, 262

EXAMPLE: Fifteen students estimated how much television they watch each week, to the nearest hour. The table shows their estimates. Show these results on a line plot. Then identify the mode of the data and any outliers that you see.

1. Title your plot. Draw a number line on grid paper. The scale of numbers should include the greatest value and the least value in the set of data.

2. For each piece of data, draw an x above the corresponding number.

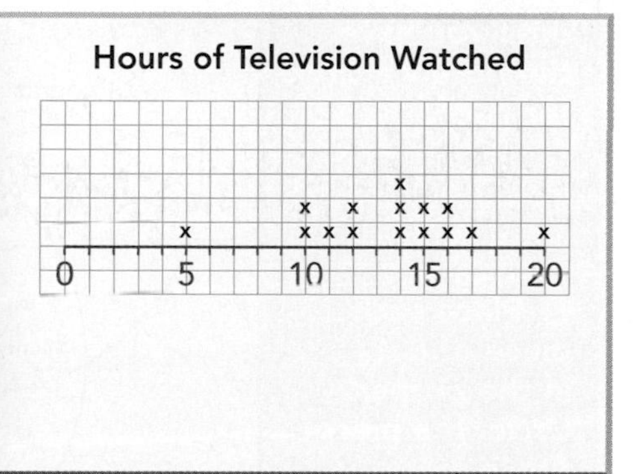

The most x's are above 14. So, the mode is 14. Five is separated from the rest of the data. So, 5 is an outlier.

283

Venn Diagrams

A **Venn diagram** is a group of intersecting circles. Each circle is named for the data in it. Data that belong in more than one circle are placed where the circles overlap.

EXAMPLE: Make a Venn diagram to display the data.

City	Warmest Weather
San Francisco	September 9–20
Seattle	July 15–August 2
Phoenix	July 11–12
Dallas	July 26–August 7
Minneapolis	July 15–August 2
St. Louis	July 22–23
Atlanta	August 3–4

(Source: National Climatic Data Center)

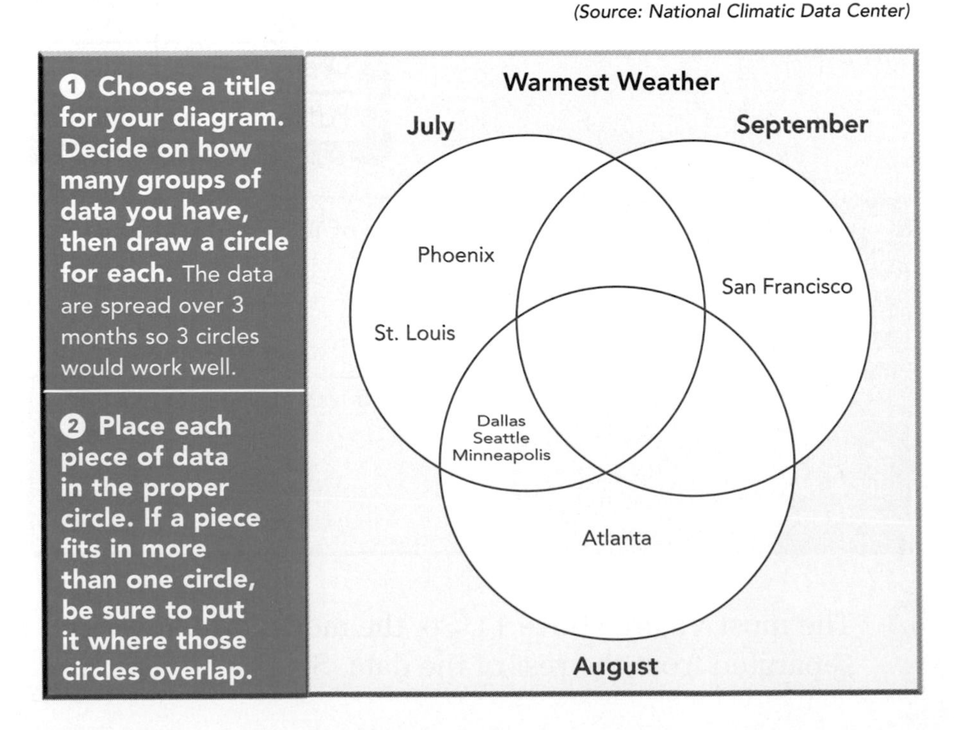

Stem-and-Leaf Plots

Stem-and-leaf plots allow you to organize the numbers in your data so that the numbers themselves make the display.

EXAMPLE: The local police department checked the speed of each car that passed its meter for one hour.

Car	1	2	3	4	5	6	7	8	9	10
Speed	35	30	32	45	30	28	42	25	33	35

Make a stem-and-leaf plot. What does the plot tell you?

❶ Title your plot.	Car Speeds
❷ Write the data in order from least to greatest.	25 28 30 30 32 33 35 35 42 45
❸ Find the least and greatest values.	25 is the least value. 45 is the greatest value.
❹ Choose the stems. Each digit in the tens place in your list is a stem.	The stems are 2, 3 and 4.
❺ Write the stems by a vertical number line, least to greatest.	Car Speeds 2 | 3 | 4 |
❻ The leaves are all the ones digits in your list. Write them next to the stems that match their tens digits.	Car Speeds 2 | 5 8 3 | 0 0 2 3 5 5 4 | 2 5
❼ Write a key that explains how to read the stems and leaves.	Key: 2 | 5 means 25 miles per hour

You can see from the plot that most cars were driving between 30 and 35 miles per hour.

285

Probability

It's helpful to know if something is likely or unlikely to happen. It's more useful if you can use a number to describe that likelihood.

Probability will help you decide how often something is likely to happen. But, it usually won't help you to know exactly when that event will happen.

286

Events and Outcomes

An **event** is something that may happen. The **probability** of an event can be any number from 0 through 1. It can be written as a fraction, a decimal, or a percent. If the probability of an event is 0, it is impossible. If an event is certain, it has a probability of 1. The more unlikely an event is, the closer its probability is to 0. The more likely an event is, the closer its probability is to 1.

Probability

less often than not			more often than not	
0	$\frac{1}{4}$	$\frac{1}{2}$	$\frac{3}{4}$	1
0.0	0.25	0.5	0.75	1.0
0%	25%	50%	75%	100%
impossible	unlikely	as likely as unlikely	likely	certain

EXAMPLE 1: When you flip a penny, two things can happen. These two things are called **outcomes**.

The penny can land with Lincoln's picture face-up. This is called **heads**.

Probability $= \frac{1}{2}$

The penny can land with the Lincoln memorial face-up. This is called **tails**.

Probability $= \frac{1}{2}$

The probability of flipping a penny and having it land with a picture of Washington face-up is 0.

EXAMPLE 2: If you roll a number cube with one of the digits 1–6 on each side, there are six things that can happen. Any one of the six digits can land face-up. These outcomes are equally likely to happen. So, the probability of rolling a 6 is $\frac{1}{6}$.

EXAMPLE 3: When you spin a spinner that's $\frac{3}{4}$ blue and $\frac{1}{4}$ green, two things can happen. The probability of spinning blue is 0.75 or $\frac{3}{4}$. The probability of spinning green is 0.25 or $\frac{1}{4}$. These two events are *not* equally likely.

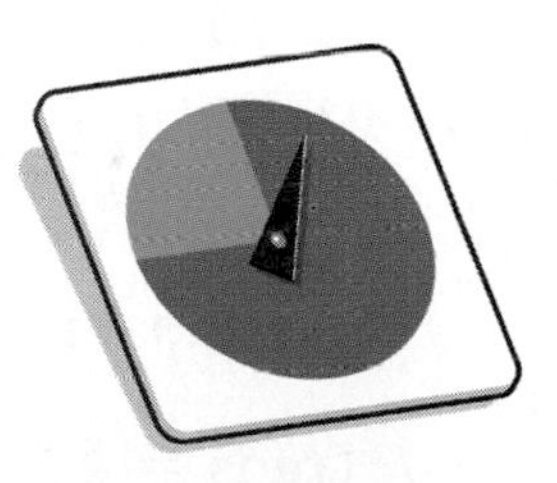

EXAMPLE 4: The weather forecaster says there's a 75% chance of rain today in Daytona Beach. This means it's more likel to rain today than not. It does not mean it *will* rain today in Daytona Beach.

EXAMPLE 5: It is certain that the sun will rise every day. So, the probability of the sun rising is 100% or 1.

287 Notation and Calculating Probability

Suppose you want to figure out just how likely it would be to spin red. We call *red* the **event**.

There are 8 equal-size sections on the spinner. The spinner is equally likely to land on any one of the 8 sections. We can say that there are 8 **possible outcomes**. They are all equally likely. There are 3 red sections. We can say that there are 3 **favorable outcomes**. When all outcomes are equally likely, use a ratio to calculate the probability of an event.

Write: $P\text{ (event)} = \frac{\text{number of favorable outcomes}}{\text{number of possible outcomes}}$

Say: *The probability of an event is the ratio of the number of favorable outcomes to the number of possible outcomes.*

MORE HELP

See 178, 286

The probability of spinning red is:

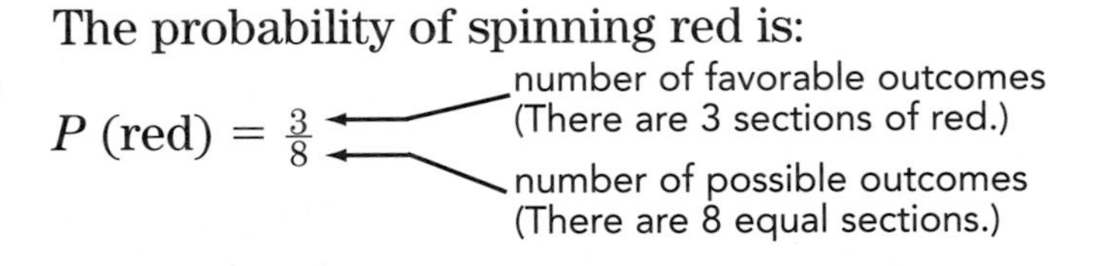

288 Sampling

When you do an experiment to get an idea about probability, you are **sampling.** Sampling may not give you exactly the same number as finding the ratio of favorable to possible outcomes. But, if you try your experiment many times, it should come close.

Surveys are also a form of sampling. Presidential candidates use sampling to try to find out what their chances are of winning. The sample will be large. It represents all the different groups of people who vote.

MATH ALERT Sampling Can Fool You

When a probability experiment has very few trials, the results can be misleading. For example, if you rolled a 1–6 number cube three times and rolled a 5 two of the times, the experiment might lead you to say that the probability of rolling a 5 would be $\frac{2}{3}$. But you know that you're not more likely to roll a 5 than any other number. The number of 5's compared to the number of rolls should drop if you continue to roll the number cube many times.

Sample Space 290

To calculate probability, you need to know all the different things that can happen. A **sample space** is a list of all the possible outcomes of an event.

EXAMPLE: Suppose you spin the spinner. Make a sample space for a spin.

This spinner can land on 12 different regions. To make the sample space, list all the possible outcomes.

The sample space is: 20, 10, 10, spin again, 50, 10, 20, 10, 100, lose turn, 50, 500.

291 Tree Diagrams

You may be able to find all possible outcomes in a sample space by drawing a tree diagram or making an organized list.

EXAMPLE: A hot dog stand offers two flavors, regular and spicy. You can put the hot dog on either a bun or bread. You can have dill relish or sweet relish as a topping. How many possible combinations of flavors, containers, and relishes are there?

Make a tree diagram to show all of the possible combinations (outcomes) in the sample space.

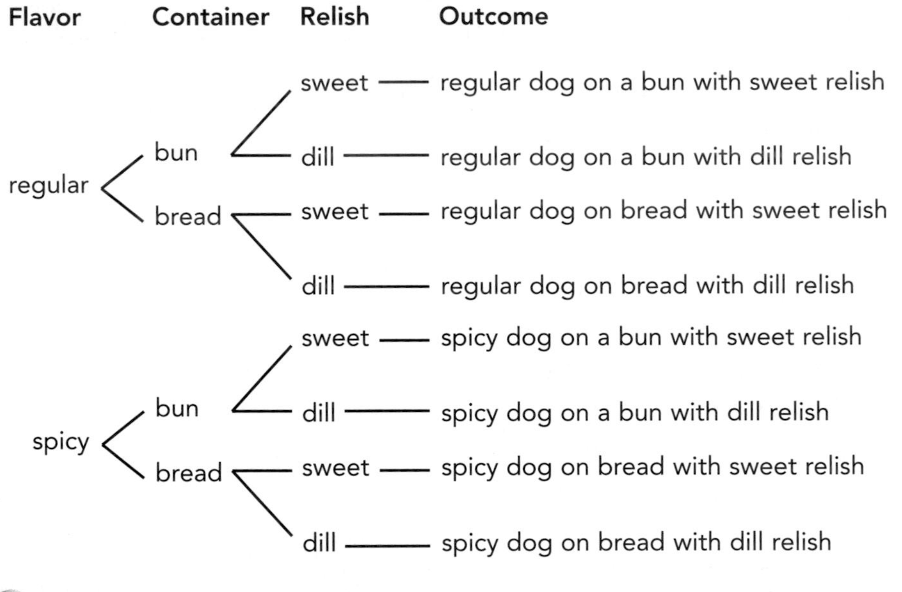

There are 8 possible combinations. The sample space contains 8 outcomes.

Counting Principle

The **counting principle** tells how you can find the number of outcomes when there is more than one way to put things together. If you start out with two choices and each of those choices leads to three choices, then you can put things together 2×3 ways.

EXAMPLE 1: A shirt comes in 4 colors, and in long or short sleeves. How many choices of shirts are there?

ONE WAY You can draw a picture.

ANOTHER WAY You can multiply lengths.

Choices = colors × sleeve
4 × 2 = 8

Either way, there are 8 choices of shirts.

EXAMPLE 2: Six people are running in the class elections. The person with the most votes will be President. The person who comes in second will be Vice President. How many different pairs of President and Vice President are possible?

Different pairs = choices for President × choices left for Vice President

$6 \times 5 = 30$

There are only 5 choices for Vice President because the same person can't be both President and Vice President!

There are 30 different President–Vice President pairs possible.

293

Measurement

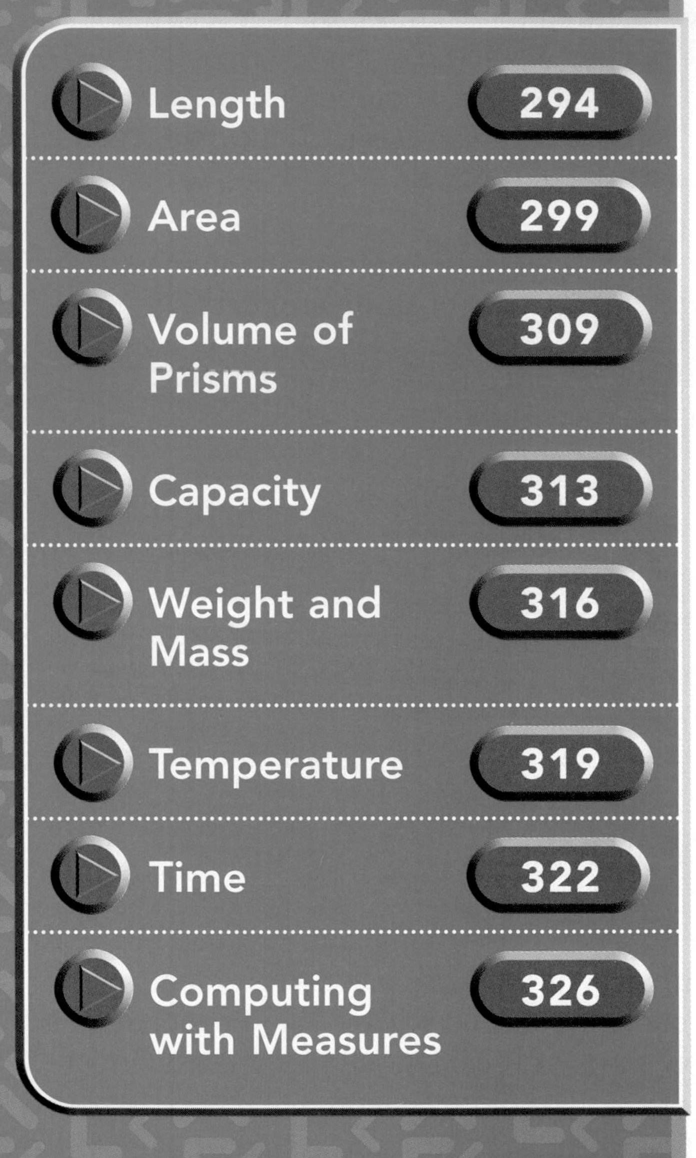

How far is it to the park? How hot is it? How long until dinner? To answer these questions, you need to know about measurement.

See 485–487

We use two systems of measurement—metric and customary. We use these systems to measure length, area, volume, capacity, weight or mass, and temperature.

294

Length

MORE HELP

See 012, 485–486

How tall are you? How far do you live from school? To answer these questions, you need to know about measuring length.

Here are some units for measuring length.

Customary Units of Length		Metric Units of Length	
inch (in.)		millimeter (mm)	1 mm = 0.001 m
foot (ft)	1 ft = 12 in.	centimeter (cm)	1 cm = 0.01 m
yard (yd)	1 yd = 3 ft	meter (m)	
mile (mi)	1 mi = 5280 ft	kilometer (km)	1 km = 1000 m

Here are some benchmarks to help you understand the size of each unit.

The diameter of a quarter is about an inch.

A doorway is about a yard wide.

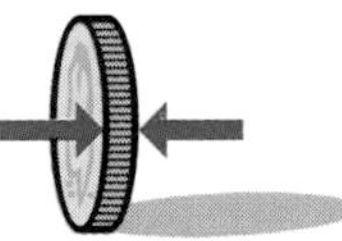

A dime is about a millimeter thick.

The width of a fingernail is about a centimeter.

295 Perimeter

Two Greek words join together to form the word *perimeter*. *Peri* means around and *metron* means measure. The distance around a figure is called the **perimeter**. To find perimeter of any shape, add the lengths of the sides.

EXAMPLE: What is the distance around this park?

To find the answer, find the perimeter of the park. Add the lengths of the sides.

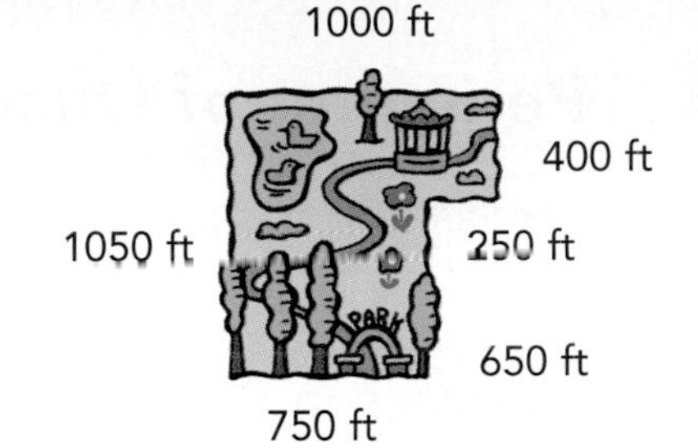

650 + 750 + 1050 + 1000 + 400 + 250 = 4100

The distance around the park is 4100 feet.

Perimeter of Triangles

CASE 1 To find the perimeter of any triangle, add the lengths of the sides.

EXAMPLE 1: Find the perimeter of the playground.

To find the perimeter, add.
30 + 40 + 50 = 120

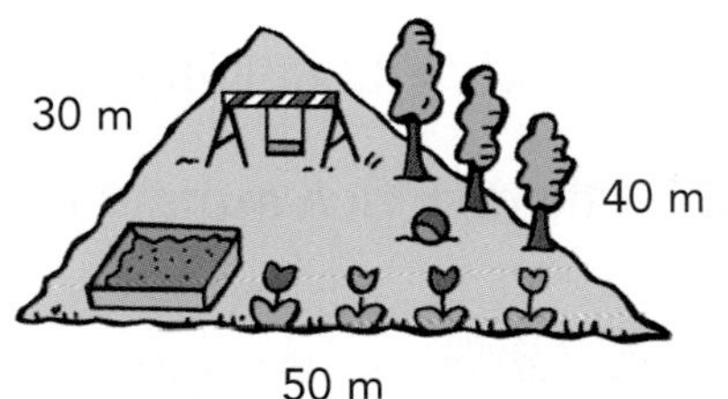

The perimeter of the playground is 120 meters.

See 362

CASE 2 If the triangle is equilateral (all sides are the same length), multiply the length of a side by 3 to find the perimeter.

EXAMPLE 2: Find the perimeter of the garden.

All of the garden's sides are the same length. To find the perimeter, you could add 20 + 20 + 20. You could also multiply the length of one side by 3. $3 \times 20 = 60$

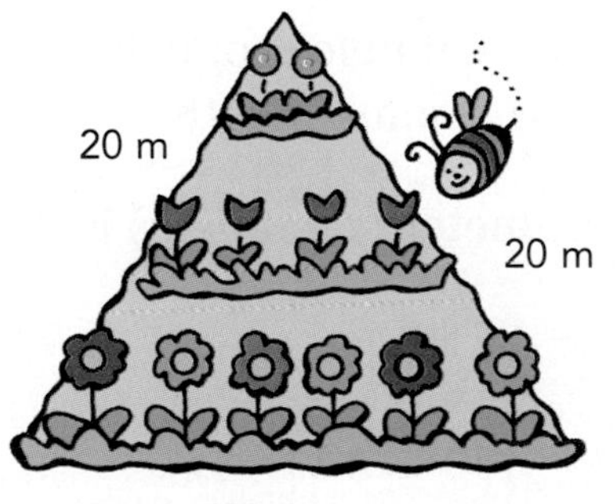

The perimeter of the garden is 60 meters.

297 Perimeter of Quadrilaterals

CASE 1 To find the perimeter of any quadrilateral, add the lengths of its sides.

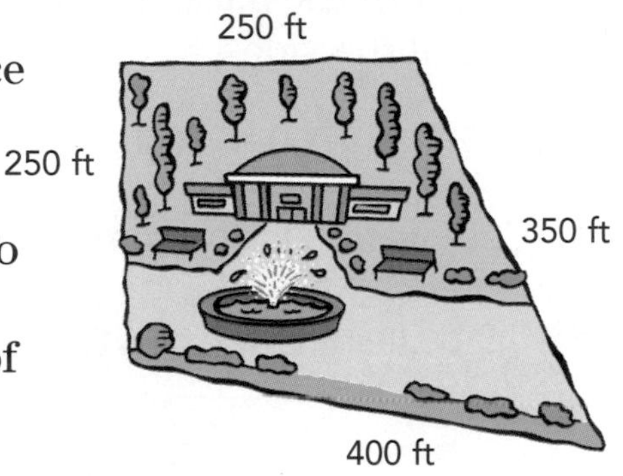

EXAMPLE 1: Find the distance around the library grounds.

The library grounds are shaped like a quadrilateral. To find the distance around the grounds, find the perimeter of the quadrilateral. Add the lengths of the sides.

$250 + 250 + 350 + 400 = 1250$

The distance around the library grounds is 1250 feet.

See 212–214, 364–366

CASE 2 For rectangles or squares, you can use a formula to find the perimeter.

EXAMPLE 2: What is the distance around this rectangular swimming pool?

In a rectangle, opposite sides are the same length.

*P*erimeter (rectangle) is 2 times *l*ength plus 2 times *w*idth

$P = 2l + 2w$
$P = 2 \times 75 + 2 \times 25$
$P = 150 + 50$
$P = 200$

Remember the order of operations. Do multiplication before addition.

The distance around the swimming pool is 200 feet.

EXAMPLE 3: Find the perimeter of the wading pool.

A square is a special rectangle. All four of a square's sides are the same length.

10 ft
10 ft
10 ft
10 ft

Perimeter (square) is 4 times length of a side

$P = 4s$

$P = 4 \times 10$

$P = 40$

The perimeter of the wading pool is 40 feet.

Circumference 298

Circles are so special that they have their own name for perimeter: **circumference**. This word comes from Latin: *circum* (around) and *ferre* (to carry).

MORE HELP

See 107, 142, 367

Circumference = pi × diameter

$C = \pi d$

OR

Circumference = 2 × pi × radius

$C = 2\pi r$

That makes sense because the diameter is twice the radius.

EXAMPLE: In many towns, rotaries, or traffic circles, are used instead of regular intersections. What is the circumference of this rotary?

Radius 100 ft
Diameter 200 ft

$C = \pi d$	OR	$C = 2\pi r$
$C \approx 3.14 \times 200$		$C \approx 2 \times 3.14 \times 100$
$C \approx 628$		$C \approx 628$

Pi is about 3.14. When numbers are approximate, use ≈ instead of =.

The circumference of the rotary is about 628 feet.

299

Area

See 364–365

Area is the number of square units needed to cover a figure. The units used to measure area are based on units of length. For example, area can be measured in square inches.

A **square inch** is the size of a square that is exactly 1 inch on each side.

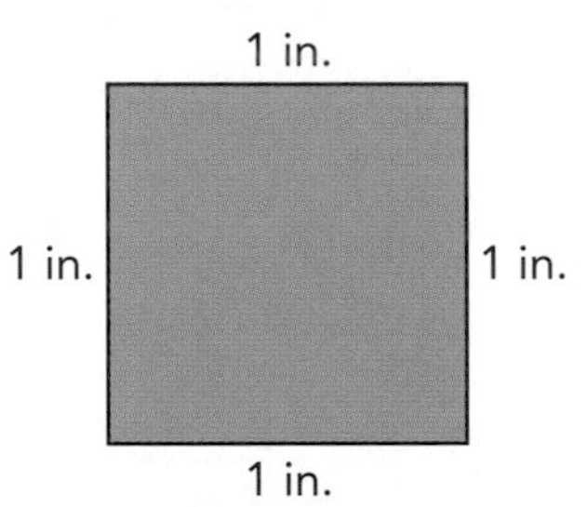

Write: 1 in.^{2}
Say: *one square inch*

Here are some units for measuring area.

Customary Units of Area	Metric Units of Area
square inch (in^2.)	square centimeter (cm^2)
square foot (ft^2)	square meter (m^2)

You can sometimes find area by counting squares.

EXAMPLE: What is the area of this room?

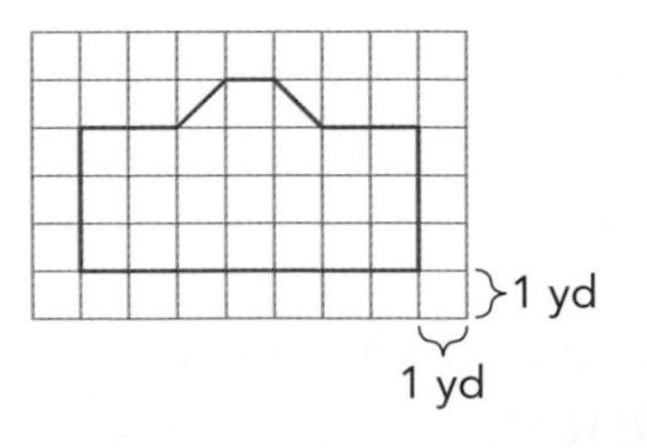

You can find the area by finding the number of whole squares that are inside the figure. In this figure, you can count all the half squares and all the whole squares.

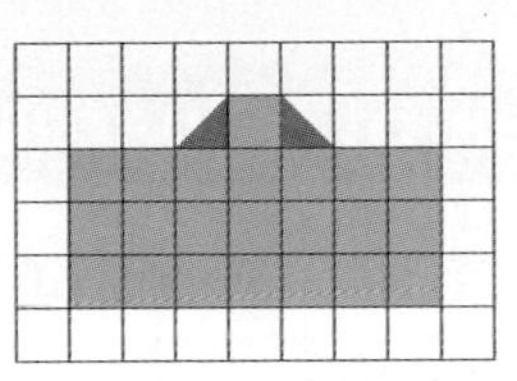

2 half squares → 1 whole square
22 whole squares → + 22 whole squares
Total 23 whole squares

ANOTHER WAY You can also find the area by finding a larger rectangle that fits around the figure. Find the area of this rectangle. Then subtract the area of the part that you must take away to get back to the original figure.

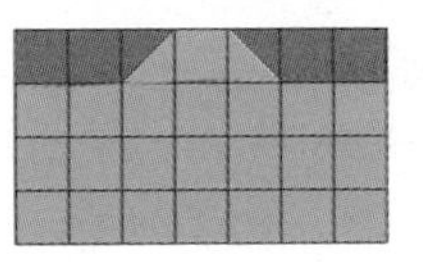

Whole rectangle → 28 whole squares
Extra pieces → − 5 whole squares
23 whole squares

Either way, the area of the floor is 23 yd^2.

Write: 23 yd^2
Say: *twenty-three square yards*

MATH ALERT A 4-Foot Square Is Not 4 Square Feet

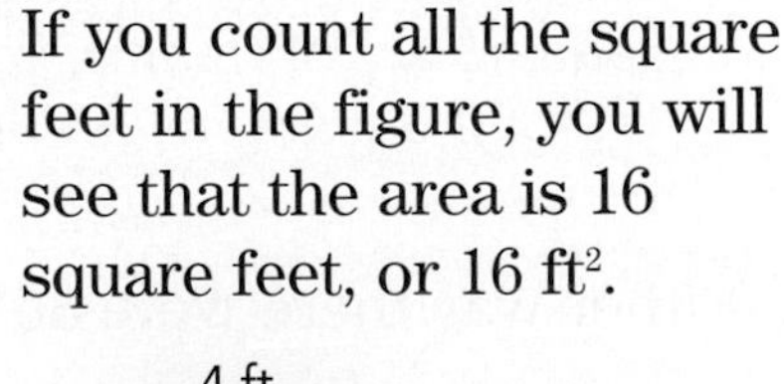

A square that is 4 feet on each side is sometimes called a 4-foot square.

If you count all the square feet in the figure, you will see that the area is 16 square feet, or 16 ft^2.

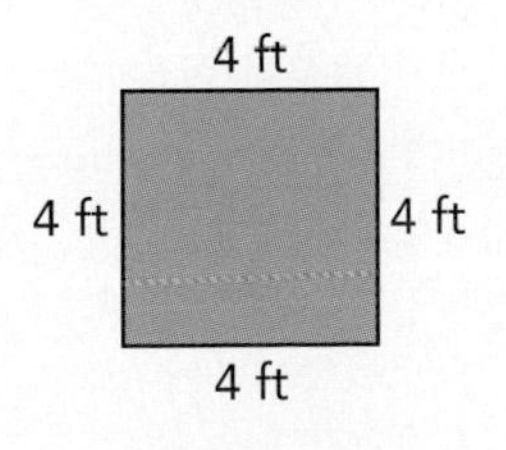

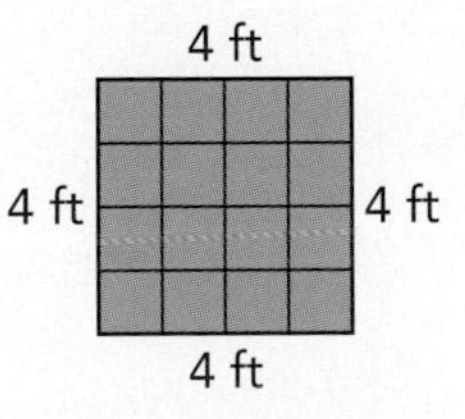

301 Area of Squares and Other Rectangles

You can find the area of a rectangle by counting squares or by multiplying.

See 246

EXAMPLE 1: In 1785, Congress established a Land Ordinance Survey System. A township was a square, 6 miles on a side. Each township was then divided into 1-square-mile sections. How many sections were in a township?

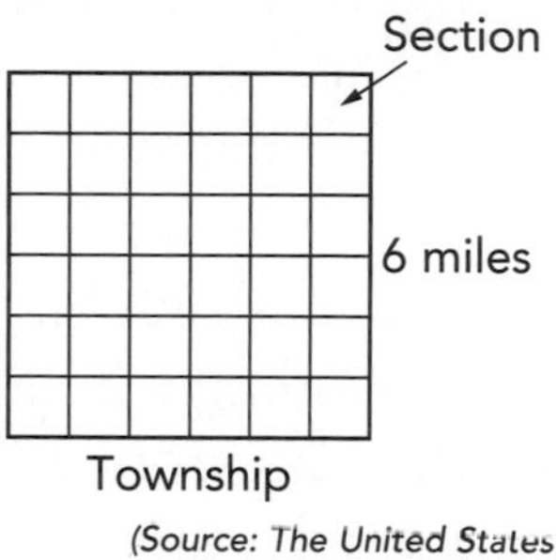

(Source: The United States and Its People)

You can find the area of a rectangle by counting the number of square units inside. The township square contains 36 squares. Each square is 1 mile on a side.

ANOTHER WAY Multiply the length by the width.

$$\text{Area of a rectangle} = \textit{length} \times \textit{width}$$
$$A = lw$$
$$6 \times 6 = 36$$

Since the length and width of a square are the same, you could say A (of a square) $= s \times s$, or s^2.

Either way there were 36 sections in a township.

Area of Parallelograms 302

Rectangles are special parallelograms. You can use what you know about rectangles to find the area of a parallelogram. If you cut up a parallelogram, you can rearrange the pieces to make a rectangle.

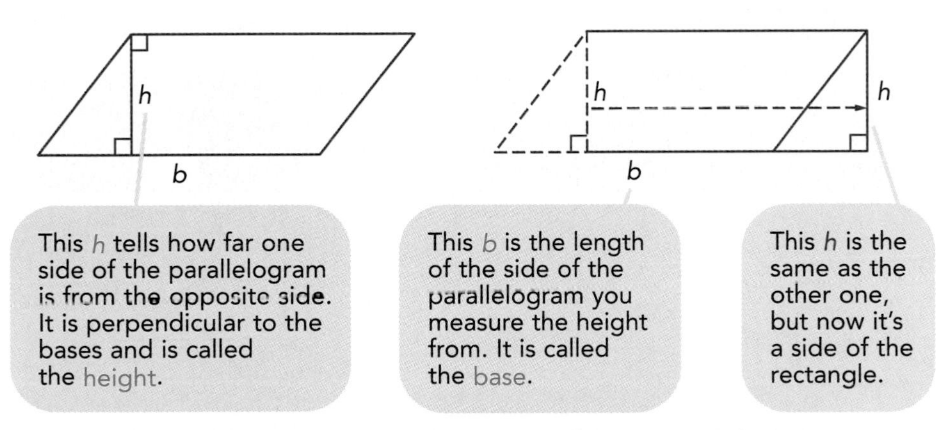

To form the rectangle, you used the whole parallelogram and nothing else. So, the area didn't change. To find the area of the parallelogram, find the area of the rectangle you formed.

MORE HELP
See 301, 364–366

Area of a rectangle $= \textit{length} \times \textit{width}$

Area of a parallelogram $= \textit{base} \times \textit{height}$

$A = bh$

EXAMPLE: Before you can buy paint, you need to know how many square feet you'll be covering. Each stripe in a crosswalk is a parallelogram. What is the area of one stripe?

10 feet

2 feet

$A = bh$

$\downarrow$

$2 \times 10 = 20$

 The area of one stripe is 20 ft^2.

303 Area of Triangles

If you can find the area of a parallelogram, you can find the area of any triangle. This is because any triangle is half of a parallelogram!

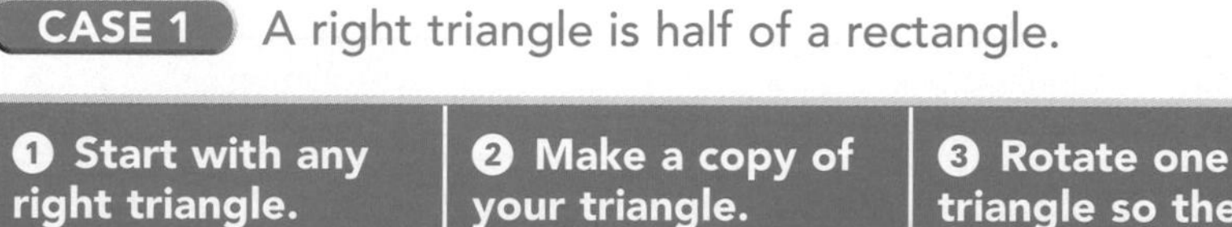

CASE 1 A right triangle is half of a rectangle.

❶ Start with any right triangle.	❷ Make a copy of your triangle.	❸ Rotate one triangle so the two form a rectangle.

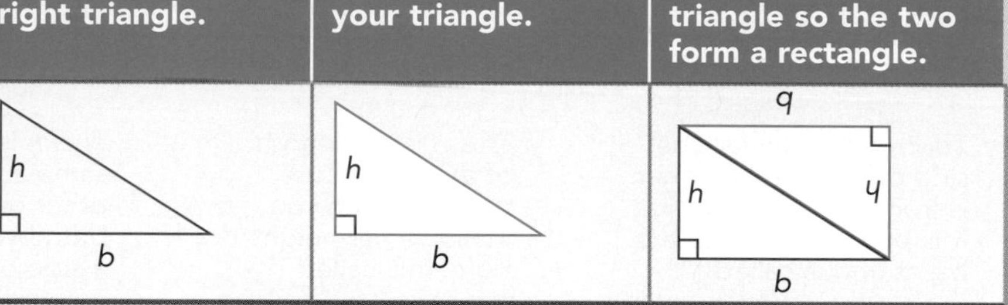

A rectangle is a special parallelogram. The area of this special parallelogram is bh. So, the formula for the area of a right triangle is $\frac{1}{2}bh$.

See 167–170, 301, 361

EXAMPLE 1: Use the formula to find the area of ΔDIM.

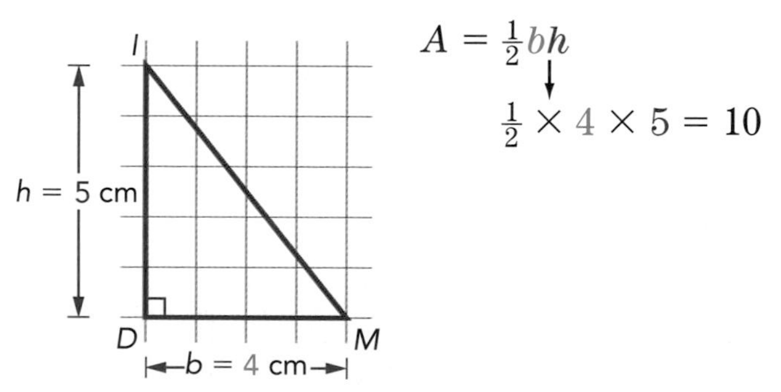

$$A = \frac{1}{2}bh$$

$$\frac{1}{2} \times 4 \times 5 = 10$$

The area of ΔDIM is 10 cm².

Say: *The area of triangle D I M is ten square centimeters.*

CASE 2 Any triangle is half of a parallelogram.

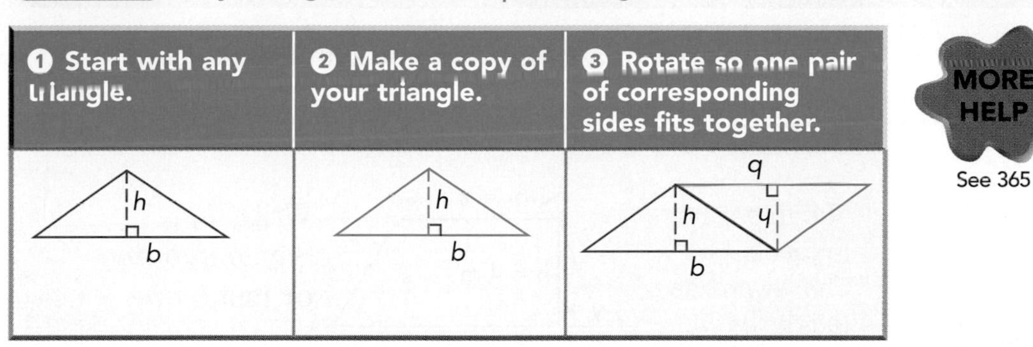

See 365

The area of any parallelogram is *bh*. So, the formula for the area of any triangle is $\frac{1}{2}bh$.

Both cases give the same formula. That makes sense because a right triangle is still a triangle.

EXAMPLE 2: Use the formula to find the area of ΔJOY.

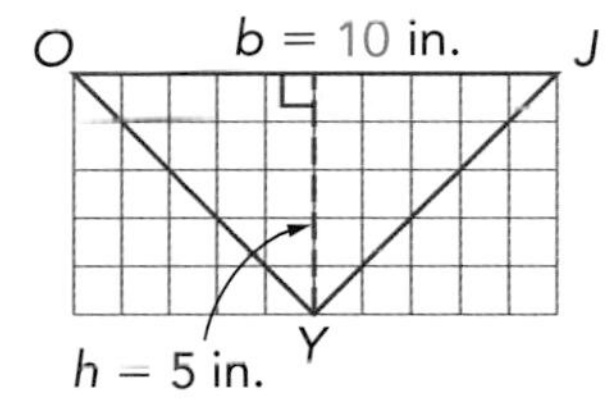

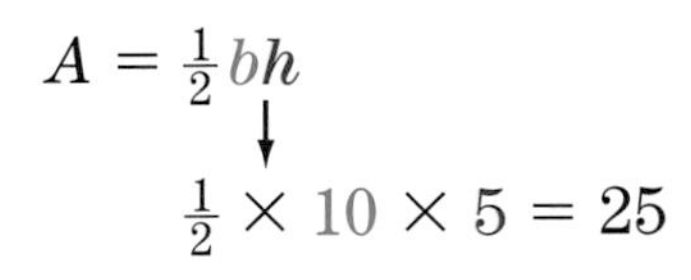

$$A = \frac{1}{2}bh$$

$$\frac{1}{2} \times 10 \times 5 = 25$$

See 167–170, 302, 359

The area of ΔJOY is 25 in.^{2}

304 Area of Trapezoids

To find the area of a trapezoid, you can break it up.

EXAMPLE: Find the area of this trapezoid.

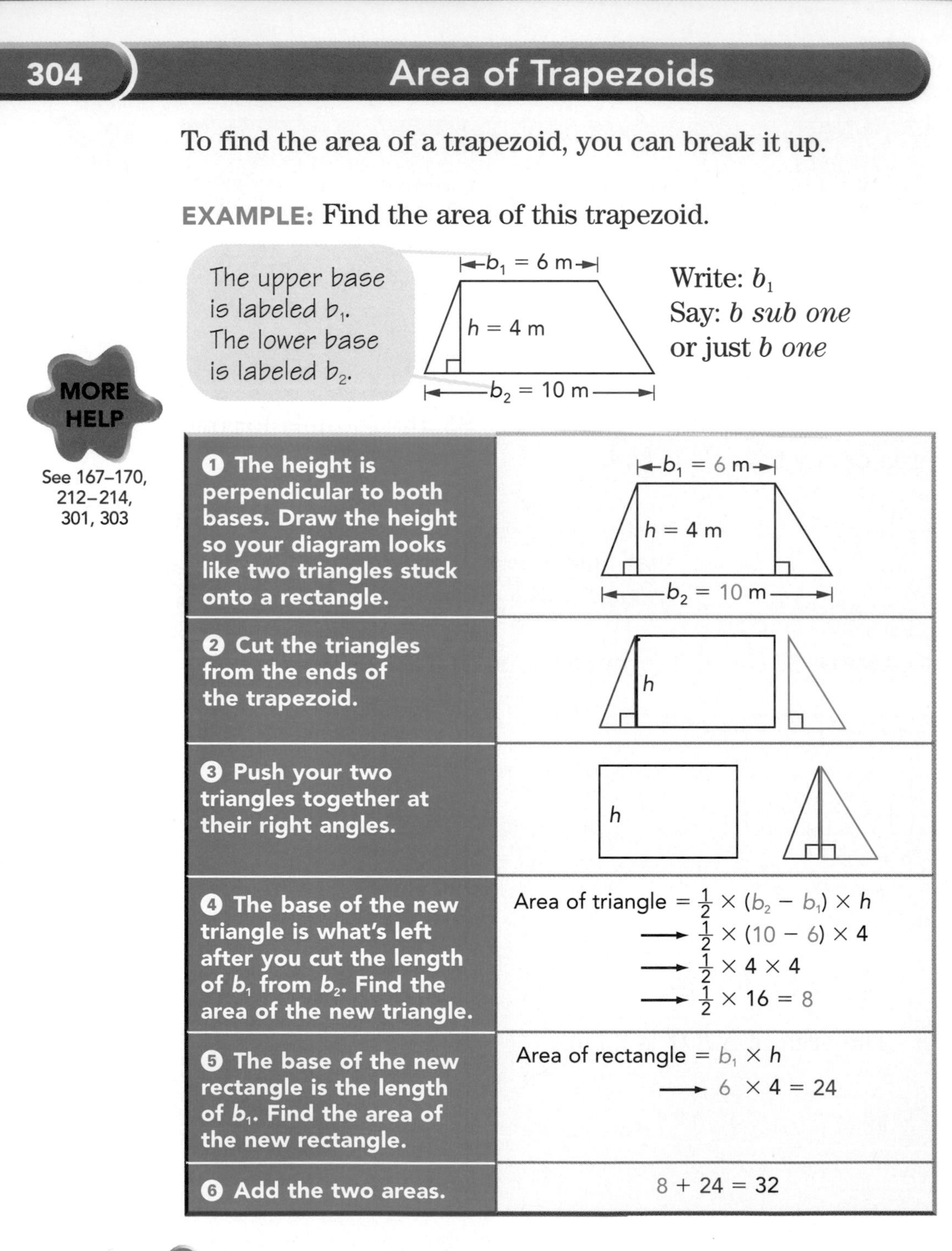

Write: b_1
Say: *b sub one* or just *b one*

MORE HELP

See 167–170, 212–214, 301, 303

❶ The height is perpendicular to both bases. Draw the height so your diagram looks like two triangles stuck onto a rectangle.	$b_1 = 6$ m, $h = 4$ m, $b_2 = 10$ m
❷ Cut the triangles from the ends of the trapezoid.	h
❸ Push your two triangles together at their right angles.	h
❹ The base of the new triangle is what's left after you cut the length of b_1 from b_2. Find the area of the new triangle.	Area of triangle $= \frac{1}{2} \times (b_2 - b_1) \times h$ $\longrightarrow \frac{1}{2} \times (10 - 6) \times 4$ $\longrightarrow \frac{1}{2} \times 4 \times 4$ $\longrightarrow \frac{1}{2} \times 16 = 8$
❺ The base of the new rectangle is the length of b_1. Find the area of the new rectangle.	Area of rectangle $= b_1 \times h$ $\longrightarrow 6 \times 4 = 24$
❻ Add the two areas.	$8 + 24 = 32$

✪ The area of the trapezoid is 32 square units.

Area of Circles 305

You can see how the formula for the area of a circle works by cutting a circle into wedges and rearranging them into a shape that is like a parallelogram.

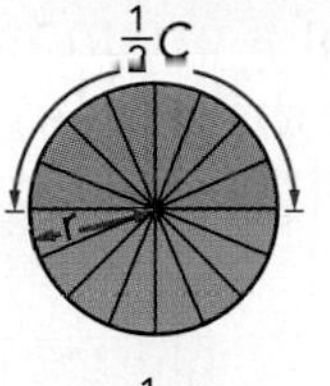

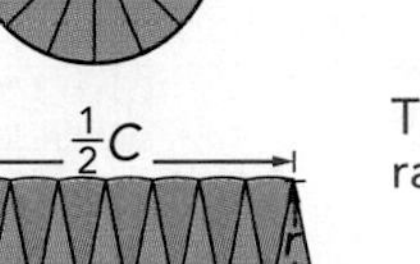

In your wedge-parallelogram, each base is half of the circumference of the circle.

The height is the radius of the circle.

See 298, 302, 367–368

Area of a parallelogram = *b*ase × *h*eight

$\longrightarrow \frac{1}{2}Cr$

But, to find the circumference, you need to use another formula!

Circumference = 2 × pi × *r*adius

$\longrightarrow 2\pi r$

Put all this information together to make one formula.

Area of a circle = $\frac{1}{2}$ × *c*ircumference × *r*adius

$\longrightarrow \frac{1}{2} \times 2\pi r \times r$

$\longrightarrow 1 \times \pi \times r^2 = \pi r^2$

EXAMPLE: Find the area of this circle.

$A = \pi r^2$

$\longrightarrow \frac{22}{7} \times 7^2$

$\longrightarrow \frac{22}{7} \times 7 \times 7 \approx 154$

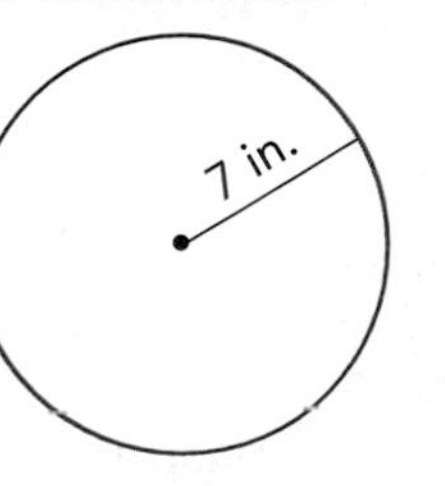

The area of the circle is about 154 in.^{2}

See 067, 168, 368

You can use the approximate value for pi that makes computing easiest for you. It's easier to multiply 49 by $\frac{22}{7}$ than by 3.14.

306 Surface Area of Prisms

The **surface area** of a prism is the sum of the areas of all the faces (including the bases).

MORE HELP
See 383–384

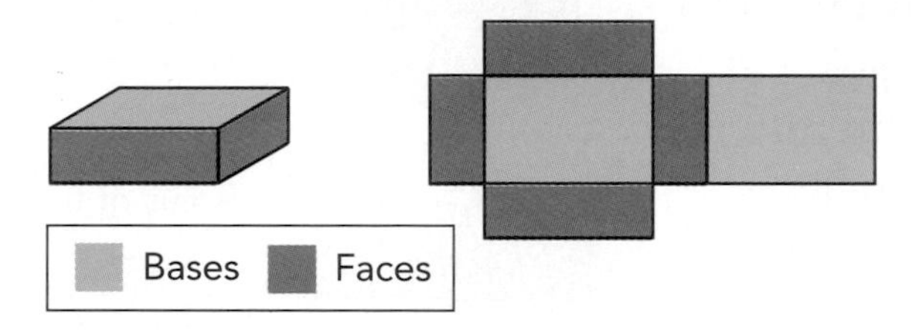

307 Surface Area of Cubes and Rectangular Prisms

The bases of rectangular prisms are rectangles. Each face of a rectangular prism has a matching parallel face.

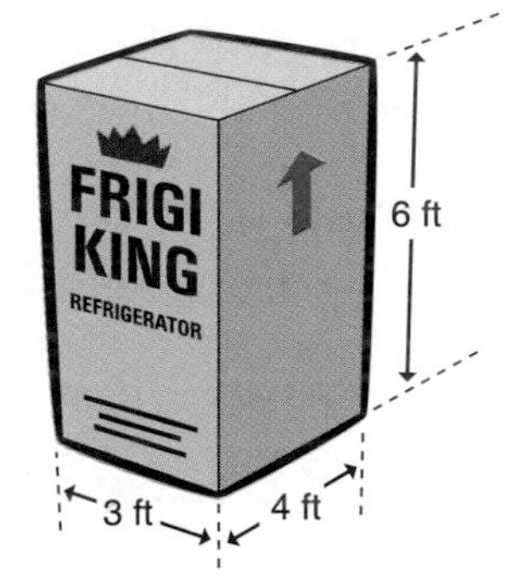

EXAMPLE: The box is a rectangular prism. Find the surface area of the box.

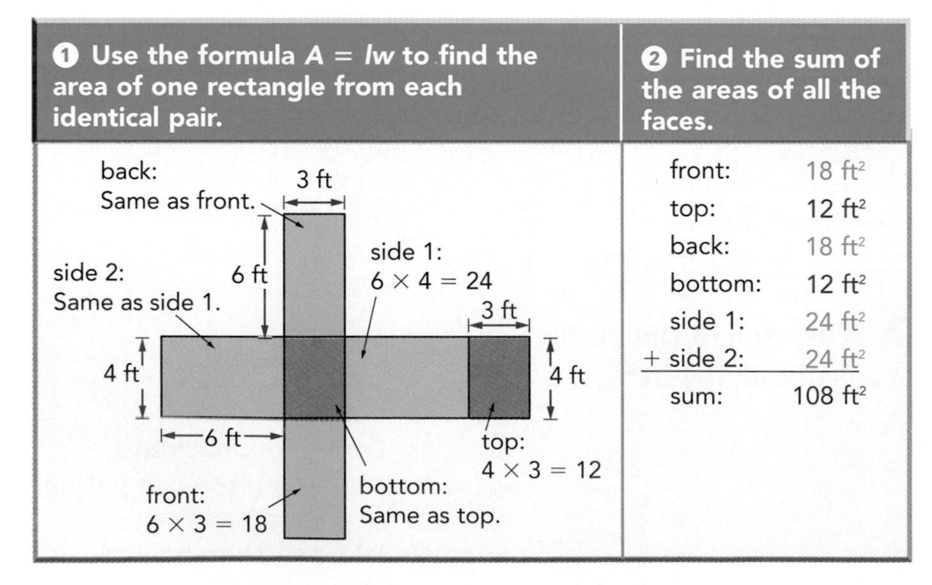

The surface area of the box is 108 ft^2.

In a cube, all six faces are identical. So, you can find surface area by multiplying the area of one face by 6. $SA = 6s^2$

Surface Area of Triangular Prisms

Triangular prisms have two bases that are triangles. Their faces are rectangles.

EXAMPLE: Some mailing boxes for posters are triangular prisms. Find the surface area of this box.

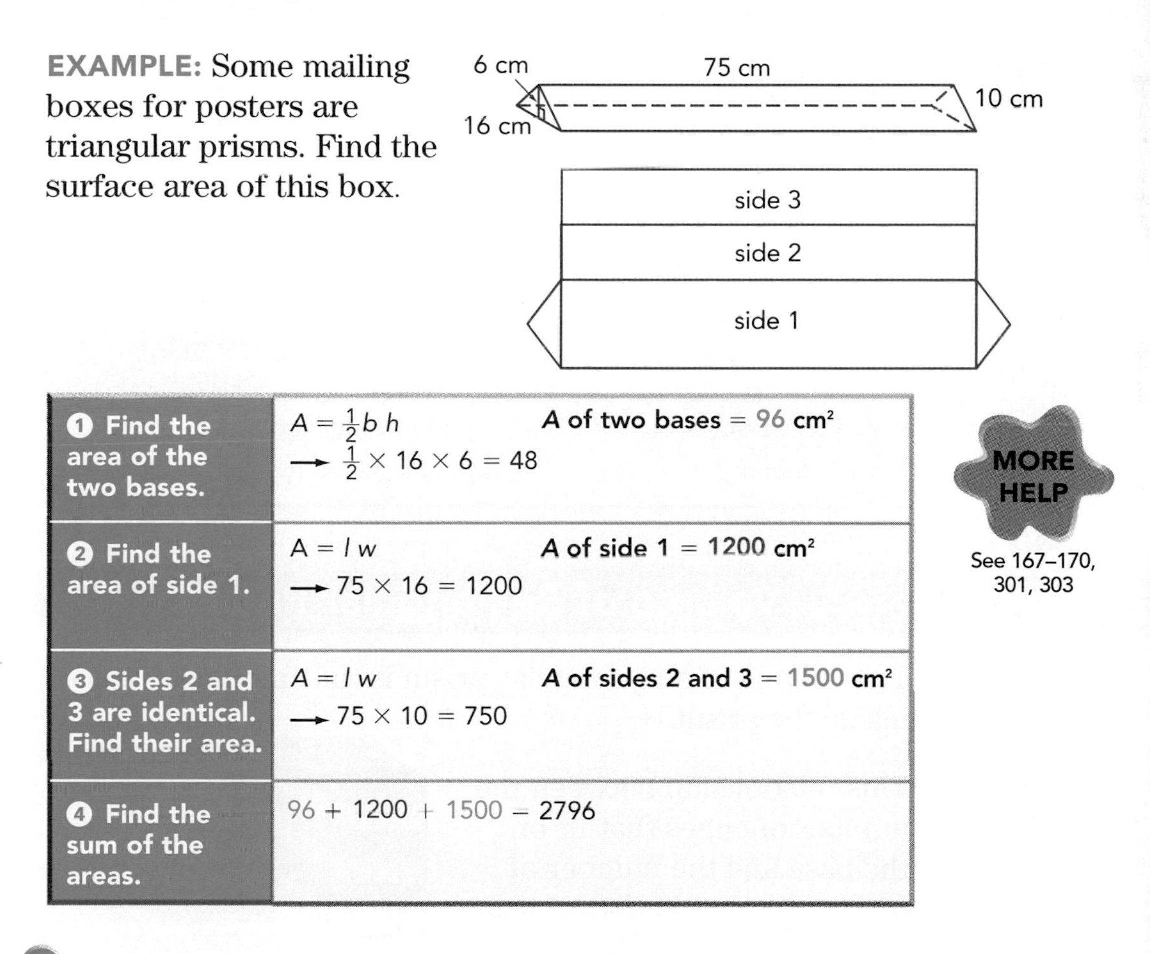

1 Find the area of the two bases.	$A = \frac{1}{2}bh$ → $\frac{1}{2} \times 16 \times 6 = 48$ **A of two bases = 96 cm²**
2 Find the area of side 1.	$A = lw$ → $75 \times 16 = 1200$ **A of side 1 = 1200 cm²**
3 Sides 2 and 3 are identical. Find their area.	$A = lw$ → $75 \times 10 = 750$ **A of sides 2 and 3 = 1500 cm²**
4 Find the sum of the areas.	$96 + 1200 + 1500 = 2796$

MORE HELP
See 167–170, 301, 303

The surface area of the box is 2796 cm².

309 Volume of Prisms

MORE HELP See 301, 383–384

The **volume** of a prism tells you how many cubes of a given size it takes to fill the prism. So, volume is measured in cubic units. The units used to measure volume are based on the units used to measure length. For example, volume can be measured in cubic inches. A **cubic inch** is the size of a cube that is exactly 1 inch on each edge.

Write: 1 in.^{3}

Say: *one cubic inch*

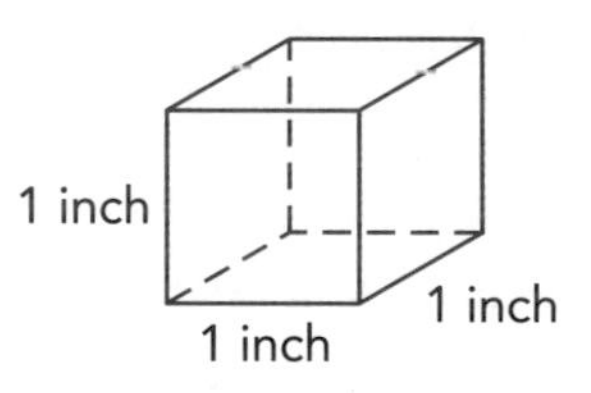

Remember: Volume is always measured in cubic units.

Here are some units for measuring volume.

Customary Units of Volume	Metric Units of Volume
cubic inch (in.^{3})	cubic centimeter (cm^3)
cubic foot (ft^3)	cubic meter (m^3)

310 Volume of Rectangular Prisms

The volume of a rectangular prism is the amount of space inside the prism.

This relationship between the number of cubes that fit on the base and the number of layers of cubes is always true.

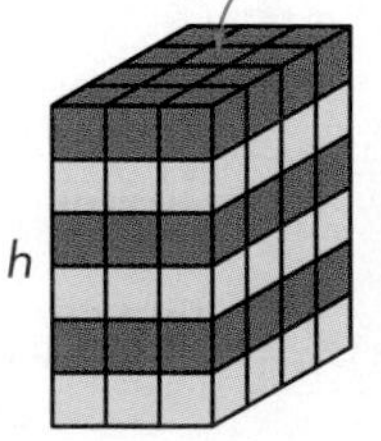

There are 6 layers of cubes. There are 12 cubes in each layer. So, the volume is 72 cubic units.

*V*olume (Prism) = Area of one *B*ase × *h*eight

$V = Bh$

$\longrightarrow lwh$

EXAMPLE: Find the volume of the space inside this cooler

The space is a rectangular prism.

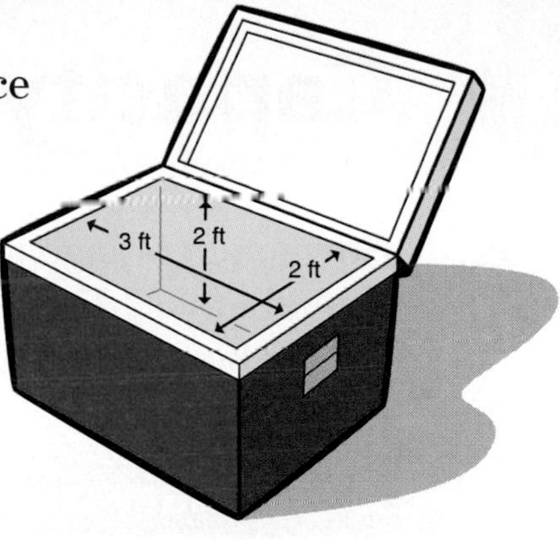

❶ **Use the formula for area of a rectangle to find the area of the base.**	$A = lw$ → $3 \times 2 = 6$
❷ **Use the formula for volume of a prism to find the volume.**	$V = Bh$ → $6 \times 2 = 12$

When you multiply ft^2 by ft, you get ft^3.

The volume inside the cooler is 12 ft^3.

Volume of Cubes 311

A cube is a special rectangular prism.
Its length, width, and height are all the same.

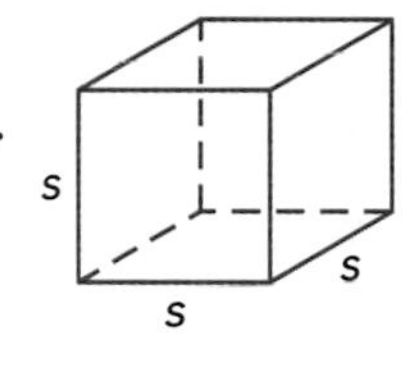

Volume of rectangular prism $= l \times w \times h$

Volume of cube $= s \times s \times s = s^3$

Multiplying the same number three times is called finding the **cube** of the number.

MATH ALERT A 2-Foot Cube Is Not 2 Cubic Feet 312

A cube that is 2 feet on each edge is sometimes called a 2-foot cube.

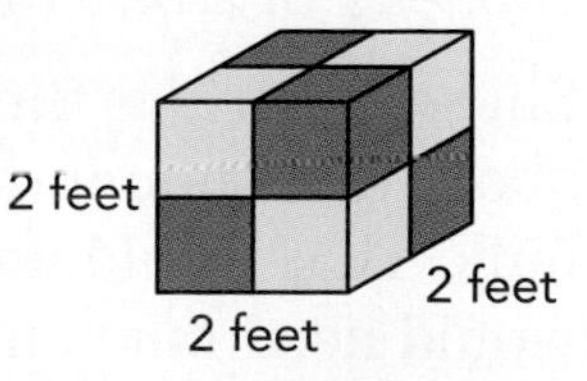

If you count all the cubic feet in the cube, you will see that the volume is 8 cubic feet, or 8 ft^3.

313 Capacity

See 485–487

How much orange juice shall we buy? How much medicine should I take? To answer these questions you need to know about capacity or liquid volume.

314 Customary System for Measuring Capacity

Here are some customary units of capacity.

Customary Units of Capacity			
ounce (oz)		quart (qt)	1 qt = 2 pt
cup (c)	1 c = 8 oz	gallon (gal)	1 gal = 4 qt
pint (pt)	1 pt = 2 c		

These containers will help you understand the size of the units in the customary system.

EXAMPLE: Which is a more reasonable measure of the capacity of a bathtub—2 quarts or 25 gallons?

Since 2 quarts is the same as half of a gallon, think about the amount of liquid in a half-gallon container of milk. That would not be enough to fill a bathtub! You would need much more.

A more reasonable measure of the capacity of a bathtub would be 25 gallons.

Metric System for Measuring Capacity 315

Here are some metric units of capacity.

These benchmarks will help you understand the size of the units in the metric system.

Metric Units of Capacity
milliliter (mL)
liter (L) 1 L = 1000 mL

See 309–312

A milliliter is about 10 drops.

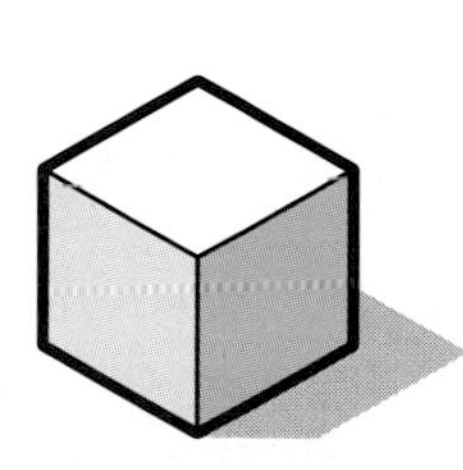

A milliliter of liquid takes up 1 cubic centimeter of space.

A liter is about 4 cups.

EXAMPLE: You want to buy a small bottle of perfume. Do you think you will buy 10 milliliters or 10 liters?

Since 1 mL is about 10 drops, 10 mL is about 100 drops. Since 1 L is about 4 cups, 10 L would be about 40 cups.

 You will be more likely to buy 10 mL than 10 L of perfume.

316 Weight and Mass

Mass and weight are similar, but they are not the same. **Mass** is a measure of the amount of matter in an object. **Weight** is a measure of how heavy an object is.

MORE HELP
See 468, 485–487

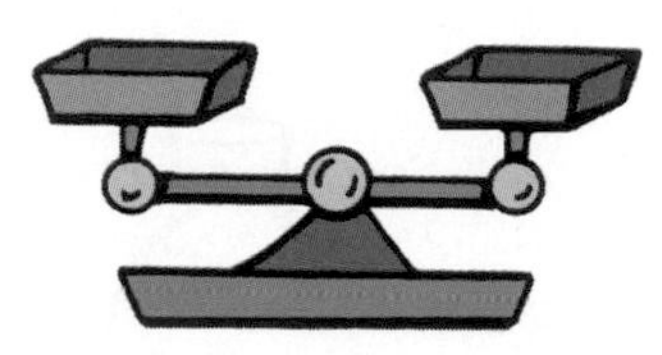

A balance is used to measure mass.

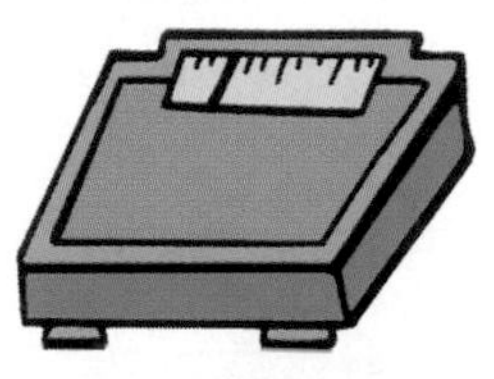

A scale is used to measure weight.

Gravity influences weight, but does not affect mass. Your weight would change if you were on another planet. Your mass would not change. Customary units are used to measure weight. Metric units are used to measure mass.

317 Weight

These benchmarks will help you understand the size of the units of weight in the customary system.

Customary Units of Weight	
ounce (oz)	
pound (lb)	1 lb = 16 oz
ton (t)	1 t = 2000 lb

A slice of bread weighs about 1 ounce.

A loaf of bread weighs about 1 pound.

A subcompact car weighs about 1 ton.

EXAMPLE: Which is the most reasonable measure of the weight of the basket of fruit—10 ounces, 10 pounds, or 10 tons?

Since 10 ounces would be about the weight of 10 slices of bread and 10 tons would be about the weight of 10 cars, the most reasonable weight must be 10 pounds.

The most reasonable measure of the weight of the fruit basket is 10 pounds.

Mass 318

These benchmarks will help you understand the size of the units of mass in the metric system.

Metric Units of Mass	
gram (g)	
kilogram (kg)	1 kg = 1000 g

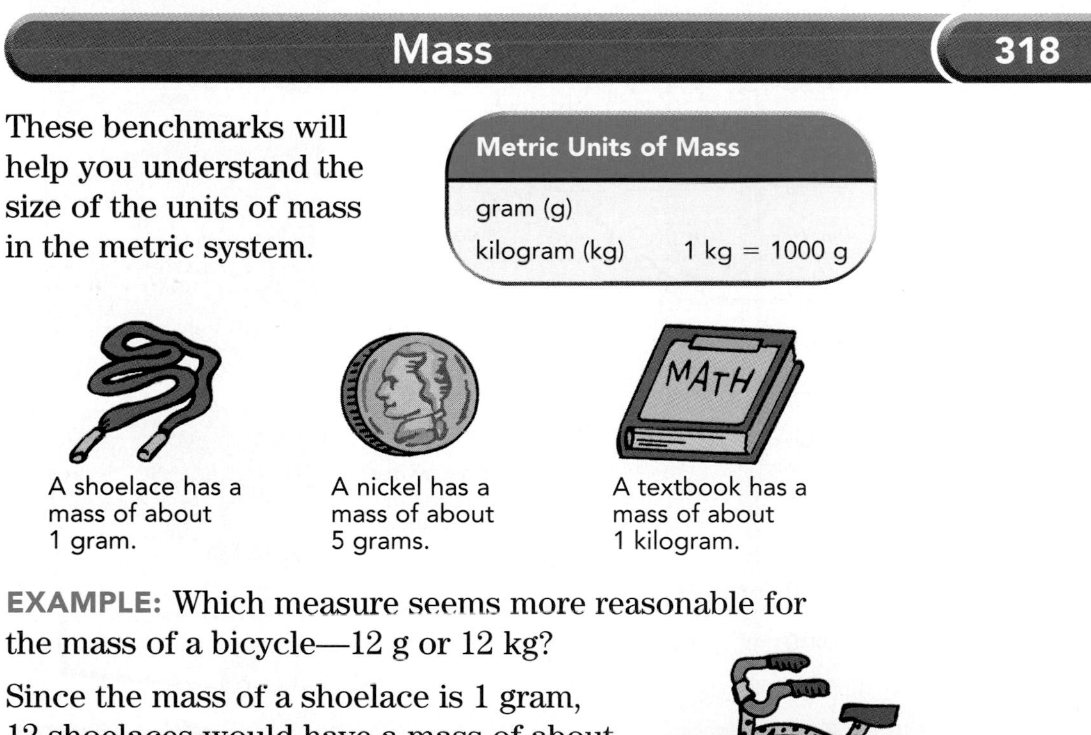

A shoelace has a mass of about 1 gram.

A nickel has a mass of about 5 grams.

A textbook has a mass of about 1 kilogram.

EXAMPLE: Which measure seems more reasonable for the mass of a bicycle—12 g or 12 kg?

Since the mass of a shoelace is 1 gram, 12 shoelaces would have a mass of about 12 grams. It makes sense that a bicycle might have a mass 1000 times that! So, 12 kg is more reasonable.

The more reasonable measure for the mass of a bicycle is 12 kg.

319

Temperature

We measure temperature with a thermometer. The colored liquid in the thermometer expands as it gets warmer, so it moves up in its tube. As it gets colder, the liquid takes up less space, so it moves down in its tube.

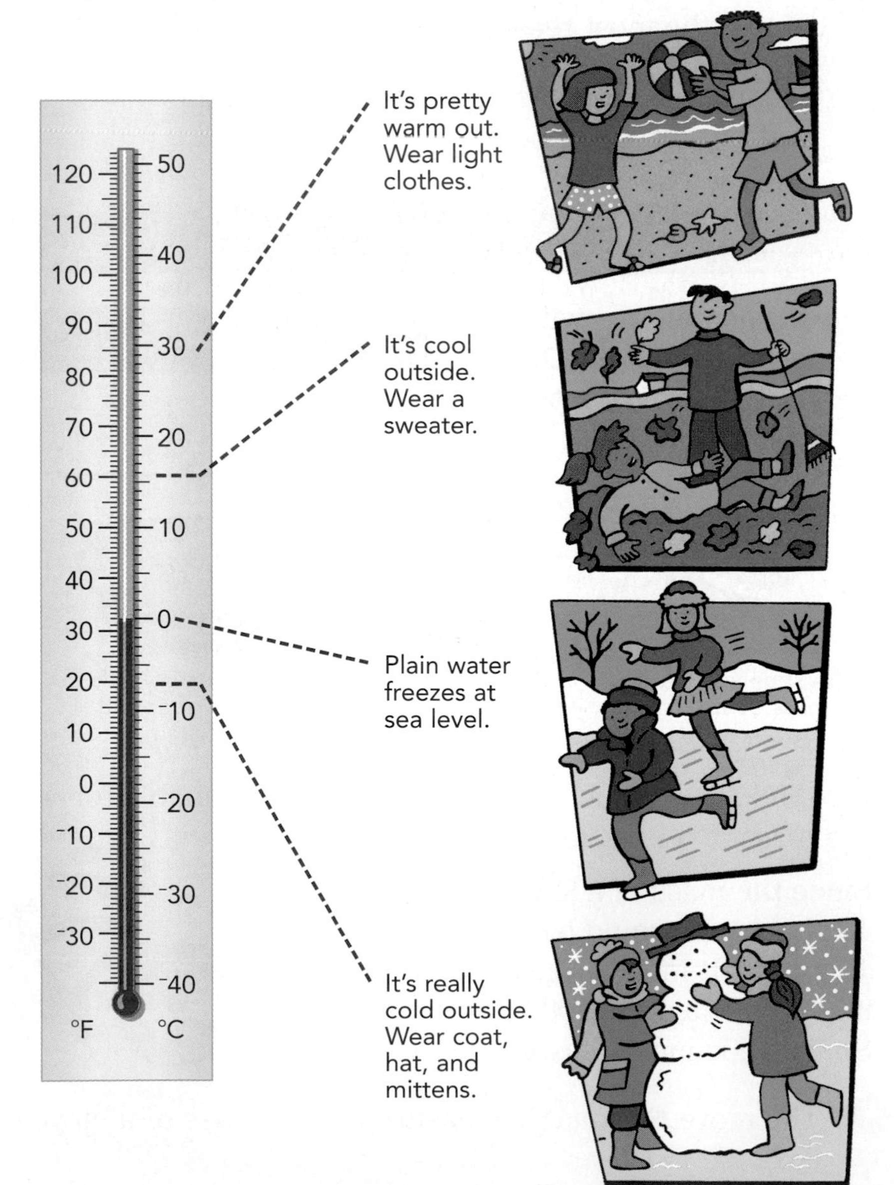

Fahrenheit 320

In the customary system of measurement, temperature is measured on the **Fahrenheit** (**Fare**-in-hite) scale. On this scale, the freezing point of plain water at sea level is 32°F.

Write: 32°F
Say: *thirty-two degrees Fahrenheit*

The United States uses the Fahrenheit scale to measure temperature.

Celsius 321

The metric system of measurement uses a scale based on 100 to measure temperature. This scale is called the **Celsius** (**Sell**-see-us) scale. It is also sometimes called the Centigrade scale. One interesting thing about this scale is that the freezing point of plain water at sea level is 0°C.

Write: 0°C
Say: *zero degrees Celsius*

Most countries, other than the U.S., measure temperature on the Celsius scale.

EXAMPLE: Which country is the weather forecast coming from?

Compare the two scales. You can see that you'd feel much better in shorts at 25° Celsius than at 25° Fahrenheit.

The weather forecast must be coming from Canada.

322

Time

We measure time just as we measure other things. The measuring tools are clocks and calendars instead of rulers and yardsticks. If you want to know how long it will take to do something or how long it's been since something happened, you may have to compute with time. If you keep the different units separate, you'll be fine.

Here are some units for measuring time.

Units of Time			
second (s)		month (mo)	1 mo = 28, 29, 30, or 31 d
minute (min)	1 min = 60 s		
hour (h)	1 h = 60 min	year (yr)	1 y = 12 mo
day (d)	1 d = 24 h		

The word *day* can be confusing. It means a 24-hour period from midnight to the next midnight. But people often use the word to mean the part of the day when the sun is out or the part of the day when they're at work or at school.

Some terms you'll use when you measure time are:

A.M. (*ante meridiem:* before midday) between midnight and noon

P.M. (*post meridiem:* after midday) between noon and midnight

Schedules

A **schedule** is an organized list. It tells, in order, what is supposed to happen at various times. A student day planner is a schedule.

EXAMPLE: Here are Ronnie's school and activity schedules. Make a weekly schedule for Ronnie.

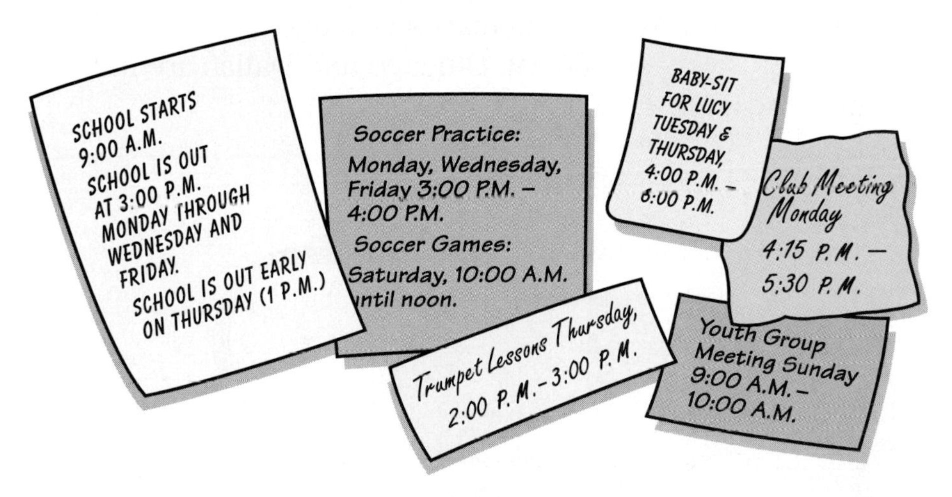

Your list should have 7 days. Each day should list, in order, the times and events for that day.

	SUN.	MON.	TUE.	WED.	THU.	FRI.	SAT.
9:00 A.M.	Youth Group	School	School	School	School	School	
10:00 A.M.		School	School	School	School	School	Soccer
11:00 A.M.		School	School	School	School	School	Soccer
12:00 noon		School	School	School	School	School	
1:00 P.M.		School	School	School		School	
2:00 P.M.		School	School	School	Trumpet	School	
3:00 P.M.		Soccer		Soccer		Soccer	
4:00 P.M.		4:15: Club	Baby-sit		Baby-sit		
5:00 P.M.		Club til 5:30	Baby-sit		Baby-sit		
6:00 P.M.							

324 Elapsed Time

The time that went by between 8:00 A.M. and 8:30 A.M. is the time that passed, or **elapsed** between 8:00 A.M. and 8:30 A.M. You can find elapsed time by counting on from the starting time to the finishing time.

See 325

EXAMPLE: You are flying from Chicago to Dallas with a stop in Atlanta. Your plane leaves Chicago at 9:15 A.M. It arrives in Dallas at 1:05 P.M. Chicago and Dallas are in the same time zone. How long is the flight?

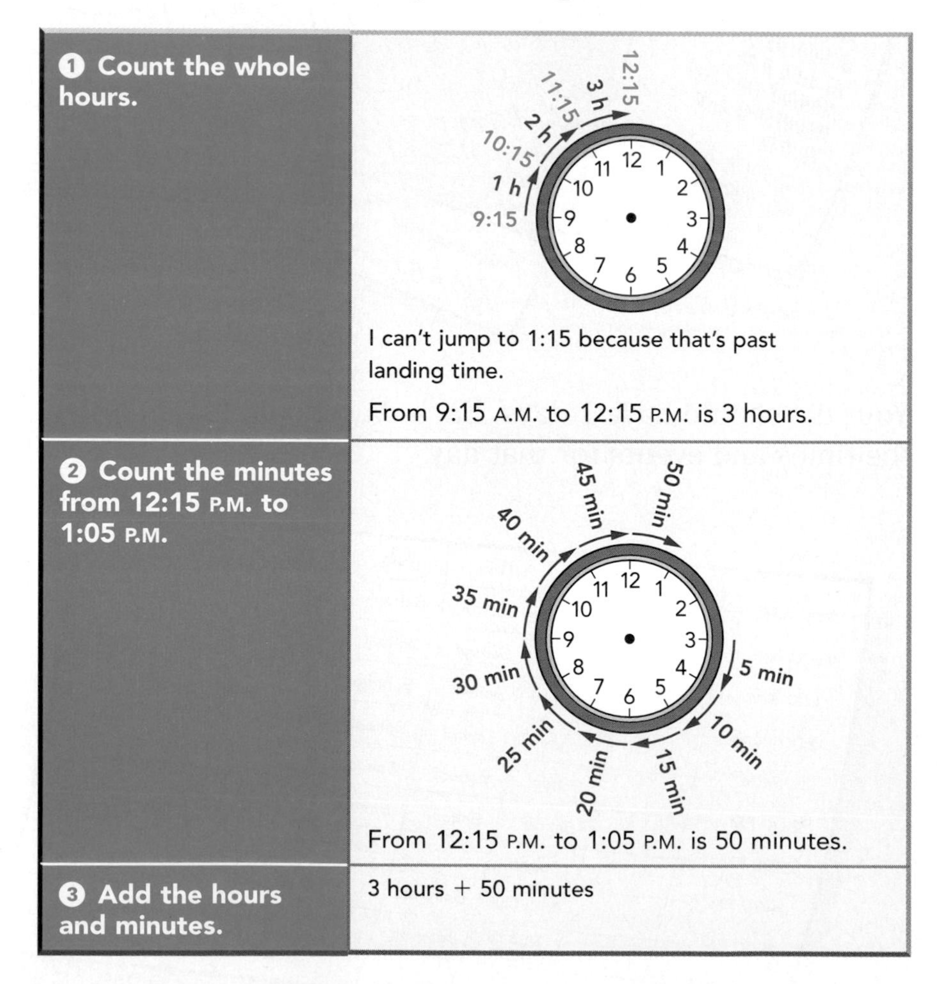

❶ Count the whole hours.	I can't jump to 1:15 because that's past landing time. From 9:15 A.M. to 12:15 P.M. is 3 hours.
❷ Count the minutes from 12:15 P.M. to 1:05 P.M.	From 12:15 P.M. to 1:05 P.M. is 50 minutes.
❸ Add the hours and minutes.	3 hours + 50 minutes

The flight is 3 hours 50 minutes long.

Time Zones

The world is divided into 24 time zones, one for each hour of the day. If you're one time zone west of friends in another zone, your time is one hour earlier than theirs. If you're one time zone east of friends in another zone, your time is an hour later than theirs.

Name of Time Zone	If it's noon in Chicago, what time is it in this time zone?
Eastern	1:00 P.M.
Central	noon
Mountain	11:00 A.M.
Pacific	10:00 A.M.
Alaska	9:00 A.M.
Hawaii-Aleutian	8:00 A.M.

EXAMPLE: If it's 8:00 A.M. in Baltimore, is it a good idea to call your friend in Seattle?

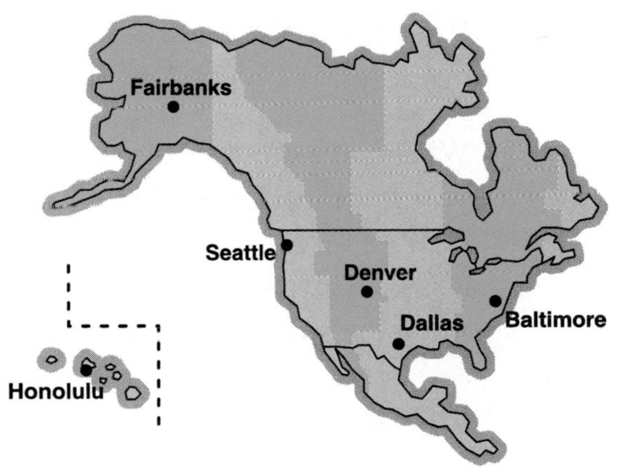

To answer the question, you need to know what time it is in Seattle. Baltimore is in the Eastern Time Zone. Seattle is in the Pacific Time Zone. Pacific time is 3 hours earlier than Eastern time. Subtract 3 hours from 8:00 A.M. It is 5:00 A.M. in Seattle.

Since it's 5:00 A.M. in Seattle, it's probably too early to call unless you know that your friend is a very early riser!

DID YOU KNOW...

Time Zones are centered on the Prime Meridian running through Greenwich near London, England? Times can be said to be a number of hours ahead of or behind Greenwich Mean Time (GMT).

326

Computing with Measures

When you add 2 feet and 24 inches, you don't get 26 feet and you don't get 26 inches. But, what do you get? There are rules that help you compute with units that are not the same.

327

Changing from One Unit to Another

If you want to add 2 feet and 24 inches, one thing you could do is change 24 inches to feet and then add.

When you change from one unit of measure to another, you need to know the relationship between the two units of measure. To change inches to feet or feet to inches, you need to know that 1 foot = 12 inches. The tables of measures in the Almanac can give you that information.

EXAMPLE 1: Change 6 yards to feet.

See 149, 485–487

Customary Units of Length		Metric Units of Length	
inch (in.)		millimeter (mm)	1 mm = 0.001
foot (ft)	1 ft = 12 in.	centimeter (cm)	1 cm = 0.01
yard (yd)	1 yd = 3 ft	meter (m)	
mile (mi)	1 mi = 5280 ft	kilometer (km)	1 km = 1000 m

Because feet are smaller than yards, you can expect that you will have more feet than you had yards. Your answer must be more than 6, so multiply 6 by 3.

 6 yards = 18 feet

EXAMPLE 2: Change 9 quarts to gallons.

Customary Units of Capacity			
ounce (oz)		quart (qt)	1 qt = 2 pt
cup (c)	1 c = 8 oz	gallon (gal)	1 gal = 4 qt
pint (pt)	1 pt = 2 c		

Because gallons are larger than quarts, you can expect that you will have fewer gallons than you had quarts. Your answer must be less than 9, so divide 9 by 4.

When you divide 9 quarts by 4, you get 2 with a remainder of 1. To find the exact number of gallons, write the remainder as a fraction.

9 quarts = $2\frac{1}{4}$ gallons

Changing units in the metric system is like changing units in the customary system. But in the metric system, we use decimals instead of fractions and we don't use mixed measures.

See 153–155

EXAMPLE 3: Change 256 centimeters to meters.

Customary Units of Length		Metric Units of Length	
inch (in.)		millimeter (mm)	1 mm = 0.001
foot (ft)	1 ft = 12 in.	centimeter (cm)	1 cm = 0.01
yard (yd)	1 yd = 3 ft	meter (m)	1 m = 100 cm
mile (mi)	1 mi = 5280 ft	kilometer (km)	1 km = 1000 m

Because meters are larger than centimeters, you can expect that you will have fewer meters than you had centimeters. Since you know your answer must be less than 256, divide 256 by 100.

256 centimeters = 2.56 meters

328 Computing with Mixed Measures

A **mixed measure** is one that has more than one unit. So, if you say you are 4 ft 8 in. tall, you are using a mixed measure for your height.

You will often need to compute with mixed measures. Sometimes there's more than one way to do this.

329 Adding Mixed Measures

EXAMPLE: Find the perimeter of the triangle.

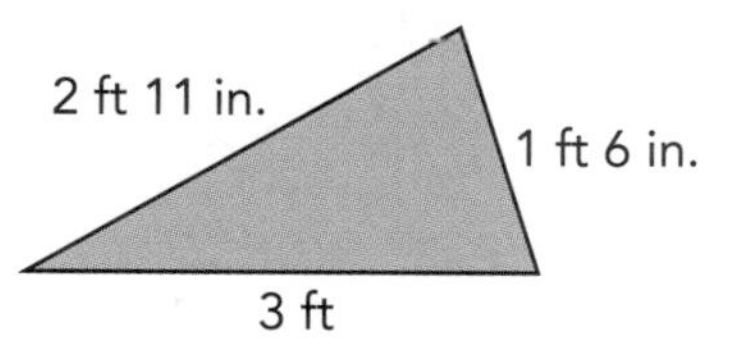

ONE WAY You can add and regroup units.

See 122, 295–298, 486

① Add inches and regroup.	② Add feet.
1 ft (regrouped) 1 ft 6 in. 3 ft 17 in. = 1 ft 5 in. + 2 ft 11 in. 5 in.	1 ft (regrouped) 1 ft 6 in. 3 ft + 2 ft 11 in. 7 ft 5 in.

ANOTHER WAY You can change to all inches or all feet and then add.

See 028–034, 486

Change to all inches.	Or, change to all feet.
1 ft 6 in. → 12 in. + 6 in. → 18 in. 3 ft → 36 in. + 2 ft 11 in. → 24 in. + 11 in. → + 35 in. 89 in. 89 ÷ 12 → 7 R5 89 in. = 7 ft 5 in. or $7\frac{5}{12}$ ft	1 ft 6 in. → $1\frac{6}{12}$ ft 3 ft → 3 ft + 2 ft 11 in. → + $2\frac{11}{12}$ ft $6\frac{17}{12}$ ft $6\frac{17}{12} = 6 + 1\frac{5}{12} = 7\frac{5}{12}$

 Either way, the perimeter is 7 ft 5 in.

Subtracting Mixed Measures

Sometimes you need to subtract mixed measures.

EXAMPLE: Mark is 4 ft 9 in. tall. His older brother Eric is 6 ft 2 in. tall. How much taller is Eric than Mark?

ONE WAY To solve the problem, subtract 4 ft 9 in. from 6 ft 2 in.

You can regroup units and then subtract. This is something like what you do when you subtract whole numbers.

See 131–132, 486

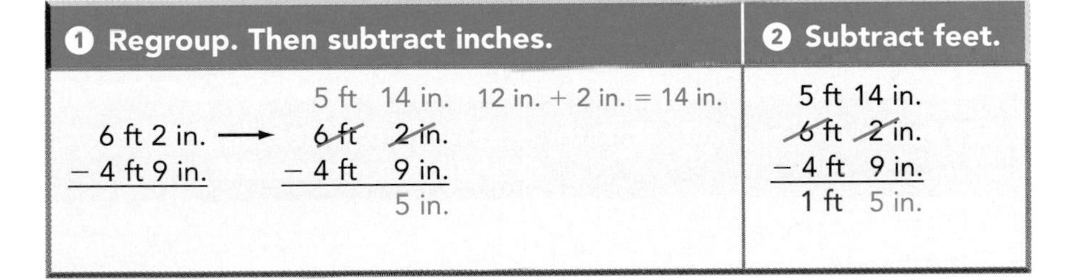

① Regroup. Then subtract inches.	② Subtract feet.
6 ft 2 in. ⟶ (5 ft 14 in.) ~~6 ft~~ ~~2 in.~~ − 4 ft 9 in. ⟶ − 4 ft 9 in. 5 in. 12 in. + 2 in. = 14 in.	5 ft 14 in. ~~6 ft~~ ~~2 in.~~ − 4 ft 9 in. 1 ft 5 in.

ANOTHER WAY You can change to all inches or all feet and then subtract.

See 165–166

Change to all inches.	6 ft 2 in. ⟶ 72 in. + 2 in. ⟶ 74 in. − 4 ft 9 in. ⟶ 48 in. + 9 in. ⟶ − 57 in. 17 in. 17 ÷ 12 ⟶ 1 R5 17 in. = 1 ft 5 in. or $1\frac{5}{12}$ ft
Or, change to all feet.	6 ft 2 in. ⟶ $6\frac{2}{12}$ ft − 4 ft 9 in. ⟶ $4\frac{9}{12}$ ft ⟶ $5\frac{14}{12}$ (~~$6\frac{2}{12}$~~) ft $-4\frac{9}{12}$ ft $1\frac{5}{12}$ ft

Either way, Eric is 1 ft 5 in. taller than Mark.

331 Multiplying and Dividing Mixed Measures

CASE 1 Sometimes you need to multiply or divide a measure by a number.

EXAMPLE 1: Find the perimeter of the square. Use the formula $P = 4s$.

2 ft 4 in.

2 ft 4 in. 2 ft 4 in.

2 ft 4 in.

To find the perimeter, multiply 2 ft 4 in. by 4.

MORE HELP

See 136–144, 295–298, 485

ONE WAY You can multiply and regroup units. This is something like what you do when you multiply whole numbers.

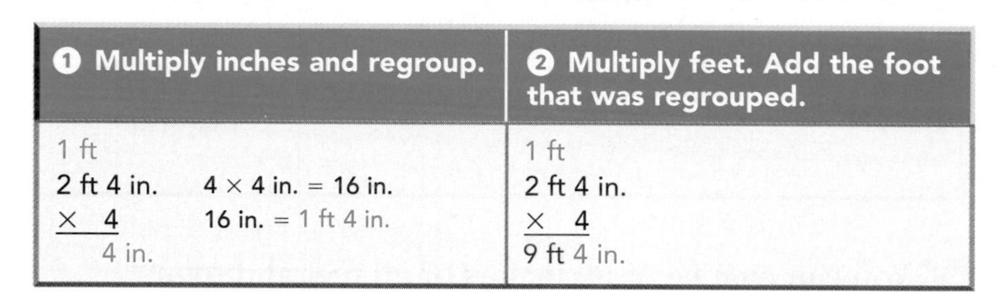

① Multiply inches and regroup.	② Multiply feet. Add the foot that was regrouped.
1 ft 2 ft 4 in. 4×4 in. = 16 in. $\times$ 4 16 in. = 1 ft 4 in. 4 in.	1 ft 2 ft 4 in. $\times$ 4 9 ft 4 in.

ANOTHER WAY You could also change to all inches or change to all feet and then multiply.

MORE HELP

See 034, 167–170, 485

Change to all inches.	Or, change to all feet.
2 ft 4 in. → 24 in. + 4 in. → 28 in. $28 \times 4 = 112$ $112 \div 12$ → 9 R4 112 in. = 9 ft 4 in., $9\frac{4}{12}$ ft, or $9\frac{1}{3}$ ft	2 ft 4 in. → $2\frac{4}{12}$ ft = $2\frac{1}{3}$ ft $2\frac{1}{3} \times 4$ ↓ $\frac{7}{3} \times \frac{4}{1} = \frac{28}{3}$ or $9\frac{1}{3}$

Either way, the perimeter of the square is 9 ft 4 in. You could also say that the perimeter is $9\frac{1}{3}$ ft.

CASE 2 To multiply or divide a measure by a measure, be sure all measures are in the same units before you begin.

EXAMPLE 2: You are making doorstops with bags of rice. You have 6 lb 4 oz of rice. You want to put 1 lb 9 oz into each bag. How many bags can you fill?

Customary Units of Weight	
ounce (oz)	
pound (lb)	1 lb = 16 oz
ton (t)	1 t = 2000 lb

To solve the problem, divide 6 lb 4 oz by 1 lb 9 oz.

Change all to ounces.	Or, change all to pounds.
6 lb 4 oz → 96 oz + 4 oz → 100 oz	6 lb 4 oz → $6\frac{4}{16}$ lb = $6\frac{1}{4}$ lb
1 lb 9 oz → 16 oz + 9 oz → 25 oz	1 lb 9 oz → $1\frac{9}{16}$ lb
100 ÷ 25 = 4	$6\frac{1}{4} \div 1\frac{9}{16}$ ↓ $\frac{25}{4} \div \frac{25}{16}$ ↓ $\frac{25}{4} \times \frac{16}{25} = \frac{400}{100}$ or 4

MORE HELP

See 034, 176, 485

Either way, you can fill 4 bags.

MATH ALERT Watch Your Units

When you multiply units of measure by other units of measure, remember two very important things:

1. Always be sure the units are the same. For example, you can multiply inches by inches or feet by feet, but don't multiply inches by feet.
2. Always be sure that your answer has the right kind of unit. For example, if you've multiplied inches by inches you get square inches. Just as $4 \times 4 = 4^2$, in. × in. = in.^{2}

333

Geometry

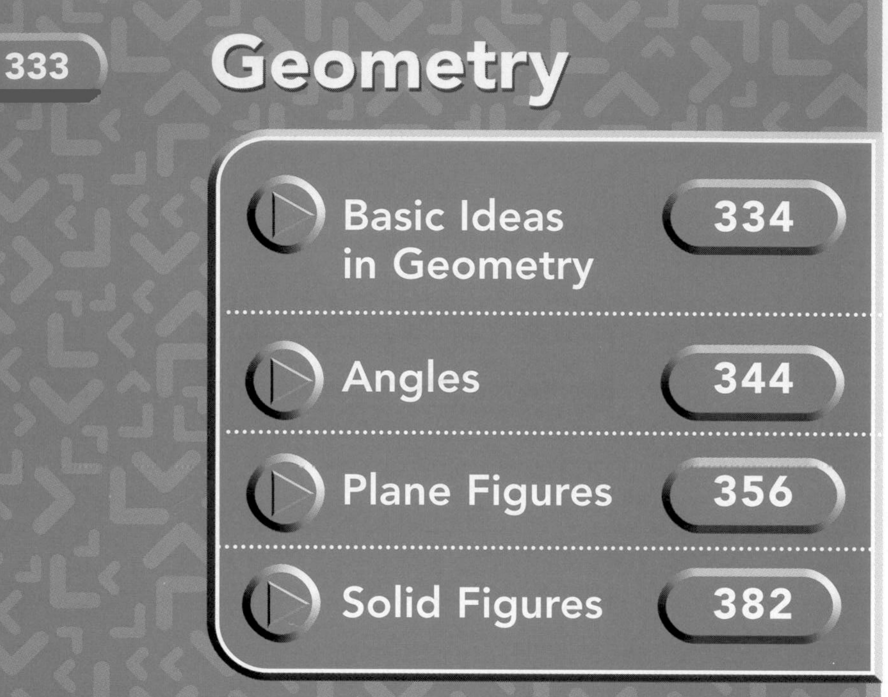

You might be surprised at how much geometry you use every day. When you give directions, play soccer, bake, or build, you use words like *straight*, *curved*, *turn*, *round*, and *corner*. Words like these help us talk about geometric ideas.

Numbers and computation help us talk about *how many* and *how big* and to make some comparisons. Geometry helps us talk about where things are, where they're going, and what shapes they have.

334

Basic Ideas in Geometry

Points, planes, lines, and curves are ideas. They help us build bridges, draw pictures, and make maps. They help us to talk about shape, position, and direction.

335

Points

In geometry, **points** are places in space. They have no length or width. You can't touch them or measure them. But, you can describe where they are. Usually, when you picture a point, you name it with a capital letter.

Where two lines **intersect** (cross), there is a point.

See 337, 340

If you name any two points, there is a line (and only one line) that will pass through both of them.

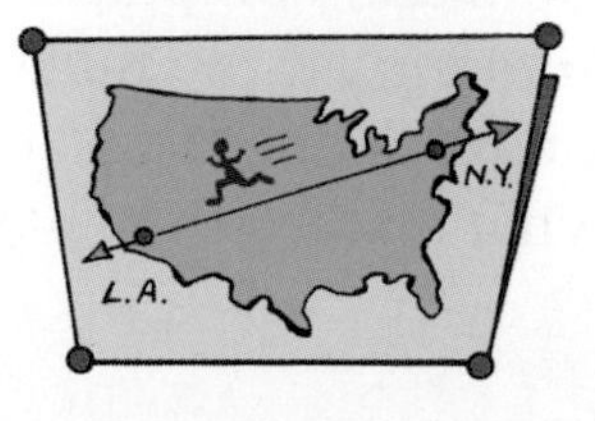

Planes 336

In geometry, planes are more like the Great Plains than like airplanes! A **plane** is a flat surface infinitely wide and infinitely long, with no thickness. This means you can't measure the length and width of a plane. Nor can you count the number of points or lines in a plane because there is an infinite number of them.

MORE HELP See 340

CASE 1 Planes can intersect each other, and where they do, there is a line.

CASE 2 Two planes can be parallel to each other.

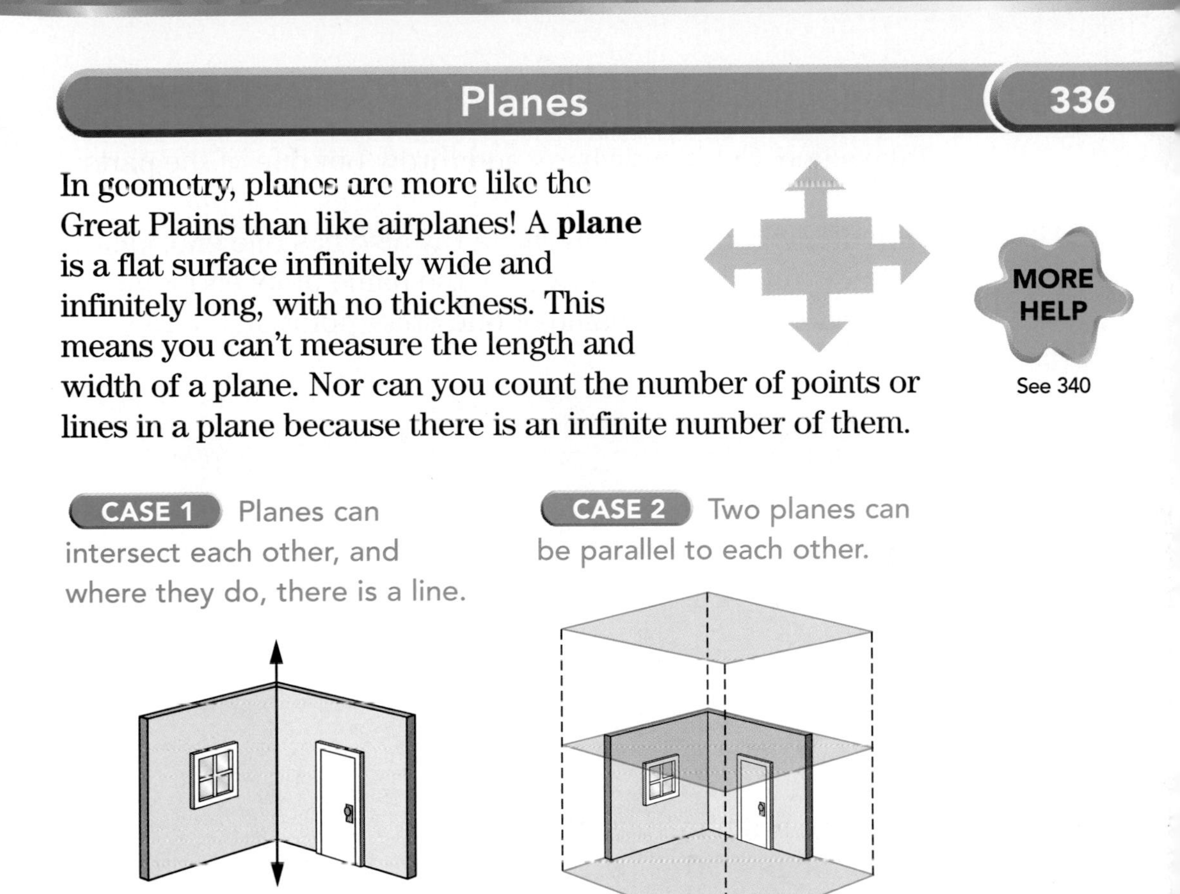

Lines 337

A **line** is a straight path of points that has no endpoints. It goes on forever in two directions. When you draw a picture of a line, you can mark two points on it and give them letter-names. You can also name it with a single letter. You should draw arrowheads to show that the line goes on in both directions without stopping.

Write: $\overleftrightarrow{FG}$ or $\overleftrightarrow{GF}$ or a
Say: *line F G, line G F, or line a*

When you see the little 2-headed arrow, say *line*.

338

Rays

Take a line. Cut it anywhere, and throw out one of the parts. What you have left is a ray. A **ray** only goes on in *one* direction without stopping. So, a ray also has one end, called an **endpoint**. To name a ray, say the name of its endpoint first and then say the name of one other point on the ray.

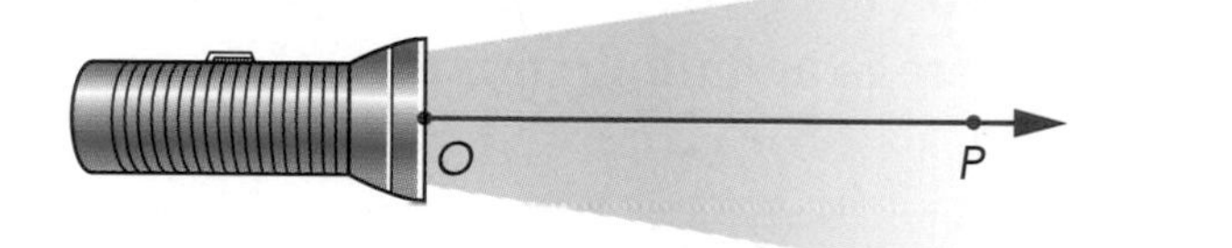

Write: $\overrightarrow{OP}$
Say: *ray O P*

When you see the little 1-headed arrow, say ray.

EXAMPLE: Name the rays with endpoint O.

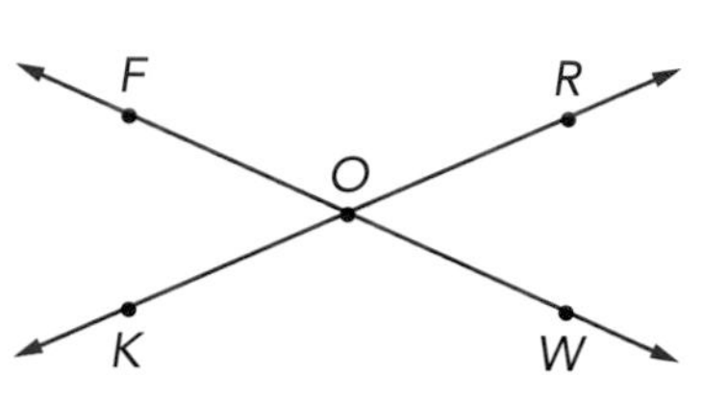

There are four rays with endpoint O: $\overrightarrow{OF}$, $\overrightarrow{OR}$, $\overrightarrow{OK}$ and $\overrightarrow{OW}$

339

Line Segments

A **line segment** is a part of a line that you can measure. It has two endpoints and includes all the points between those endpoints. To name a line segment, use the endpoints.

EXAMPLE: Name the line segments that form this rectangle.

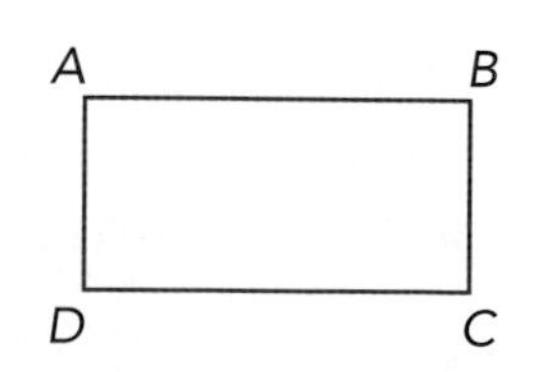

The rectangle has 4 sides. Each side is a line segment. They are $\overline{AB}$, $\overline{BC}$, $\overline{CD}$, and $\overline{DA}$.

Write: $\overline{AB}$
Say: *line segment A B*

When you see the little segment over the letters that name the endpoints, say *line segment.*

Parallel and Intersecting Lines

MORE HELP
See 336

Parallel lines are lines that are in the same plane. They will never cross because they are always the same distance apart. They don't share any points. Segments of parallel lines are parallel, too.

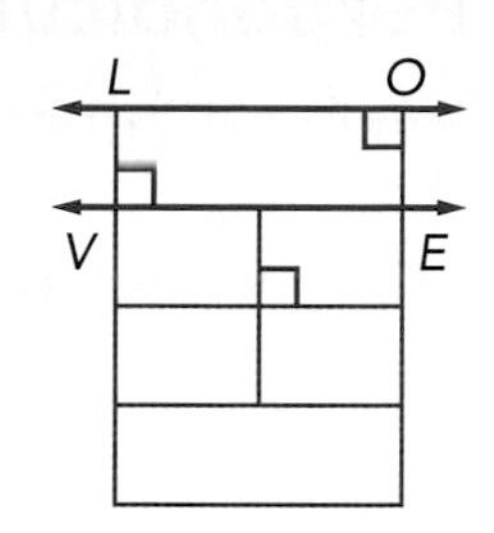

Write: $\overleftrightarrow{LO} \parallel \overleftrightarrow{VE}$
Say: *line LO is parallel to line VE*

When you see the little parallel line symbol, say *is parallel to*.

Intersecting lines are lines that do cross. They have one point in common. Line segments can also intersect.

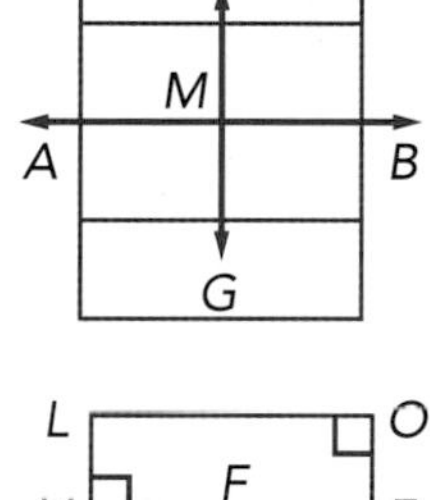

Segment *AB* intersects segment *FG*. The point of intersection is *M*.

EXAMPLE: Name one other pair of parallel segments and one pair of intersecting segments in this diagram of a singles tennis court.

A tennis court is a rectangle and all its markings make rectangles.

- $\overline{LO}$ and $\overline{AB}$ are parallel. So are $\overline{VE}$ and $\overline{AB}$.
- Segments $\overline{LA}$ and $\overline{VE}$ intersect. So do $\overline{LA}$ and $\overline{AB}$, $\overline{VE}$ and $\overline{FG}$, $\overline{VE}$ and $\overline{OB}$, $\overline{LO}$ and $\overline{OB}$, $\overline{AB}$ and $\overline{OB}$, and $\overline{AB}$ and $\overline{FG}$.

parallel

intersect

You can remember what these terms mean by thinking about the way they look.

Perpendicular Lines

Perpendicular lines are special intersecting lines. They form right angles where they intersect. Segments of perpendicular lines can also be perpendicular.

MORE HELP

See 331, 337

Perpendicular

Not Perpendicular

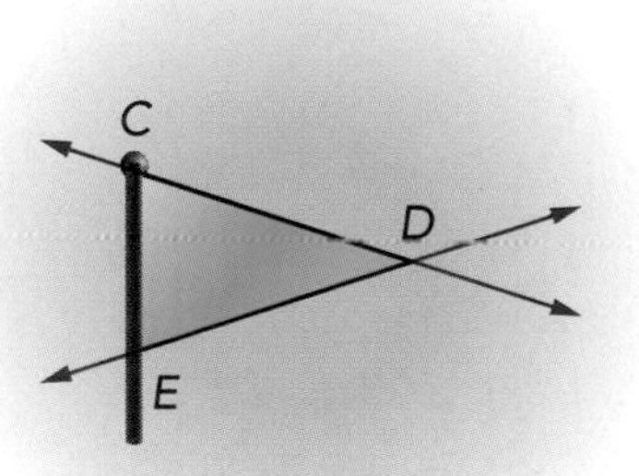

The angles formed where $\overleftrightarrow{CD}$ and $\overleftrightarrow{DE}$ intersect are not right angles.

Write: $\overleftrightarrow{OT} \perp \overleftrightarrow{IN}$
Say: *line O T is perpendicular to line I N*

When you see the little upside down T, say *is perpendicular to*.

EXAMPLE: Name the perpendicular segments in this figure.

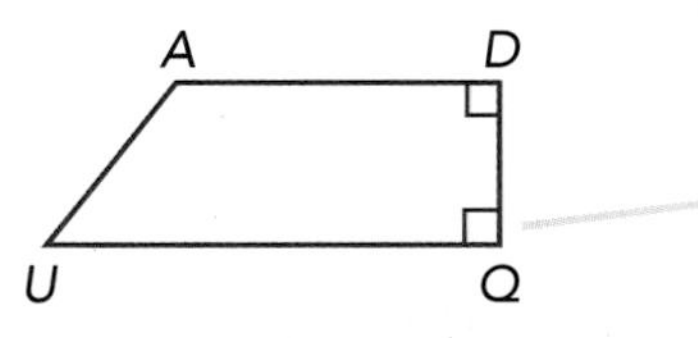

The little square corner tells you that this is a right angle. It measures 90°.

This figure is a trapezoid. It has two right angles: Q and D.

Angle	Intersecting line segments
Angle Q	$\overline{DQ} \perp \overline{QU}$
Angle D	$\overline{QD} \perp \overline{DA}$

MATH ALERT Don't Assume Perpendicular Lines

342

If you see a square corner drawn at the intersection of two lines or segments, then you can be sure they are perpendicular. If you don't see the little square corner and no other clue tells you that the lines are perpendicular, then don't assume they are. Clues, other than the symbol, that you might find in problems are:

- *right* triangle or *right* angle
- *square* or *rectangular* figure

Curves

343

Imagine smoothly bending a line or a line segment without forming a corner. You would have a **curve**. A curve can be open or closed.

Open Curves

Closed Curves

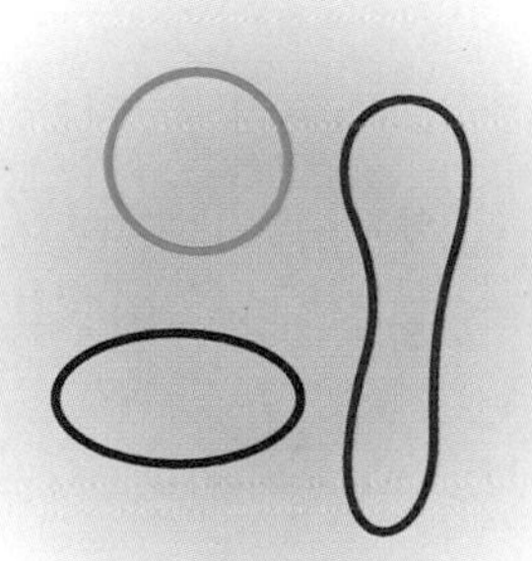

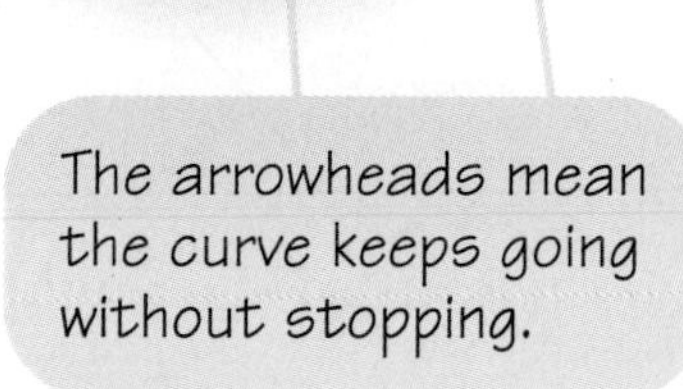

344

Angles

MORE HELP
See 467

You can think of angles as turns. Or, you can think of them as corners. Look in the Almanac to find out how to use a protractor to measure angles.

345 Parts of Angles

Angles are formed by two rays that have the same endpoint. This endpoint is called the **vertex**. You can find angles anywhere lines and line segments intersect. You can name an angle in three different ways.

- Use three letters *in this order*: a point on one ray, the vertex, a point on other ray.

  Write: $\angle BAG$ or Write: $\angle GAB$
  Say: *angle B A G* Say: *angle G A B*

- Use one letter at the vertex.

  Write: $\angle A$
  Say: *angle A*

- Use a number written inside the rays of the angle.

  Write: $\angle 1$
  Say: *angle 1*

When you see the little sharp corner, say *angle*.

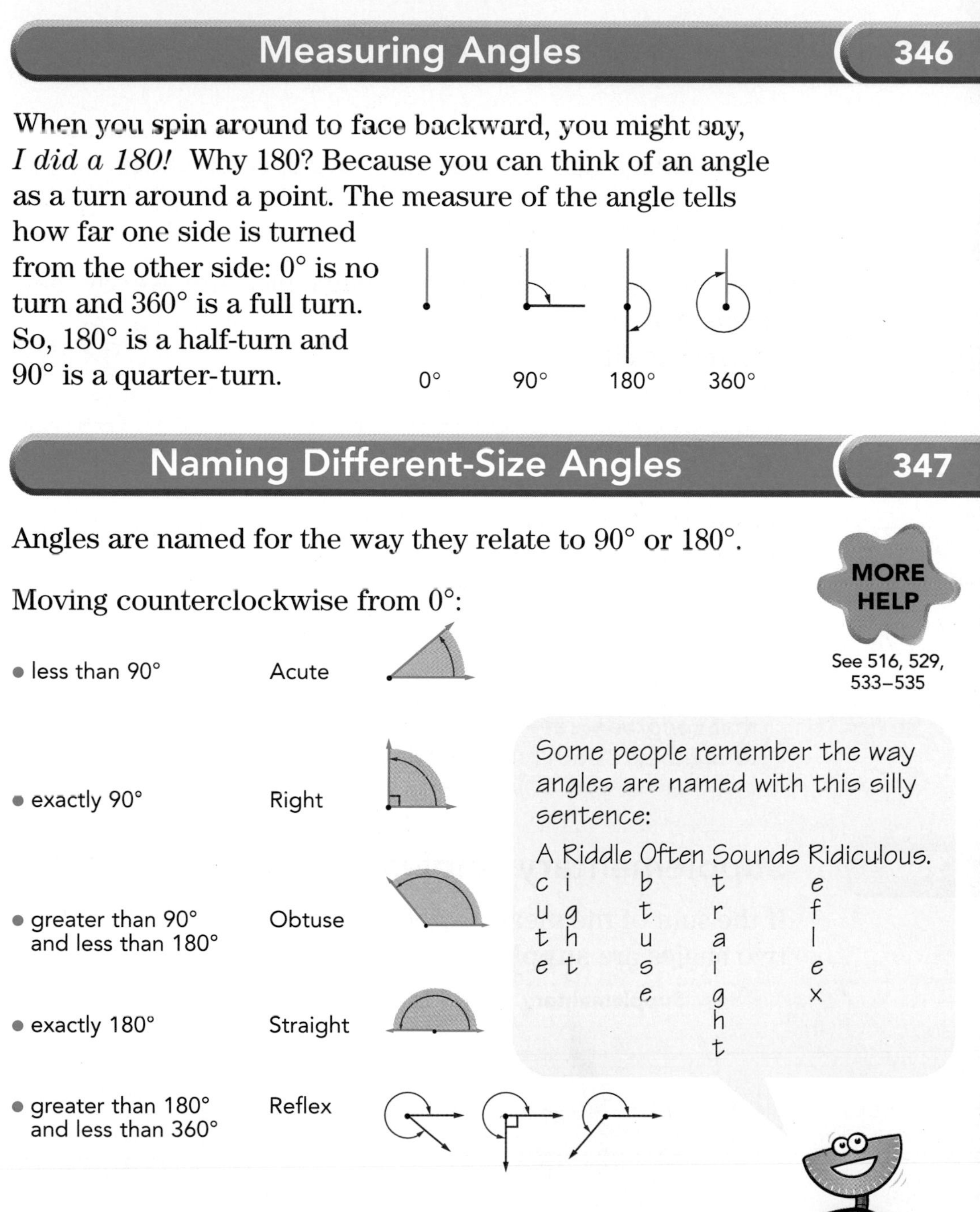

Measuring Angles 346

When you spin around to face backward, you might say, *I did a 180!* Why 180? Because you can think of an angle as a turn around a point. The measure of the angle tells how far one side is turned from the other side: 0° is no turn and 360° is a full turn. So, 180° is a half-turn and 90° is a quarter-turn.

0° 90° 180° 360°

Naming Different-Size Angles 347

Angles are named for the way they relate to 90° or 180°.

MORE HELP
See 516, 529, 533–535

Moving counterclockwise from 0°:

- less than 90° — Acute
- exactly 90° — Right
- greater than 90° and less than 180° — Obtuse
- exactly 180° — Straight
- greater than 180° and less than 360° — Reflex

Some people remember the way angles are named with this silly sentence:

A Riddle Often Sounds Ridiculous.

Acute Right Obtuse Straight Reflex

348 Angle Relationships

Angles can fit together in specific ways, or have special sums. When angles do this they have special names.

349 Congruent Angles

When two angles have the same angle measure, we say they are **congruent**. This means either angle can fit exactly over the other angle.

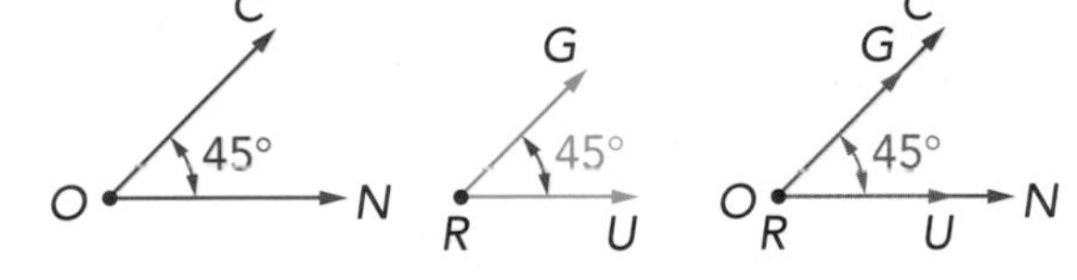

Write: ∠CON ≅ ∠GRU
Say: *angle C O N is congruent to angle G R U*

When you see the equals sign with the little wavy segment over it, say *is congruent to*.

DID YOU KNOW...

that *congruent* refers to geometric figures and *equal* refers to numbers? Both mean *the same as*.

350 Supplementary Angles

If the sum of the measures of two angles is 180°, then the two angles are **supplementary**.

Supplementary

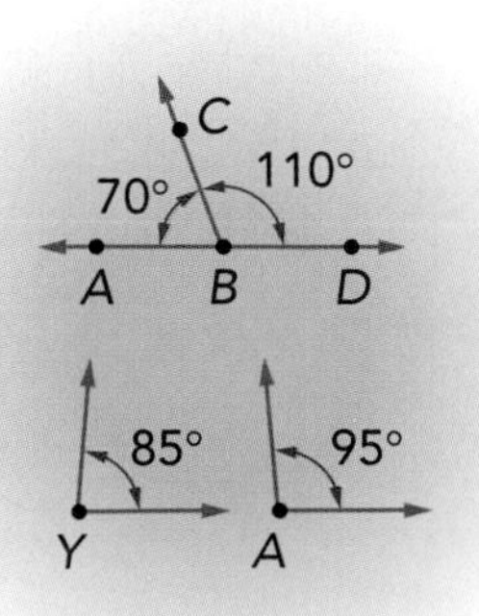

Not Supplementary

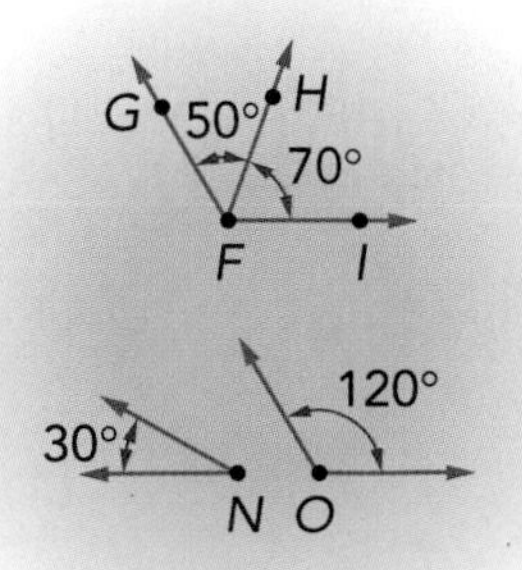

Complementary Angles

351

If the sum of the measures of two angles is 90°, then the angles are **complementary**.

Complementary | **Not Complementary**

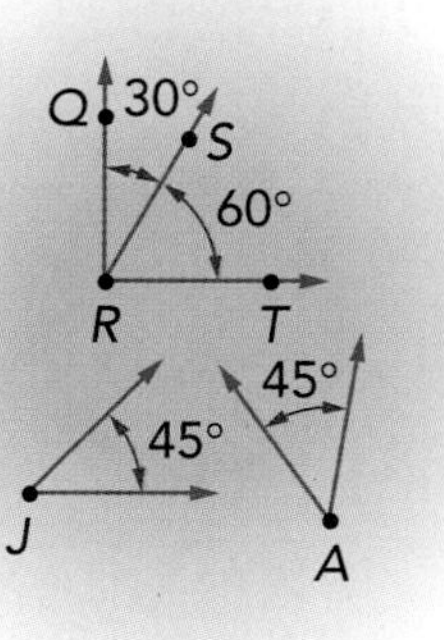

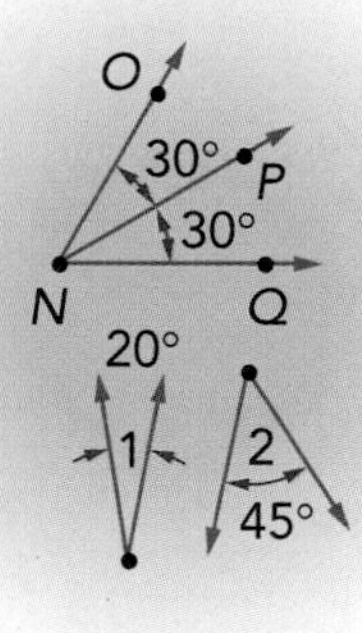

Angle Bisectors

352

An **angle bisector** is a ray that separates an angle into two congruent angles.

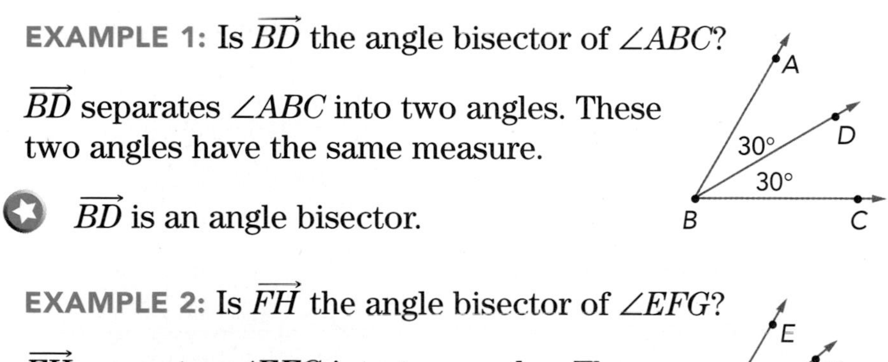

EXAMPLE 1: Is $\overrightarrow{BD}$ the angle bisector of $\angle ABC$?

$\overrightarrow{BD}$ separates $\angle ABC$ into two angles. These two angles have the same measure.

$\overrightarrow{BD}$ is an angle bisector.

EXAMPLE 2: Is $\overrightarrow{FH}$ the angle bisector of $\angle EFG$?

$\overrightarrow{FH}$ separates $\angle EFG$ into two angles. These two angles do not have the same measure.

$\overrightarrow{FH}$ is not an angle bisector.

353

Angles Formed by Intersecting Lines

When two lines intersect, they form four angles that measure less than 180°. In this case, two angles next to each other are called **adjacent angles**. Also in this case, two angles *not* next to each other are called **vertical angles**. Look at how these angles are related to each other.

The four angles go all the way around a point, so the sum of their measures must be 360°.

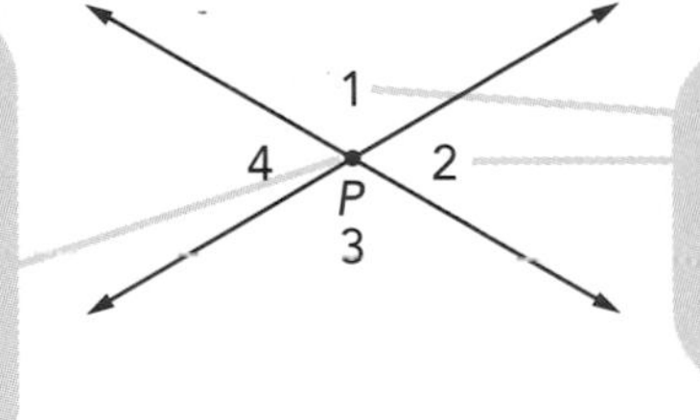

Adjacent angles form a straight angle, so they must be supplementary.

If you add the measures of angle 1 and angle 4, the sum is 180°. If you add the measures of angle 1 and angle 2, the sum is also 180°. This means that angle 2 and angle 4 must have the same measure. Vertical angles always have the same measure.

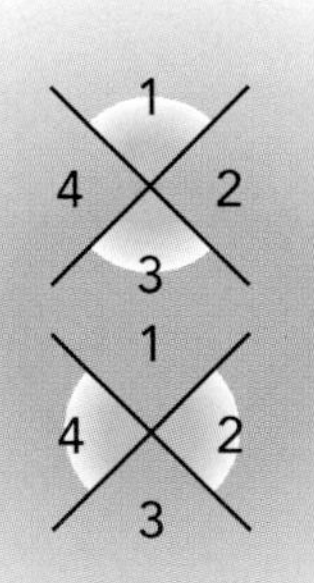

Supplementary Angles

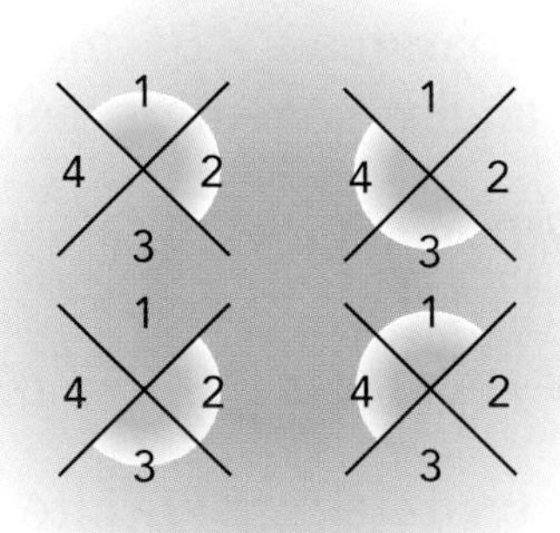

Sums of Interior Angles of Triangles

Look at the inside of a triangle. There are three angles. These are called the **interior angles**. Now look what happens when you tear off the corners of the triangle and line up its angles.

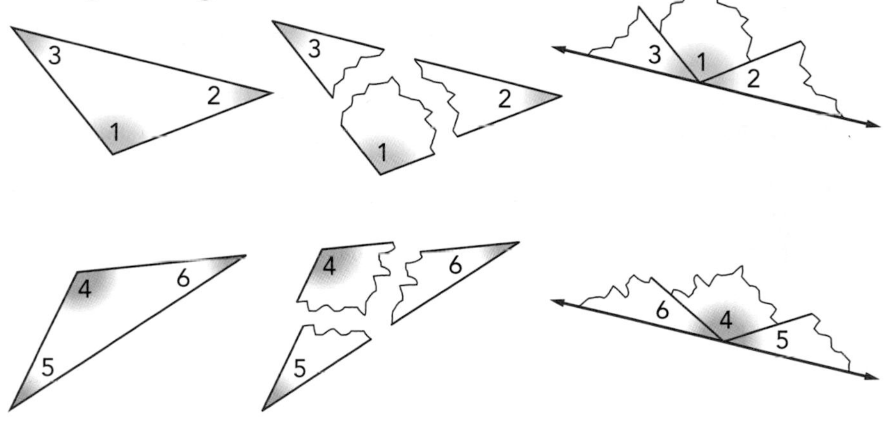

For any triangle, if you put the interior angles together, you get a straight angle. So, the sum of the measures of the interior angles of any triangle is 180°.

Central Angles of Circles

If you draw radii out from the center of a circle, you are forming **central angles** in that circle. Remember, turning completely around is turning 360°. So, adding the measures of all the central angles in any circle gives a sum of 360°.

MORE HELP
See 367

$90 + 150 + 60 + 60 = 360$

Mathematicians chose 360 as the number of degrees in a circle because it could be divided easily by lots of numbers.

356

Plane Figures

MORE HELP

See 336

Have you ever made shadow puppets on a wall? Those images you project onto the wall are flat. Squares, triangles, and other plane figures are flat, too. They are called **plane figures** because they lie in a single plane.

In geometry, a plane is a 2-dimensional surface that is perfectly flat and infinitely large. Even though they are perfectly flat, planes and plane figures are useful in our 3-dimensional world. Think about city maps, diagrams of basketball plays, and computer games.

Plane figures come in a variety of shapes and sizes. We name them according to the number, size, and position of their sides and angles. Here are some terms you will use when you talk about plane figures.

When you talk about more than one vertex, say *vertices* (**ver** tuh sees).

side: a curve or segment that's part of a figure

vertex: a corner where two line segments meet

Polygons 357

A **polygon** is a closed figure whose sides are all line segments. In a regular polygon all the sides are the same length.

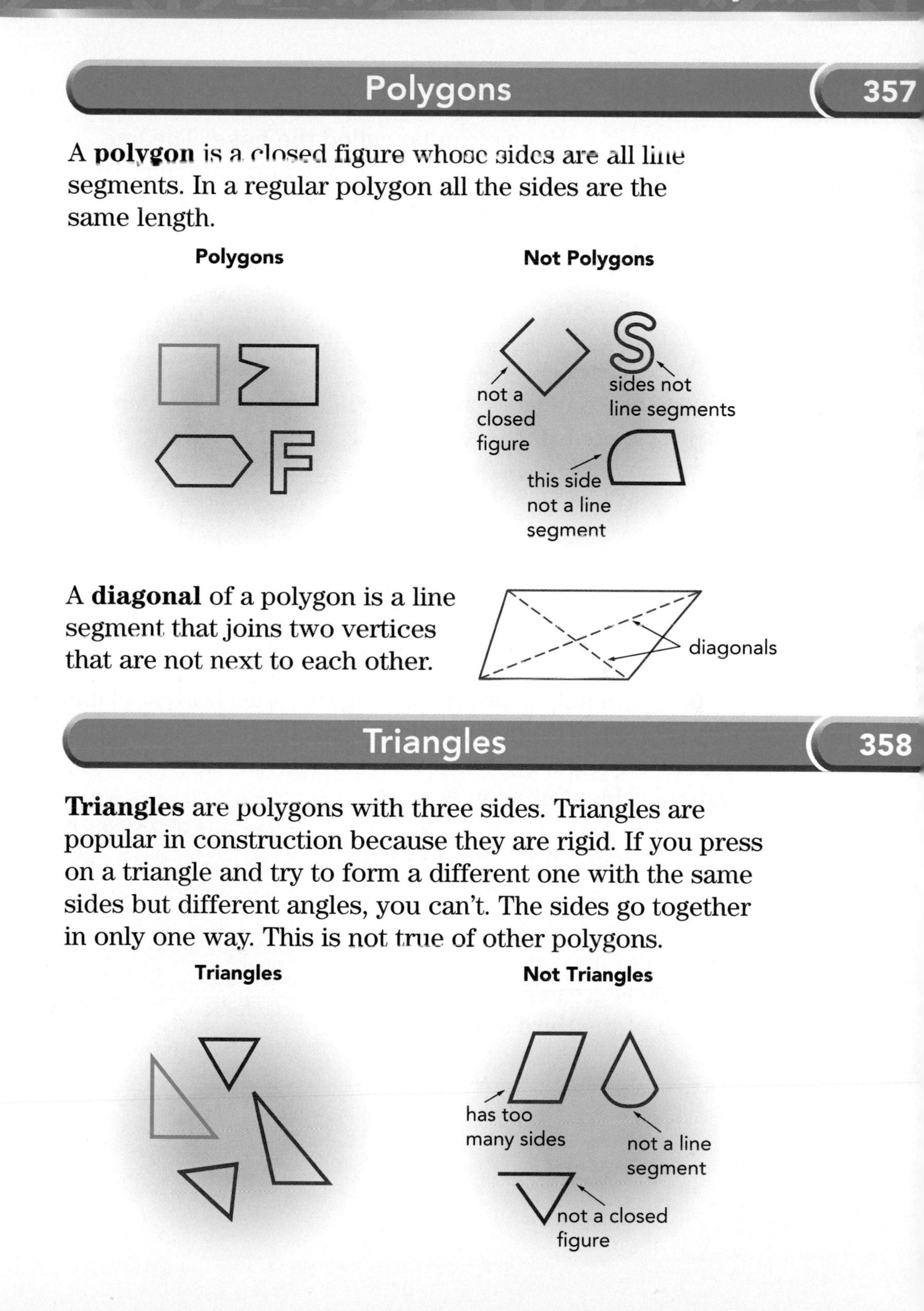

A **diagonal** of a polygon is a line segment that joins two vertices that are not next to each other.

Triangles 358

Triangles are polygons with three sides. Triangles are popular in construction because they are rigid. If you press on a triangle and try to form a different one with the same sides but different angles, you can't. The sides go together in only one way. This is not true of other polygons.

359

Parts of Triangles

MORE HELP

See 341

Any side of a triangle can be called the **base**. The line segment that starts at a vertex and is perpendicular to the base is called the **height**, or altitude.

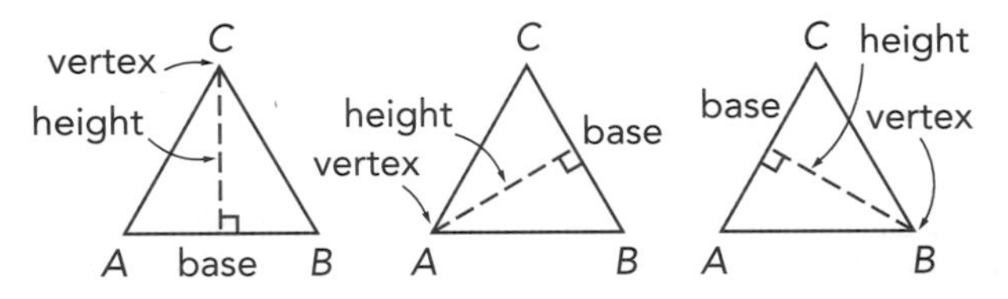

Note: The same triangle can have 3 different pairs of heights and bases. But the product of height and base will be the same no matter which pair you choose!

To name a triangle or any other figure, name the vertices in either clockwise or counterclockwise order. There are 6 ways to name this triangle: ΔABC, ΔBCA, ΔCAB, ΔACB, ΔBAC, and ΔCBA.

Write: ΔABC
Say: *triangle A B C*

360

Lengths of Sides of Triangles

You've probably heard the saying that the shortest distance between any two points is a straight line. The saying really should say *line segment*, not *line*. You can use a triangle to see that this statement is true.

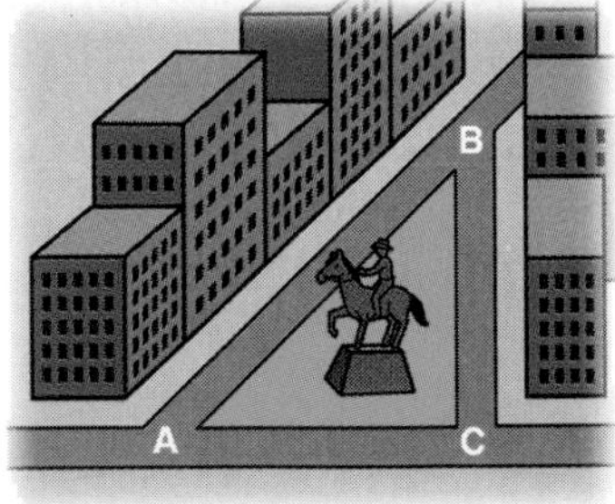

The shortest route from A to C is from A to C, not from A to B to C.

You can see that the sum of the lengths of any two sides of a triangle is greater than the length of the third side.

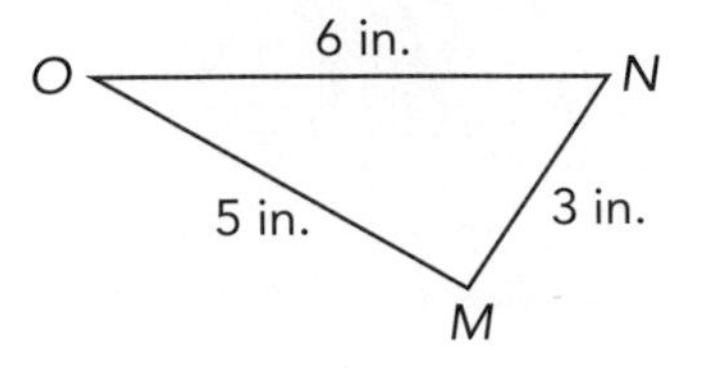

3 in. + 6 in. > 5 in.
3 in. + 5 in. > 6 in.
5 in. + 6 in. > 3 in.

Using Angle Size to Classify Triangles

361

In any triangle, the sum of the angle measures is always 180°. Here are some triangle names that are based on the angles.

Name	Description	Example
Acute Triangle	All angles measure less than 90°.	65°, 35°, 80°
Equiangular Triangle An equiangular triangle is also equilateral!	All angles measure exactly 60°.	60°, 60°, 60°
Obtuse Triangle	One angle measures more than 90°.	40°, 110°, 30°
Right Triangle	One angle measures exactly 90°.	60°, 30°, 90°

Using Side Length to Classify Triangles

362

Here are some triangle names that are based on the lengths of sides.

Name	Description	Example
Equilateral Triangle An equilateral triangle is also equiangular!	All three sides are the same length.	4 cm, 4 cm, 4 cm
Isosceles Triangle	Two sides are the same length.	5 in., 5 in., 6 in.
Scalene Triangle	No side is the same length as any other.	6 ft, 3 ft, 4 ft

363 Overlapping Classifications of Triangles

You can name any triangle by the way its sides are related and by the way its angles are related. This means that every triangle can be described in more than one way.

	Equilateral	Isosceles	Scalene
Acute	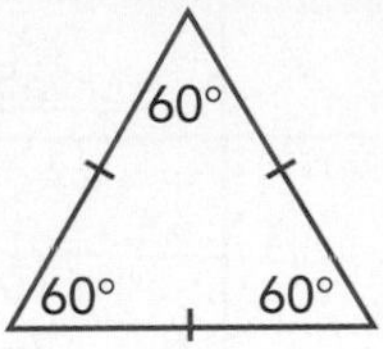 60°, 60°, 60°	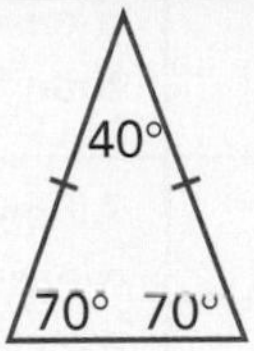 40°, 70°, 70° Sides marked the same way are the same length.	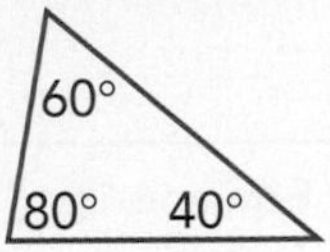 60°, 80°, 40°
Right	Equilateral triangles are always equiangular. You can't make a triangle with 3 right angles, so you can't have an equilateral right triangle.	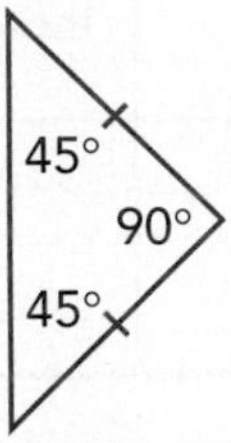 45°, 90°, 45°	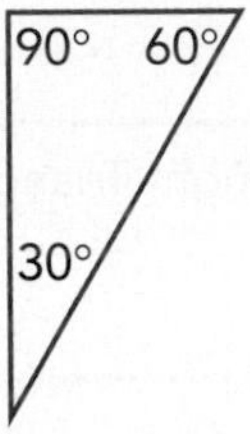 90°, 60°, 30°
Obtuse	Equilateral triangles are always equiangular. You can't make a triangle with 3 angles greater than 90°, so you can't have an obtuse equilateral triangle.	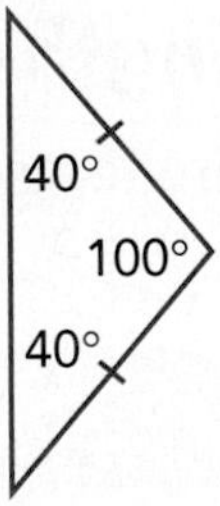 40°, 100°, 40°	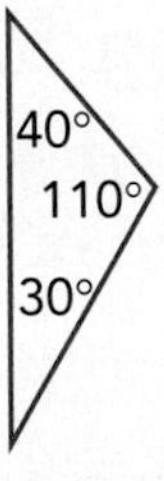 40°, 110°, 30°

To read the table above, think about the phrases *is always*, *may be*, or *is never*. For example, a right triangle *may be* either scalene or isosceles but *is never* obtuse.

Quadrilaterals

364

A **quadrilateral** is a polygon. Its four sides are line segments.

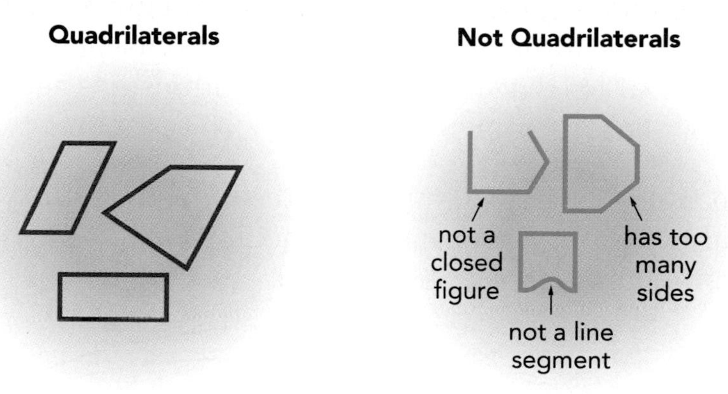

Classifying Quadrilaterals

365

Some quadrilaterals have special traits, so they have special names.

See 340

Name	Description	Example
Trapezoid	Exactly one pair of parallel sides	Sides *AB* and *CD* are parallel.
Parallelogram	Opposite sides the same length and parallel	Sides *JK* and *LM* are parallel. Sides *KL* and *MJ* are parallel.
Rectangle	Parallelogram with 4 right angles	Sides *RS* and *TU* are parallel. Sides *ST* and *UR* are parallel.
Rhombus	Parallelogram with all sides the same length	Sides *ZA* and *BC* are parallel. Sides *AB* and *CZ* are parallel.
Square	Rectangle with all sides the same length	Sides *HI* and *JK* are parallel. Sides *IJ* and *KH* are parallel.

366 Overlapping Classifications of Quadrilaterals

Think about a family tree. Joe is Ann's *husband.* He's also Jake's *father* and Shauna's *grandfather*. Joe is just one person in the family, but there are many family names for him. Many quadrilaterals can be described using more than one name. This diagram shows the relationships among different quadrilaterals.

To read this diagram, think about the phrases *is always*, *may be*, and *is never*. For example, a rhombus *may be* a square, but a rhombus *is never* a trapezoid.

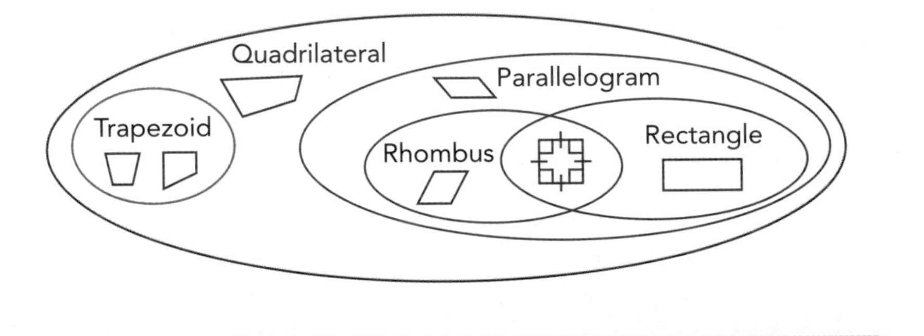

367 Circles

A **circle** is a set of points in a plane, all of which are the same distance from a given point. That point is the **center** of the circle.

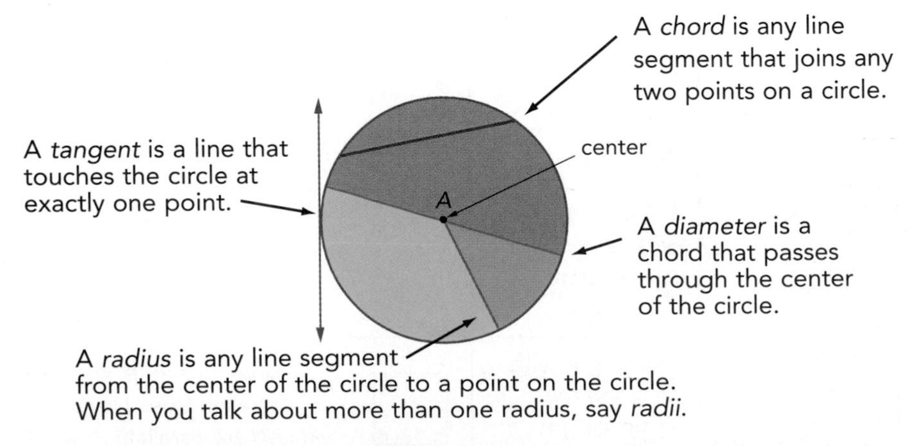

Pi

For all circles, the ratio of the circumference to the diameter (C/d) is always the same. This ratio is called pi (π). The value of π is *approximately* 3.14, or $\frac{22}{7}$. That means the circumference of a circle is a little more than three times as long as its diameter.

MORE HELP

See 178, 298

Similarity

Usually, when people say that two things are similar, they mean that the things are alike. In geometry, **similar figures** are alike in a very specific way.

- They have the same shape.
- They may or may not be the same size.

Here are three drawings: the original drawing, an enlargement of the original, and a reduction of the original. Because all three drawings show exactly the same shape, the figures are similar.

These two polygons are similar.

Two similar polygons have corresponding angles that have the same measure, and corresponding sides that are proportional.

The ratios of lengths of corresponding sides are equal.

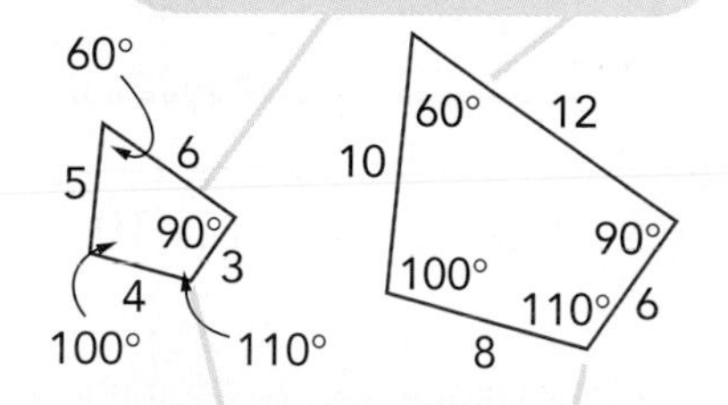

See 178, 181

370 Scale Drawings

A **scale drawing** is a drawing that is the same shape but *not* the same size as the object it shows. Blueprints and maps are examples of scale drawings.

EXAMPLE: The bedroom in this blueprint is rectangular. What are its real length and width?

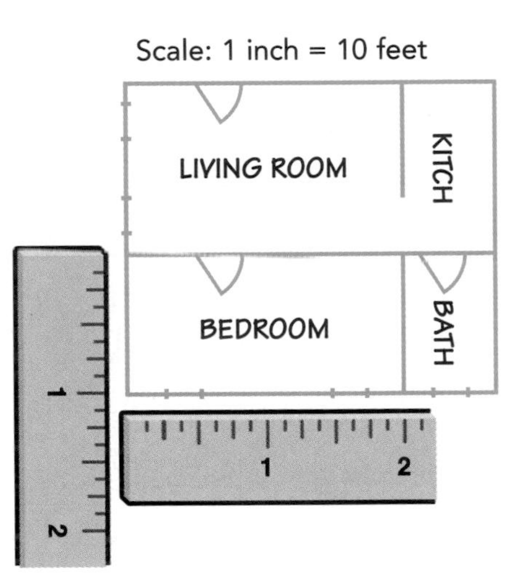

The scale says one inch on the drawing stands for ten feet in real life. In the drawing, the bedroom is one inch wide. So, the bedroom must really be ten feet wide. To find the length, use a proportion.

See 184, 365

1 Measure the drawing.	The bedroom is 2 inches long.
2 Write a proportion.	$\frac{\text{scale length}}{\text{real length}} \rightarrow \frac{1 \text{ in.}}{10 \text{ ft}} = \frac{2 \text{ in.}}{\blacksquare} \leftarrow \frac{\text{length in drawing}}{\text{length in real life}}$
3 Solve the proportion.	$\frac{1 \times 2 = 2}{10 \times 2 = 20}$
4 Decide which units to use.	The scale is inches to feet. You measured in inches, so your answer must be in feet.

★ The bedroom is 10 feet wide and 20 feet long.

When you work with ratios you don't have to make all your units the same, but you do have to use the correct units in your answer.

MATH ALERT Changing the Scale

371

EXAMPLE: Make scale drawings of a square with sides 8 cm long. First, use a scale of 1 unit = 2 cm. Then, use a scale of 1 unit = 1 cm. How does changing the scale change the area of the square in the drawing?

See 301

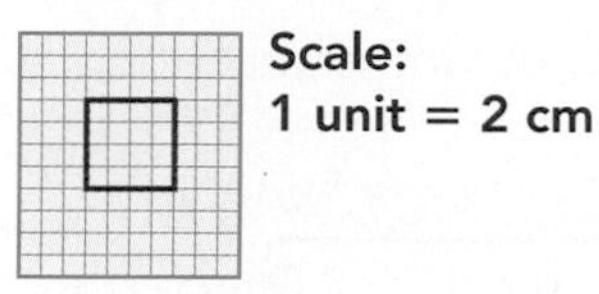

Scale:
1 unit = 2 cm

Count by twos as you draw. Each side will be 4 units long. The area of the scale drawing is 16 square units.

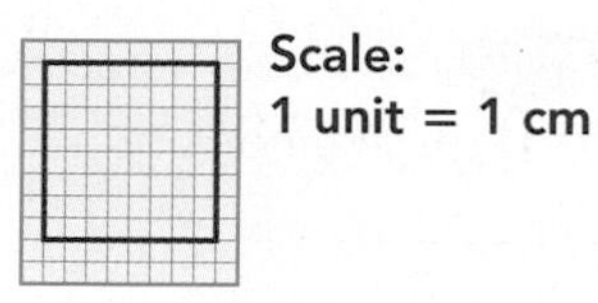

Scale:
1 unit = 1 cm

With this scale, to show sides 8 cm long, draw sides 8 units long. The area of this scale drawing is 64 square units.

When you changed the scale from 1 unit = 2 cm to 1 unit = 1 cm, you multiplied the area of the square in the scale drawing by four.

Congruence

372

Figures that have the same shape and size are **congruent.** You can put one over the other so they will match exactly.

Congruent

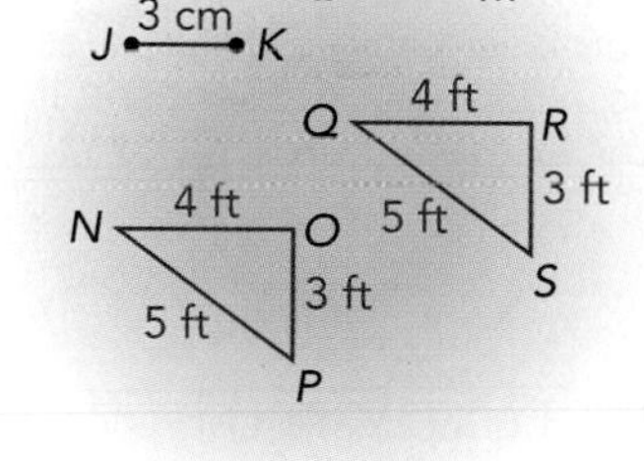

Not Congruent

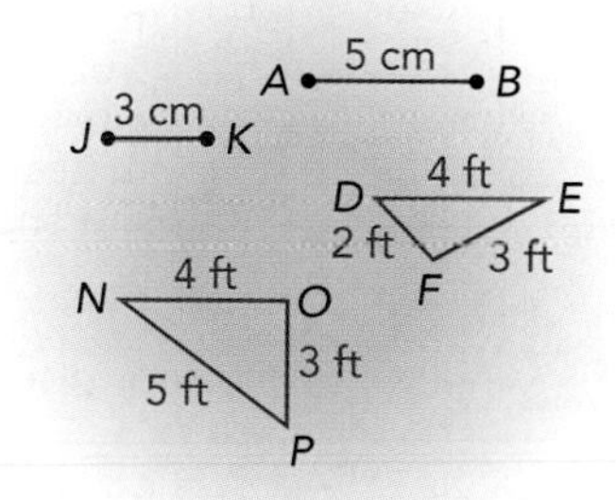

Write: $\Delta NOP \cong \Delta QRS$
Say: *triangle N O P is congruent to triangle Q R S*

When you see the equals symbol with the little wavy line above it, say is congruent to.

373 Congruent Segments and Angles

When you compare segments and angles on their own, you just look at their size to decide whether they are congruent. There are three ways to decide whether segments or angles are congruent.

CASE 1 You can measure segments or angles to decide whether they are congruent.

A 3 cm B

C 4 cm D

Segment *AB* is shorter than segment *CD*. So, $\overline{AB}$ and $\overline{CD}$ are not congruent.

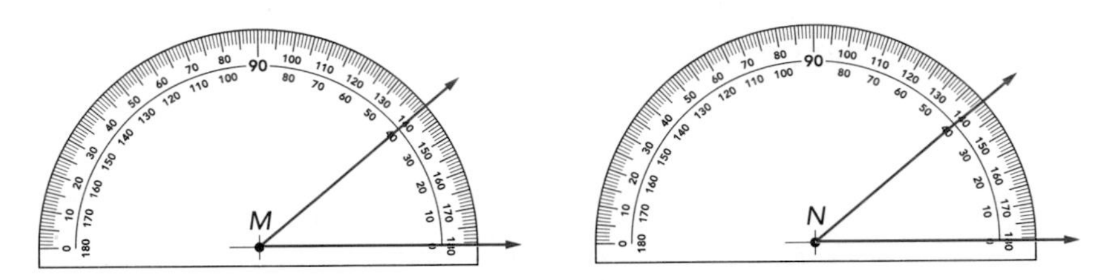

Angle *M* has the same measure as angle *N*. So, $\angle M \cong \angle N$.

See 365

CASE 2 If the measures are given, you can compare them.

A 8 cm B, D C; E 8 cm F, H G

Side *AB* and side *EF* are marked with the same measure. You can say that $\overline{AB} \cong \overline{EF}$. If the problem tells you that *ABCD* and *EFGH* are squares, then you also know that all sides and angles of one square are congruent to all sides and angles of the other square.

Segments marked the same way are congruent.

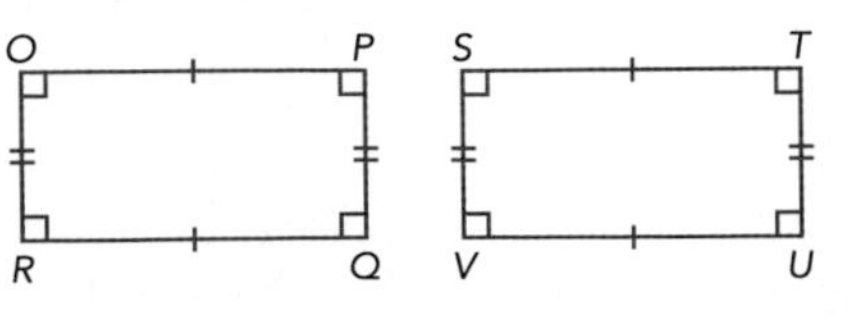

$\overline{OP}$ and $\overline{QR}$ and $\overline{ST}$ and $\overline{UV}$ are all marked with one tick mark. They are all congruent. $\overline{PQ}$ and $\overline{RO}$ and $\overline{TU}$ and $\overline{VS}$ all have two tick marks. They are all congruent. All of the angles in the two figures are marked as right angles. So, all the angles are congruent.

Congruent Figures

When two figures are congruent, corresponding sides and corresponding angles are all congruent.

See 369

EXAMPLE: These two triangles are congruent. What is the length of side *OP*?

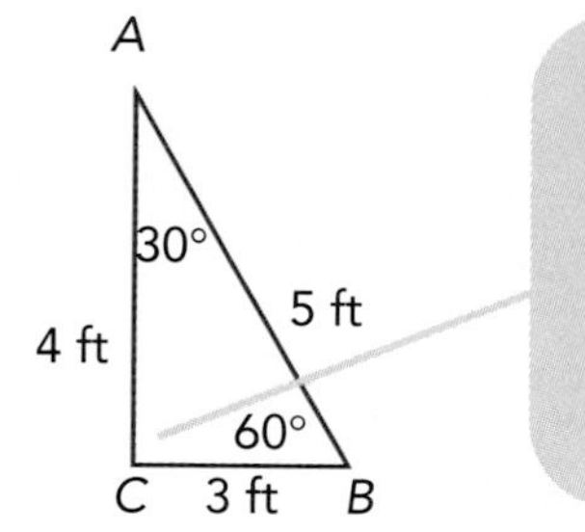

Since the sum of the measures of angles in a triangle must be 180°, these angles must both measure 90°.

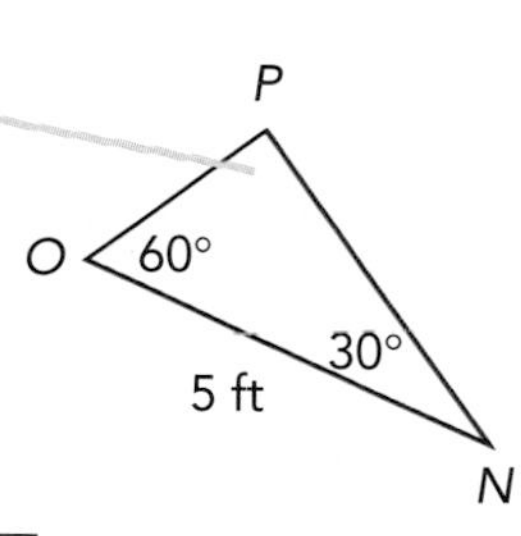

The longest sides correspond. $\overline{AB}$ and $\overline{NO}$ are the longest sides. If $\overline{AB}$ and $\overline{NO}$ correspond, then $\overline{BC}$ and $\overline{OP}$ correspond. $\overline{CA}$ and $\overline{PN}$ also correspond.

$\overline{OP}$ has the same measure as $\overline{BC}$. It is 3 feet long.

Transformations

When you transform something, you change it. In geometry, when you move a figure, you make a **transformation** of it.

CASE 1 You can **slide** a figure. This is a *translation.*

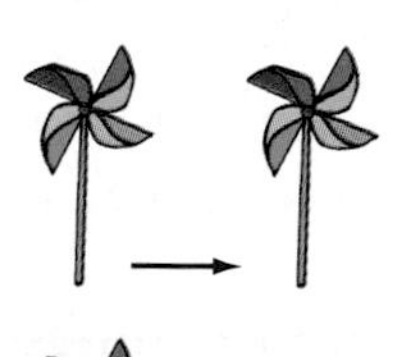

CASE 2 You can **turn** a figure around a point. This is a *rotation.*

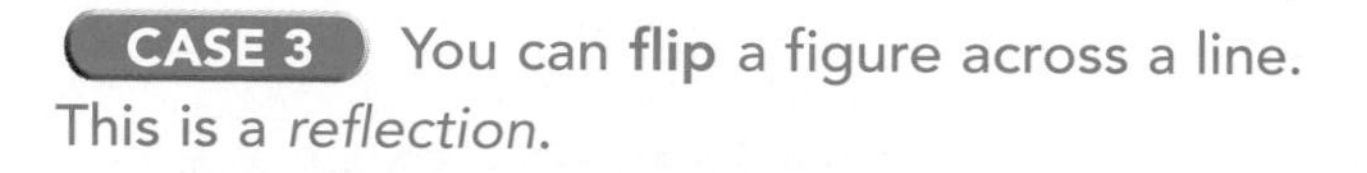

CASE 3 You can **flip** a figure across a line. This is a *reflection.*

376 Translations

A translation is also called a **slide**. That's because in a **translation**, every point in the figure slides the same distance in the same direction.

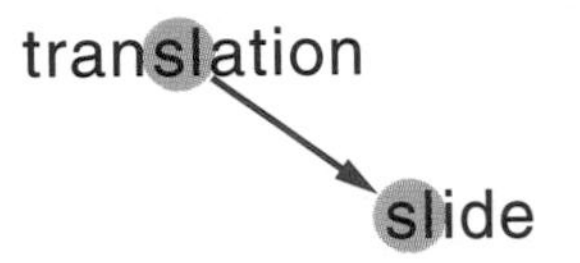

EXAMPLE: Does this diagram show a translation?

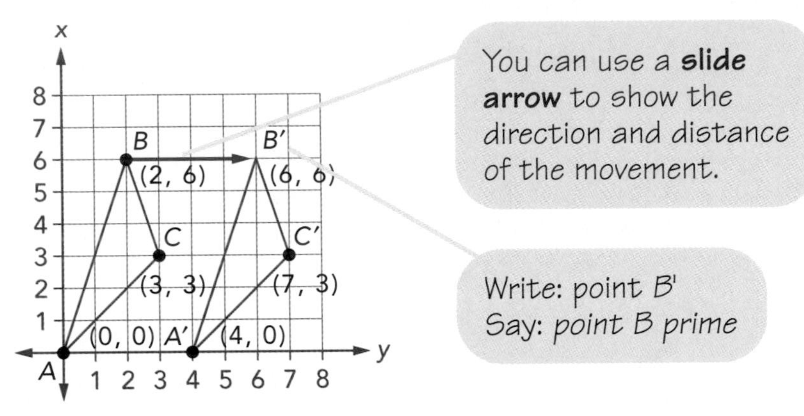

See 265–266

Look at the slide arrow. It shows that point B moves 4 units along the x axis and 0 units along the y axis to point B'. Check the other vertices.

Point A moves from 0 to 4 (4 units) along the x axis and doesn't move at all along the y axis.

Point C moves from 3 to 7 (4 units) along the x axis and doesn't move at all along the y axis.

Since every vertex moves the same distance in the same direction, you can be sure that every point inside the figure does, too. The diagram does show a translation.

Rotations

377

When something rotates, it turns. In geometry, rotating a figure means turning the figure around a point. The point can be on the figure or it can be some other point. This point is called the **turn center**, or **point of rotation**.

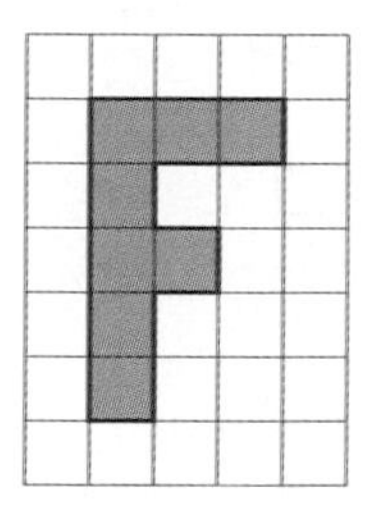

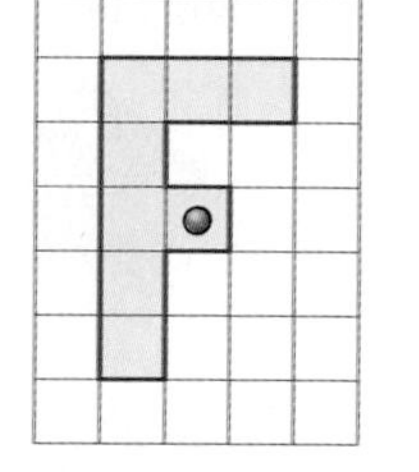

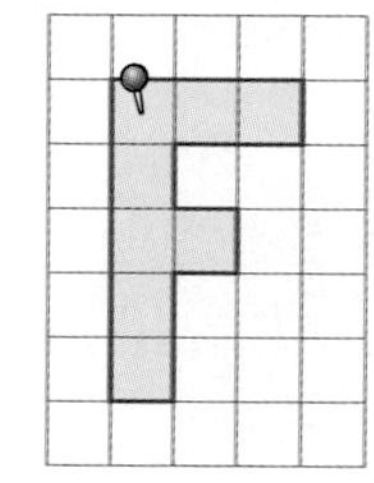

When you turn a figure around a point the point just spins in place. All of the rest of the points in the figure move.

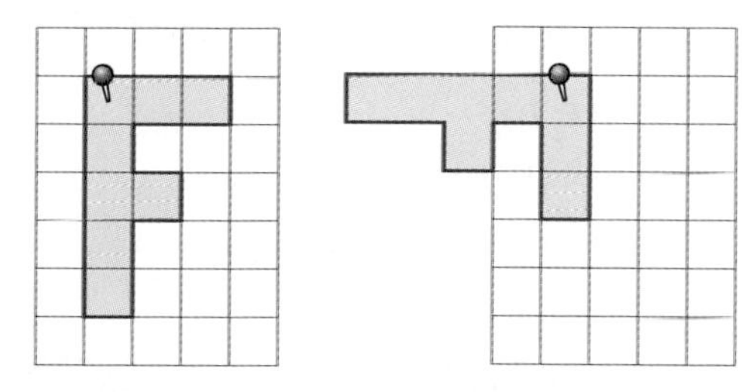

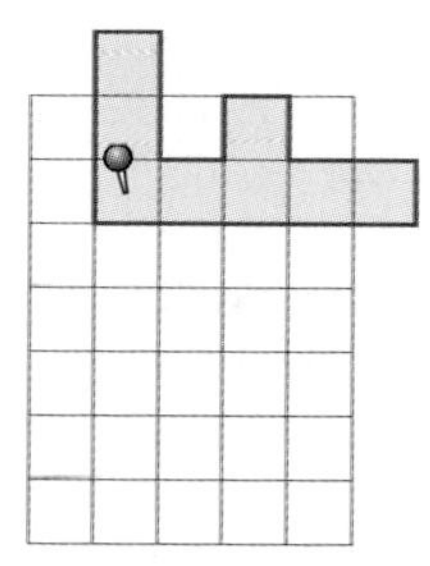

When you rotate a figure, you can describe the rotation by telling two things:

1. the direction
2. the angle that the figure is rotated around (or *about*) the point of rotation.

Imagine a clock. The way the hands move is **clockwise.** The wrong way around a clock is **counterclockwise.**

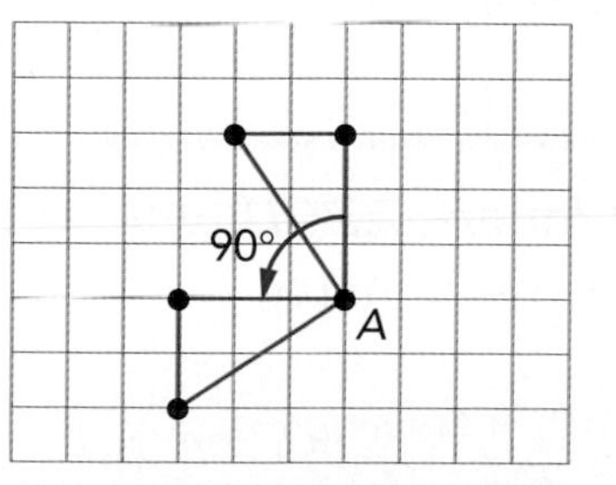

This figure was rotated 90° counterclockwise about point *A*.

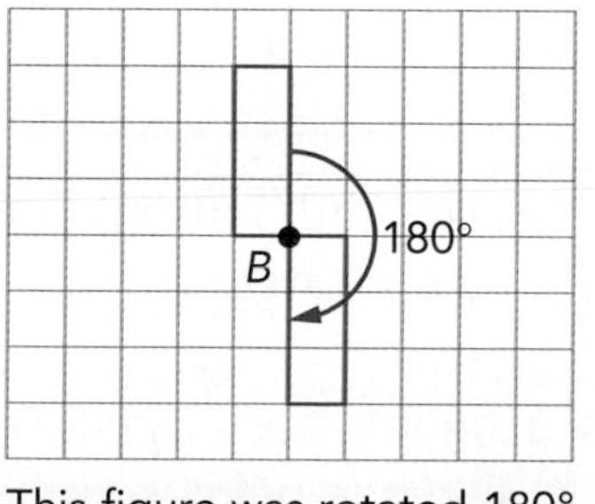

This figure was rotated 180° clockwise about point *B*.

378

Turn Symmetry

If you rotate any figure 360°, you get back to where you started. No big deal. But with some figures, it looks like you're back where you started *before* you make a full turn. Figures that can do that have **turn symmetry**.

Turn Symmetry **No Turn Symmetry**

EXAMPLE: Does rectangle *ABCD* have turn symmetry?

A good way to tell whether a figure has turn symmetry is to trace it and see how far you have to turn the tracing before it matches the original exactly.

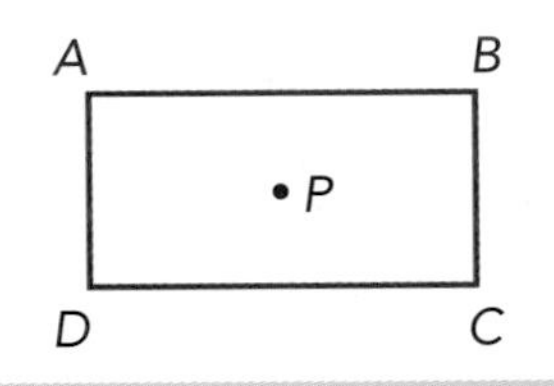

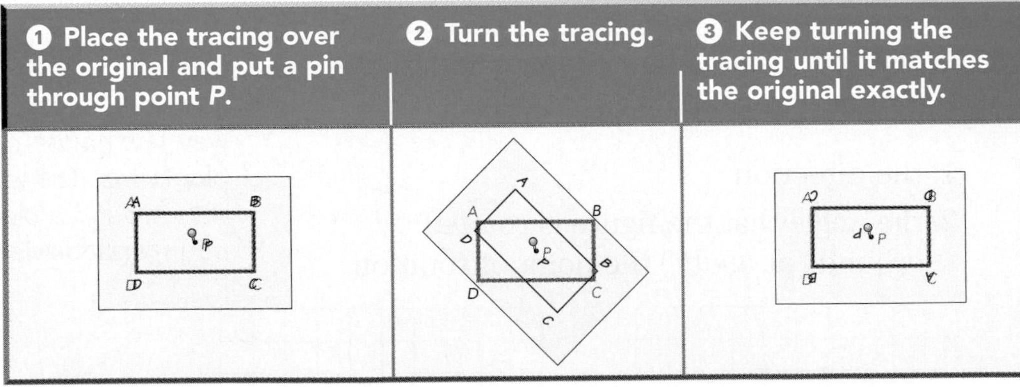

A half-turn (180°) about point *P* rotates *ABCD* onto itself. So, rectangle *ABCD* has turn symmetry.

When a figure rotates onto itself after turning exactly 180° it also has **point symmetry**.

Reflections

Stand in front of a mirror. Put your right hand on your right ear.

Now look at a friend. Each of you put your right hand on your right ear.

What you see in a mirror image is a reverse image of the object you are looking at! Right becomes left and left becomes right in a mirror image. In geometry, a **reflection** is a transformation in which a figure is flipped over a line. That's why it's also called a **flip**. Each point in a reflection image is the same distance from the line as the corresponding point in the original figure.

EXAMPLE: Does the diagram show a reflection?

This is the line of reflection. It acts as a mirror to reflect a figure.

❶ Check the red segments.	❷ Check the green segments.	❸ Check the blue segments.
All red points are 5 units from the line of reflection. Both segments are 5 units long. They are straight across from each other.	All green points are 2 units from the line of reflection. Both segments are 5 units long. They are straight across from each other.	Both blue segments connect the top of the red segment with the bottom of the green segment.

The diagram shows a reflection.

380

Line Symmetry

If you can fold a figure so that it has two parts that match exactly, that figure has **line symmetry**, or just plain symmetry. The fold line is along the **line of symmetry**. A figure can have no lines of symmetry, one line of symmetry, or more than one line of symmetry.

EXAMPLE: Does figure *KITE* have line symmetry?

A good way to tell whether a figure has line symmetry is to trace it and fold the tracing, looking for a way to match the two folded parts exactly.

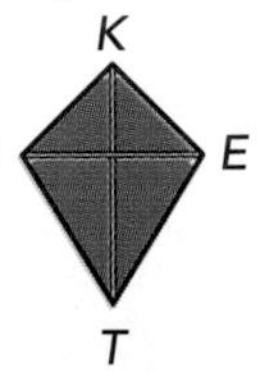

❶ Trace *KITE*. Fold along vertical axis.	❷ Fold along horizontal axis.	❸ Fold randomly.
Folding along the diagonal from *K* to *T* makes the two parts match.	Folding along the diagonal from *I* to *E* does not make the two parts match.	There doesn't seem to be any other way to fold the kite so the two parts match.

 The kite has one line of symmetry, line *KT*.

Tessellations

In a **tessellation**, a figure or pattern of figures is repeated to cover a flat surface. The figures must fit together so that none of them overlap and there are no gaps. You can make tessellations from many different polygons.

Shape	Example
Rectangle	Rectangles fit together with no overlaps and no gaps. So, rectangles tessellate.
Equilateral Triangle	Equilateral triangles fit together with no overlaps and no gaps. So, equilateral triangles tessellate.
Regular Pentagon	Regular pentagons do not fit together with no gaps. So, regular pentagons do not tessellate.
Regular Octagon with Square	Some tessellations use more than one shape.

You can also make your own tessellation shapes.

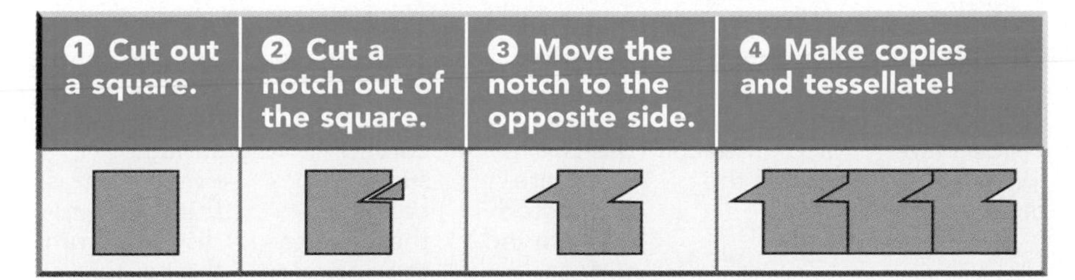

❶ Cut out a square.	❷ Cut a notch out of the square.	❸ Move the notch to the opposite side.	❹ Make copies and tessellate!

382

Solid Figures

Imagine a world where people built things without paying attention to shape. What would it be like if wheels were shaped like prisms instead of cylinders? What would it be like to keep a spherical cereal box on the shelf? Would there be very much room upstairs in a house shaped like a cone?

Because we live in a three-dimensional world, not a flat one, solid figures are important. When we say **solid figure**, we don't mean solid in the usual sense; we mean the figure is not flat, it's three-dimensional.

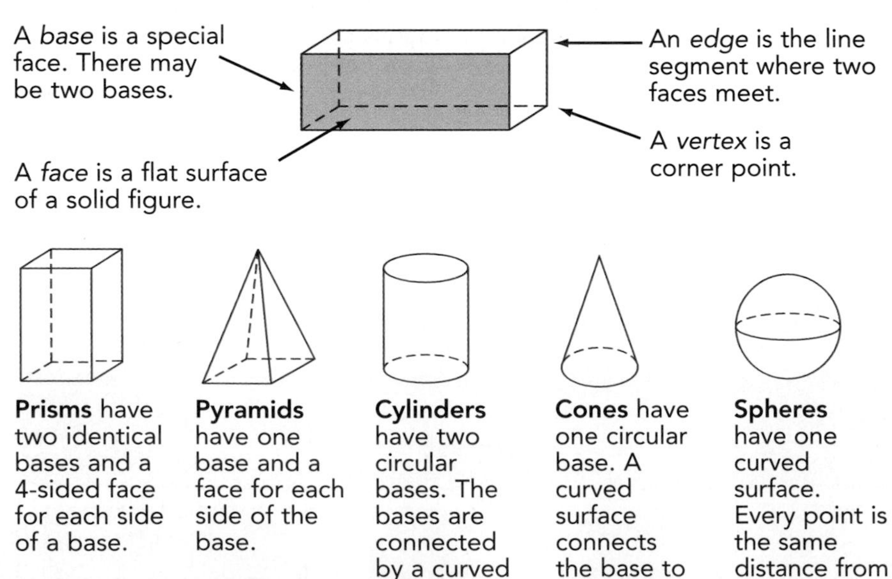

Prisms

Prisms have two identical parallel bases and faces that are polygons.

Prisms **Not Prisms**

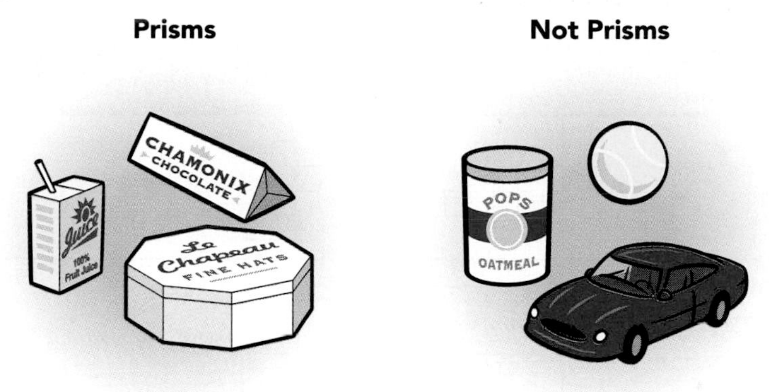

A prism is named according to the shape of its bases.

Name	Shape of Bases	Number of Other Faces	Example
Rectangular Prism	Rectangle	4	
Square Prism or Cube	Square	4	
Pentagonal Prism	Pentagon	5	

MORE HELP

See 340, 374

This solid figure has two identical parallel bases and faces that are polygons. So, it's a *prism*. Its bases are triangles, so it's a *triangular* prism.

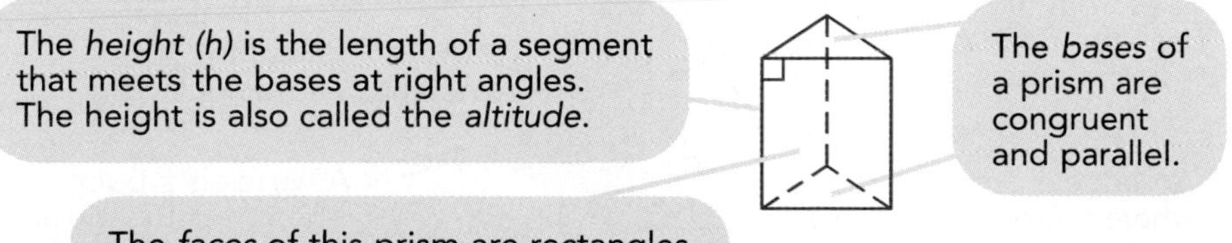

384

Nets of Prisms

If you could cut a prism along some of its edges and fold it out flat, you would have a **net** of the prism. There is often more than one possible net for a prism.

Solid	Nets
Cube	
Rectangular Prism	
Rectangular Prism	
Triangular Prism	

385

Pyramids

Pyramids have one base and a triangular face for each side of the base. This pyramid is a square pyramid. Its apex is directly above the center of its base, so its altitude meets the base at the center.

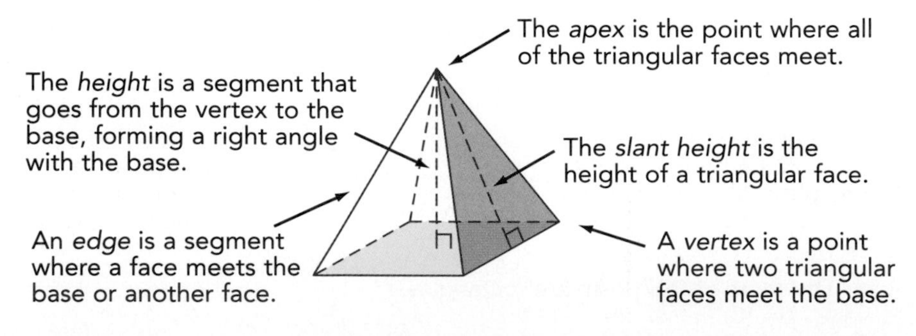

386

Types of Pyramids

A pyramid is named according to the shape of its base.

Name	Shape of Base	Number of Other Faces	Example
Triangular Pyramid	Triangle	3	
Square Pyramid	Square	4	
Rectangular Pyramid	Rectangle	4	

387

Nets of Pyramids

You can unfold a pyramid so that you show all its faces. This unfolded figure is called a **net** of the pyramid.

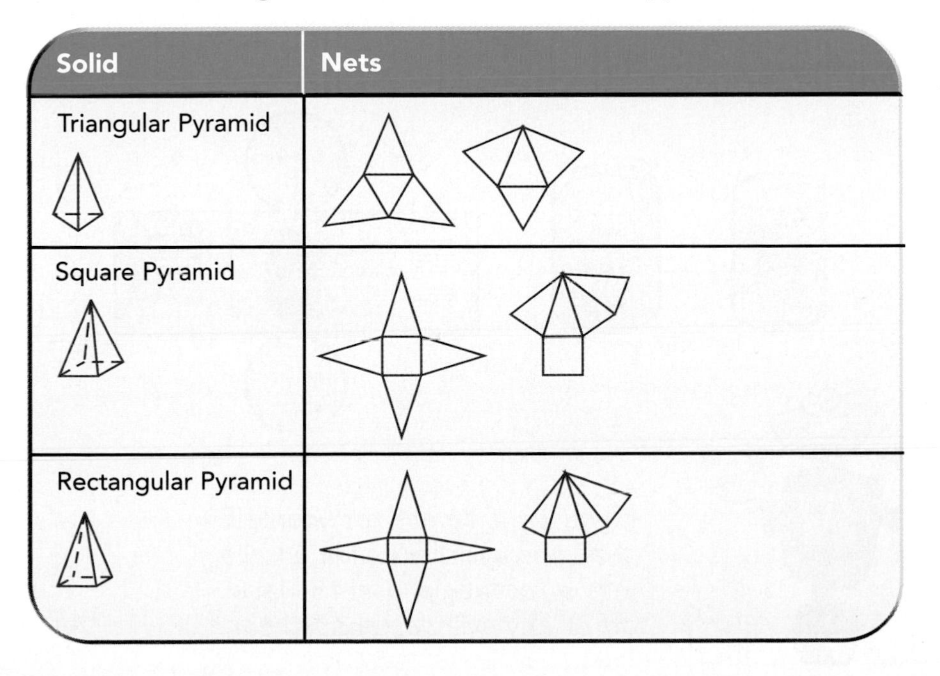

Solid	Nets
Triangular Pyramid	
Square Pyramid	
Rectangular Pyramid	

388 Cylinders

Most cans are cylinders. So are paper rolls, mailing tubes, and drain pipes.

Cylinders have two bases that are circles. As in a prism, the two bases are congruent and parallel. But since a circle has only one *side*, a cylinder has only one other face.

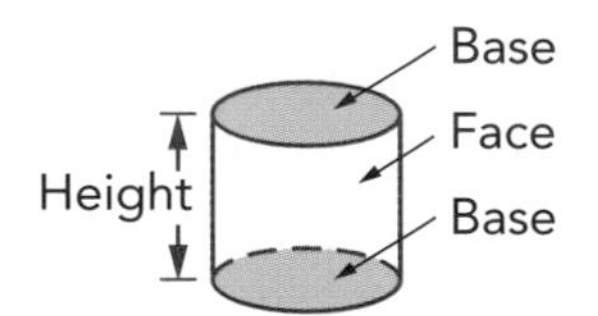

The *height* of a cylinder is a segment that meets both bases at right angles.

389 Nets of Cylinders

If you cut around the bases and down the face, you can unroll a cylinder to make a net. The face unrolls into a rectangle!

If you want to see for yourself that the cylinder's side unrolls into a rectangle, peel the label off of any cylinder-shaped can.

Cones 390

You've probably seen lots of objects that are shaped like cones.

Cones have one base that is a circle. A curved surface connects the base to the apex.

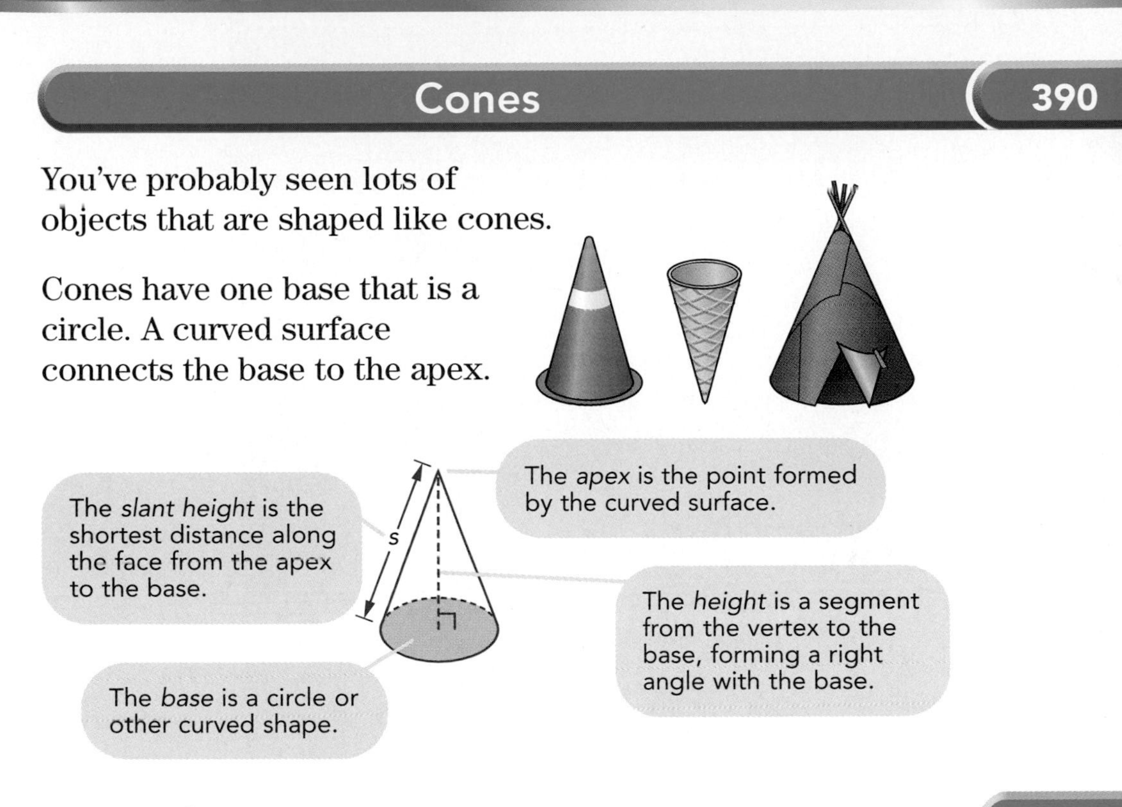

Nets of Cones 391

If you could cut around the circle and up the face to the apex, you could unroll a cone into a net.

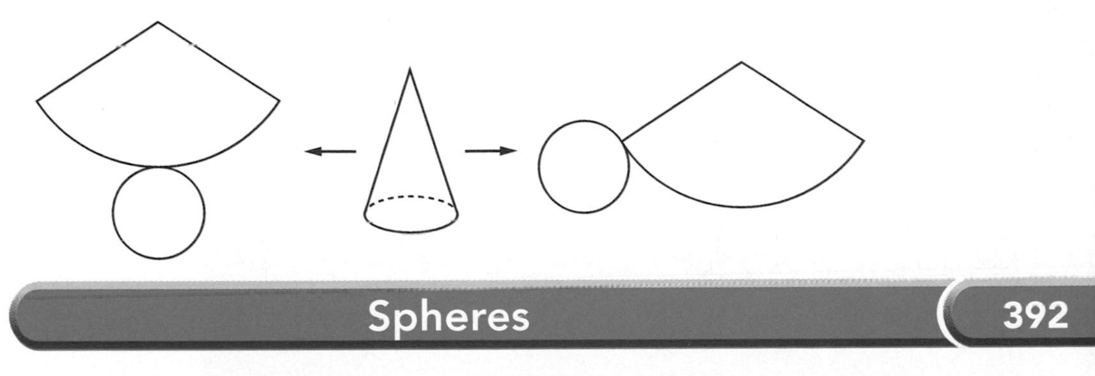

Spheres 392

A **sphere** is one smooth curved surface where every point is the same distance from the center.

center
radius
r

The center of a sphere is the same distance from every point on the sphere.

393

Problem Solving

You may not realize how many problems you solve every day. Maybe you figure out how to get a higher score in your new video game, how to fix your bicycle, what to pack when you sleep over at your friend's house, or when to start your chores so you'll have time to watch your favorite show. You just seem to be able to figure things out. Well, you can just figure out math problems, too. Of course, as with any problem, to solve a math problem it helps to know what's going on and what you're supposed to find out. It helps to know methods that have worked for you before. It helps to care about finding a solution that works. You may be surprised to see how much of what you already do can be used to help solve math problems.

394

Problem-Solving Tips

Keep trying

Imagine a baseball player who swings at and misses the first pitch and then walks away and never tries again. We would never have seen some of the greatest players in the game if they gave up so quickly. When you work on a problem, your first try may not work out. Don't worry. And don't give up. Just look at what you've done and give it another try. One thing that makes good problem solvers so good is that they see each wrong turn as just another step to success.

Take chances

Think about baseball again. You can't get a hit without swinging the bat. In the same way, you can't solve a problem without giving it a try. Take your best shot. You've got nothing to lose. If it doesn't work, you get up to the plate and swing again. In one way, problem solving is easier than baseball: you get as many strikes as you want.

Stay positive

Remember how relieved you felt when you found your lost homework after searching for it all day? Or how satisfied you felt when you fixed your computer, scored a goal in soccer, or did anything that you didn't think you could do? Next time you're working on a tough problem, remind yourself how good you'll feel when you find a solution.

Use what you know

Once you've gotten soaked on a rainy camping trip, you'll probably always remember to pack a poncho or raincoat on future trips. When you solve problems, use your experience. Look for things that are familiar. Maybe you'll remember a method that you've seen or used before.

Practice

Whether you play the piano, play soccer, or do any other activity, the more you practice, the better you get. It's the same with problem solving. With each problem you work on, you learn something new. You become better prepared for solving problems that lie ahead.

Watch what you do

As you work on a problem, keep an eye on the method you're using. Are you getting closer to a solution? Or are you just doing lots of calculations that won't help? If you think you're on the right track, stay with it. If you think you're on a road to nowhere, stop and look around. You may want to try a different path.

Take a break

Sometimes you're working on a really tough problem and you feel you've tried everything you can think of. It may be time to get away from the problem for a while. Give your brain a chance to refresh itself. Throughout history, scientists, mathematicians, generals, people in all fields, have come up with great solutions after taking a rest.

395

A Four-Step Problem-Solving Plan

When you get stuck on a problem, there's a four-step plan that acts like a map. It won't solve the problem for you, but it can help you find a way to solve it.

Imagine going to the store without knowing what you're supposed to buy. There's not much chance you'll buy the right things. In problem solving, too, you need to understand the situation. Ask yourself, *What do I know? What do I need to find out*?

If you don't understand the problem, use these hints.

- Read the problem again, slowly. Take notes or draw pictures to help.
- Study any illustrations, charts, or diagrams.
- Look up any words or symbols you don't know.

When you're not sure how to solve a problem, take a little time to think about what you might do. You can try one or more strategies.

Carry out your plan. Work carefully and keep thinking while you work. And don't forget, if your first attempt doesn't work, try something else.

After you find an answer, go over what you've done.

- Does your answer make sense?
- Does it really answer what the problem is asking?
- Did you compute correctly?
- If you try another way, do you get the same answer?

Problem-Solving Strategies

There are lots of different ways to get from one place to another: walking; biking; skating; riding in a car; taking a bus, train, or plane. Solving a problem is like traveling. You go from not knowing an answer to knowing it, and there are lots of different ways to get there. Methods for getting to a solution are often called **problem-solving strategies**.

Keep in mind a few things about strategies.

- You can use more than one strategy on a single problem.
- You can use a strategy that's not on the list (even one that you make up yourself).
- You can use whatever strategy works for you. (Strategies are not right or wrong; people have different strengths and different tastes.)

397 Act It Out or Make a Model

Sometimes you can use objects to try things out. Acting out a problem with coins, counters, paper, marbles, connecting cubes, or other handy items can help you see what's going on.

EXAMPLE 1: How many folds do you need to make to divide a sheet of paper into 16 equal sections?

UNDERSTAND 1
- You want to fold the paper to form 16 equal sections.
- How can you do that with the fewest folds?

PLAN 2
It would help to be able to see how many sections each fold makes. Try using a sheet of paper to act out the problem.

TRY 3

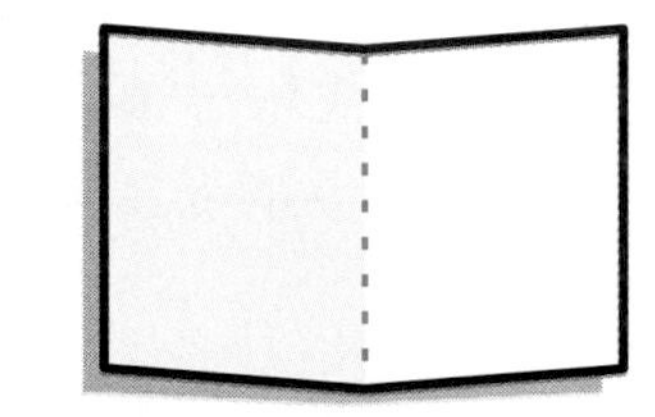

1 fold makes 2 equal sections.

2 folds make 4 equal sections.

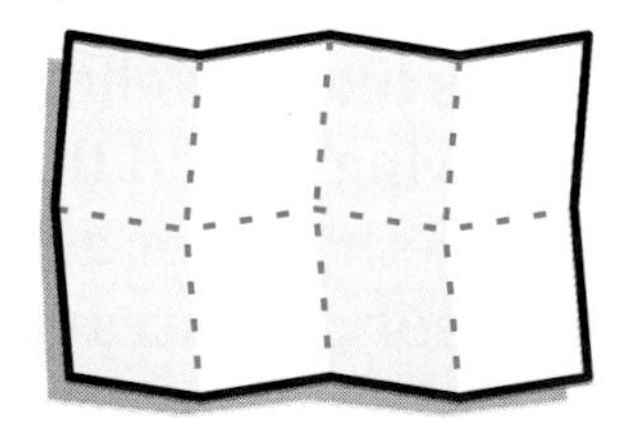

3 folds make 8 equal sections.

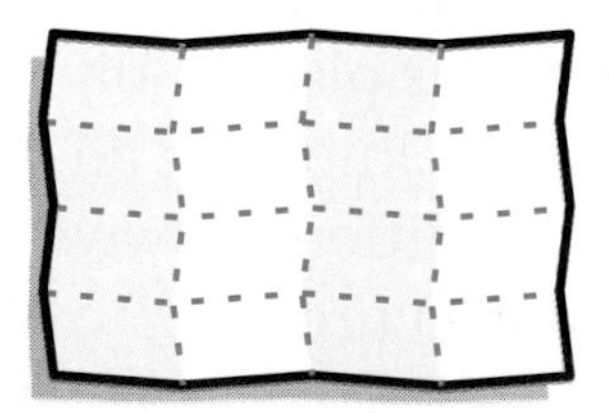

4 folds make 16 equal sections.

LOOK BACK 4
Can you make 16 equal sections with fewer than 4 folds? No, the best you can do with each fold is double the number of sections: $2 \times 2 \times 2 \times 2 = 16$

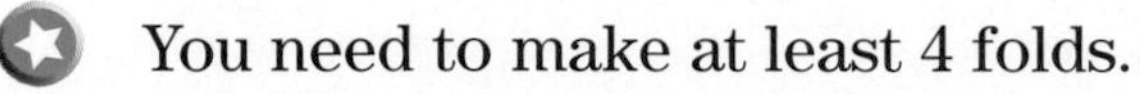

You need to make at least 4 folds.

EXAMPLE 2: A cube has the letters *A*–*F* with one letter on each face. The diagram shows the cube in two different positions. What letter would be on top when the letter *F* is on the bottom?

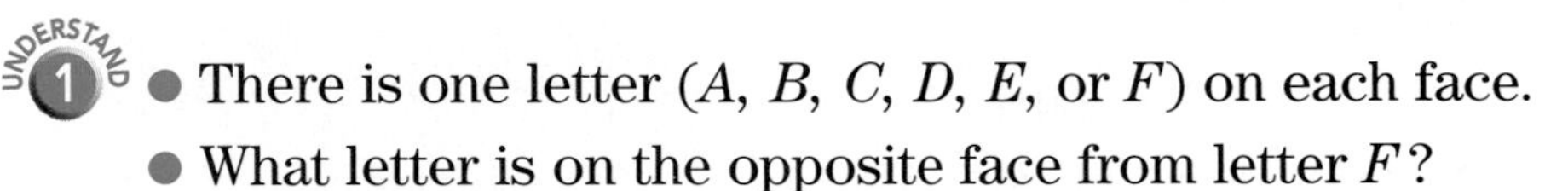

UNDERSTAND 1

- There is one letter (*A*, *B*, *C*, *D*, *E*, or *F*) on each face.
- What letter is on the opposite face from letter *F*?

PLAN 2

If you had the cube in your hand it would be easier to see what happens when you turn it. Try using any solid with 6 sides to make a model.

TRY 3

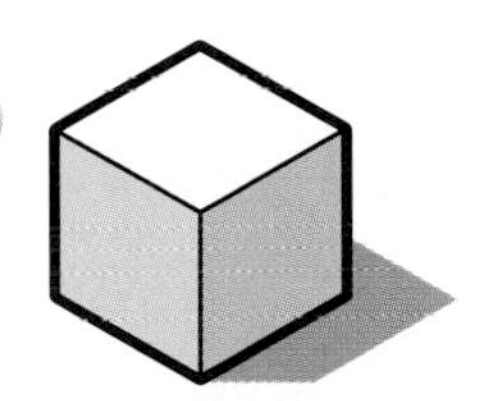

Get a box or make one with paper and tape.

Write *A*, *B*, and *C* as shown.

Turn the box this way.

Write *D* and *E* as shown.

There's only one face left for the *F*. Write an *F* on it. Turn the cube to see which letter is on the opposite side.

LOOK BACK 4 Check your model. Make sure it matches the pictures.

The letter *C* is on the top when the letter *F* is on the bottom.

398

Make a Diagram

If you are explaining a play in soccer or football, planning a parade, or designing a bulletin board display, you might draw a diagram to help you. You don't have to be an artist to draw useful diagrams. If you include the important information, diagrams can help you solve lots of different kinds of problems.

EXAMPLE 1: The Little League™ is planning to put up a straight fence to protect the players in the dugout. Each section of fence is 6 feet long. How many fence posts will they need if the fence is 30 feet long?

- Each section is 6 feet long; the fence will be 30 feet long.
- There are fence posts at each end.
- There is a post between sections.
- How many posts do they need?

A diagram would be simple to make and would show what is needed.

Draw and label a diagram.

Count the posts. There are 6 posts.

6 ft 6 ft 6 ft 6 ft 6 ft

Make sure that you correctly drew and labeled the diagram.

They will need 6 fence posts.

A diagram doesn't have to look like the problem, as long as it helps you see what's going on.

EXAMPLE 2: Five basketball teams have entered a tournament. Each team will play every other team once. How many games will be played in the tournament?

- There are 5 teams.
- Each team plays the other 4 teams once.
- How many games will there be?

Maybe a diagram can show how to match up teams with other teams.

Use a dot for each team. Connect dots to show teams playing each other.

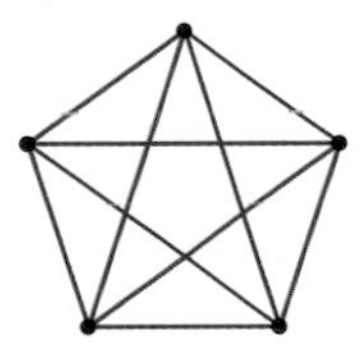

There are 10 line segments.

- Did you connect all the dots?
- Did you count correctly?

There will be 10 games in the tournament.

399 Guess, Check, and Revise

Guess, Check, and Revise is the same strategy you use with puzzles. The key is to make thoughtful guesses.

EXAMPLE: How many hardcover books and how many paperback books did the customer buy?

UNDERSTAND 1

- Hardcover books cost \$9.00; paperbacks cost \$4.00.
- The customer bought 12 books and spent \$73.00.
- How many of each kind did the customer buy?

PLAN 2

Make a guess and check it. If it's right, great. If it's not right, make a better guess. Use the first guess as a benchmark. Then decide whether to go higher or lower.

TRY 3

See 212

Guess (sum must be 12)	Check (total must be 73)	Evaluate and revise
hardcover: 6 paperback: 6 (6 + 6 = 12)	$(6 \times 9) + (6 \times 4)$ ↓ $54 + 24 = 78$	Too high! Make a guess that lowers the total.
hardcover: 2 paperback: 10 (2 + 10 = 12)	$(2 \times 9) + (10 \times 4)$ ↓ $18 + 40 = 58$	Too low! Make a guess that raises the total.
hardcover: 5 paperback: 7 (5 + 7 = 12)	$(5 \times 9) + (7 \times 4)$ ↓ $45 + 28 = 73$	Just right!

LOOK BACK 4

- Is the total cost \$73? Yes, $(5 \times 9) + (7 \times 4) = 73$.
- Is the total number of books 12? Yes, $5 + 7 = 12$.

The customer bought 5 hardcover books and 7 paperbacks.

Make a Table 400

Putting information in rows and columns makes it easier to keep track of numbers in a math problem.

EXAMPLE: A recipe for jelly squares calls for $1\frac{1}{2}$ tablespoons of butter for 12 squares. Suppose you need to make 60 squares. How much butter will you need?

UNDERSTAND 1

- You need $1\frac{1}{2}$ tablespoons of butter for 12 squares.
- You want to make 60 squares.
- How much butter do you need?

PLAN 2

If you need $1\frac{1}{2}$ tablespoons for 12 squares, you need another $1\frac{1}{2}$ tablespoons for another 12 squares. You can keep adding $1\frac{1}{2}$ tablespoons until you have enough for 60 squares. A table will help you keep track.

TRY 3

Make a table. Label it and fill in the first column. Fill in the second column to show that you can make 12 more squares if you have another $1\frac{1}{2}$ tablespoons of butter. Keep filling in the table until you reach 60 squares.

$+1\frac{1}{2}$

Tablespoons of butter	$1\frac{1}{2}$	3	$4\frac{1}{2}$	6	$7\frac{1}{2}$		
Jelly Squares	12	24	36	48	60		

$+12$

See 161, 267

LOOK BACK 4

- Did you find enough for 60 squares? Yes.
  $12 + 12 + 12 + 12 + 12 = 60$
- Did you compute correctly? Yes.
  $1\frac{1}{2} + 1\frac{1}{2} + 1\frac{1}{2} + 1\frac{1}{2} + 1\frac{1}{2} = 7\frac{1}{2}$

 You need $7\frac{1}{2}$ tablespoons of butter to make 60 squares.

401 Look for a Pattern

Imagine you have a dog that often digs up the neighbor's flowers. Let's say you notice that he does this only when you let him out before you feed him. You're noticing a pattern in his behavior. You can use the pattern to solve the problem. Feed him before you let him out! Looking for patterns can be a powerful tool in solving math problems, too.

EXAMPLE: In the lobby of the space museum, a life-size model of a rocket turns clockwise at the same rate, all day and night. It makes a complete turn every hour. At 9:00 A.M. it is facing north. In which direction will it face at 6:45 P.M.?

- The rocket faces north at 9:00 A.M.
- It makes a complete turn every hour.
- Which way will it face at 6:45 P.M.?

Since the rocket keeps turning in the same way, maybe there's a pattern you can use.

TRY

3 Find out what the rocket does for the first couple of hours. See if there's a pattern.

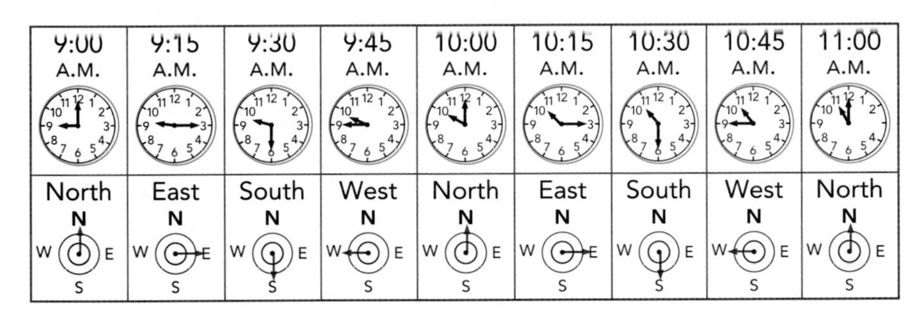

The rocket faces west every time it's 45 minutes after an hour.

LOOK BACK

4 Make sure the pattern you used will really continue. In this case, it will continue because the rocket keeps turning at the same rate.

The rocket will be facing west at 6:45 P.M.

MATH ALERT Not All Patterns Continue

Just because you see a pattern doesn't mean it will continue. Suppose a baseball team scores 1 run in the first inning, 3 runs in the next inning, and 5 runs in the inning after that. You might see a pattern in the numbers 1, 3, 5. Does that mean the team will score 7 runs in the next inning? They might, but probably won't. There's no reason why the pattern has to keep going.

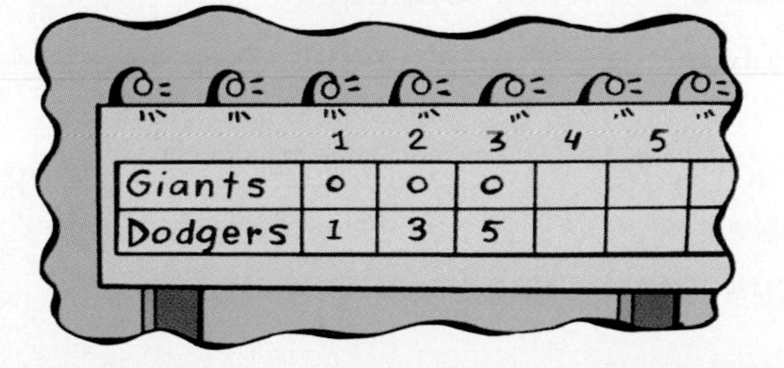

Whenever you use a pattern to solve a problem, ask yourself: Is there a reason for the pattern to continue?

403 Make an Organized List

If you've ever collected stamps or baseball cards, you know that keeping them organized makes it easier to find a stamp or card that you're looking for. Organizing also makes it easier to see if you have duplicates or any missing stamps or cards. When you make a list to solve math problems, you should organize it for the same reasons.

Sometimes, if you put things in a different order, you haven't changed the final result. So, you need to count carefully.

EXAMPLE: At Pi's Pizza Place, how many different Double-Top Specials are there?

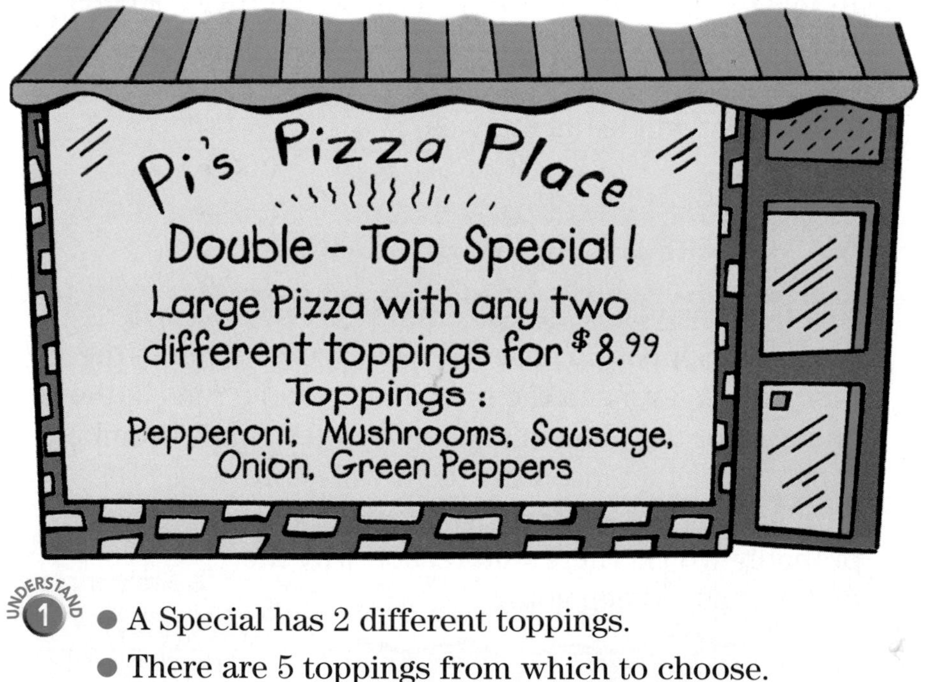

UNDERSTAND 1

- A Special has 2 different toppings.
- There are 5 toppings from which to choose.

PLAN 2

Since you need to count possible pairs, try making a list. Use a pattern to be sure you don't miss any possibilities and don't use the same pair twice.

TRY

3 Use letters to stand for toppings. First, list all the Specials with pepperoni, then all the Specials with mushrooms, and so on, until you've listed them all.

with Pepperoni	with Mushrooms	with Sausage	with Onion	with Green Peppers
■	MP	SP	OP	GP
PM	■	SM	OM	GM
PS	MS	■	OS	GS
PO	MO	SO	■	GO
PG	MG	SG	OG	■

It looks as if there are 20 pairs on the list.

4 Are mushroom-pepperoni and pepperoni-mushroom really any different? No, they're the same pizza.
You need to cross out one of each pair that is listed twice.

Remember, each Special must have two different toppings.

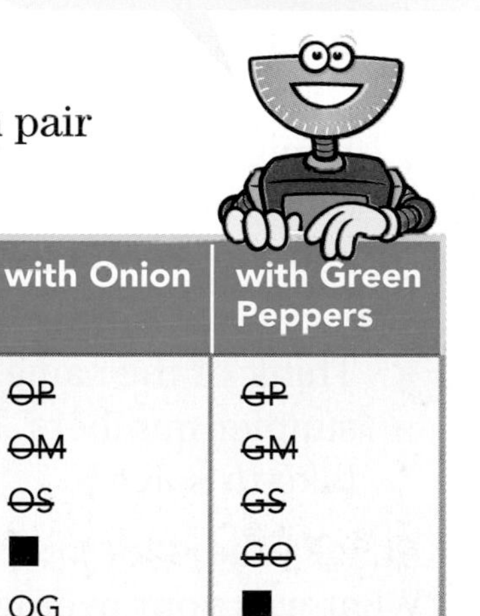

with Pepperoni	with Mushrooms	with Sausage	with Onion	with Green Peppers
■	~~MP~~	~~SP~~	~~OP~~	~~GP~~
PM	■	~~SM~~	~~OM~~	~~GM~~
PS	MS	■	~~OS~~	~~GS~~
PO	MO	SO	■	~~GO~~
PG	MG	SG	OG	■

(PM and MP: same)

There are 10 different Double-Top Specials.

404 Use Simpler Numbers

Sometimes things can seem more complicated than they really are. If large or messy numbers are making you confused, you can try making them simpler. The easier numbers may help you come up with a plan for solving the problem.

MORE HELP

See 185, 260, 324

EXAMPLE: You are riding in the car with your family. At 11:45 A.M., the odometer shows 11,963 miles. You continue riding, and at 1:15 P.M. the odometer shows 12,041. What was your average speed during that time?

UNDERSTAND 1

- From 11:45 A.M. to 1:15 P.M. the odometer changed from 11,963 to 12,041.
- What was the average speed?

Average speed of a car is measured in miles per hour.

PLAN 2

It might be easier to see how the numbers relate to one another if they were smaller and more rounded.

TRY 3

Think of the same kind of problem, but with simpler numbers. Keep track of the steps you take to solve it.

At 1:00, the odomoter shows 100. At 3:00, it shows 500. What was your average speed during that time?

① Subtract to find the distance you traveled.	500 miles − 100 miles 400 miles
② Find how much time passed.	From 1:00 to 3:00 is 2 hours.
③ Divide to find the average speed.	$400 \div 2 = 200$ 400 miles ÷ 2 hours = 200 miles per hour

Of course, that answer is not realistic. That's OK, because you made up the numbers. Remember, you're not going to use the answer; you're just going to use the steps. Follow those same steps with the actual numbers in the problem.

See 132, 176, 324

① Subtract to find the distance you traveled.	12,041 miles − 11,963 miles 78 miles
② Find how much time passed.	From 11:45 A.M. to 1:15 P.M. is $1\frac{1}{2}$ hours.
③ Divide to find the average speed.	$78 \div 1\frac{1}{2} = 78 \div \frac{3}{2}$ ↓ $78 \times \frac{2}{3} = 52$ 78 miles ÷ $1\frac{1}{2}$ hours = 52 miles per hour

LOOK BACK 4

- Is the answer reasonable? Yes, if there are no stops or slow-downs.
- Is your computation correct? Yes. You subtracted and divided correctly. You computed miles per hour correctly.

Your average speed from 11:45 A.M. to 1:15 P.M. was 52 miles per hour.

405 Work Backward

You probably know how to get from your house to school. To go back, you can follow the same directions backward. Of course, if you make a *right* turn at the corner of Elm and 7th Street on the way to school, you would make a *left* turn at that corner on the way home.

You can work backward to solve some math problems. If you know the end and you know the steps in between, you can find the beginning by going step by step in reverse order.

EXAMPLE: Elizabeth bought a T-Shirt with her first name on it. The shirt usually costs $15.95 plus a certain amount for each letter. The store was having a sale: half off the total cost of the shirt. Elizabeth paid only $11.80. How much does it usually cost for each letter?

- She paid $11.80 for a T-shirt with 9 letters on it.
- The regular price of the shirt is $15.95 plus some amount per letter.
- She paid only half of what the shirt would usually cost.
- How much does each letter usually cost?

PLAN 2

Since you know the end (what she paid) and all the steps (how the cost is calculated), you can work backward to find the beginning (the cost for each letter).

TRY

3 First write out the steps to show how the cost of Elizabeth's T-shirt was calculated:

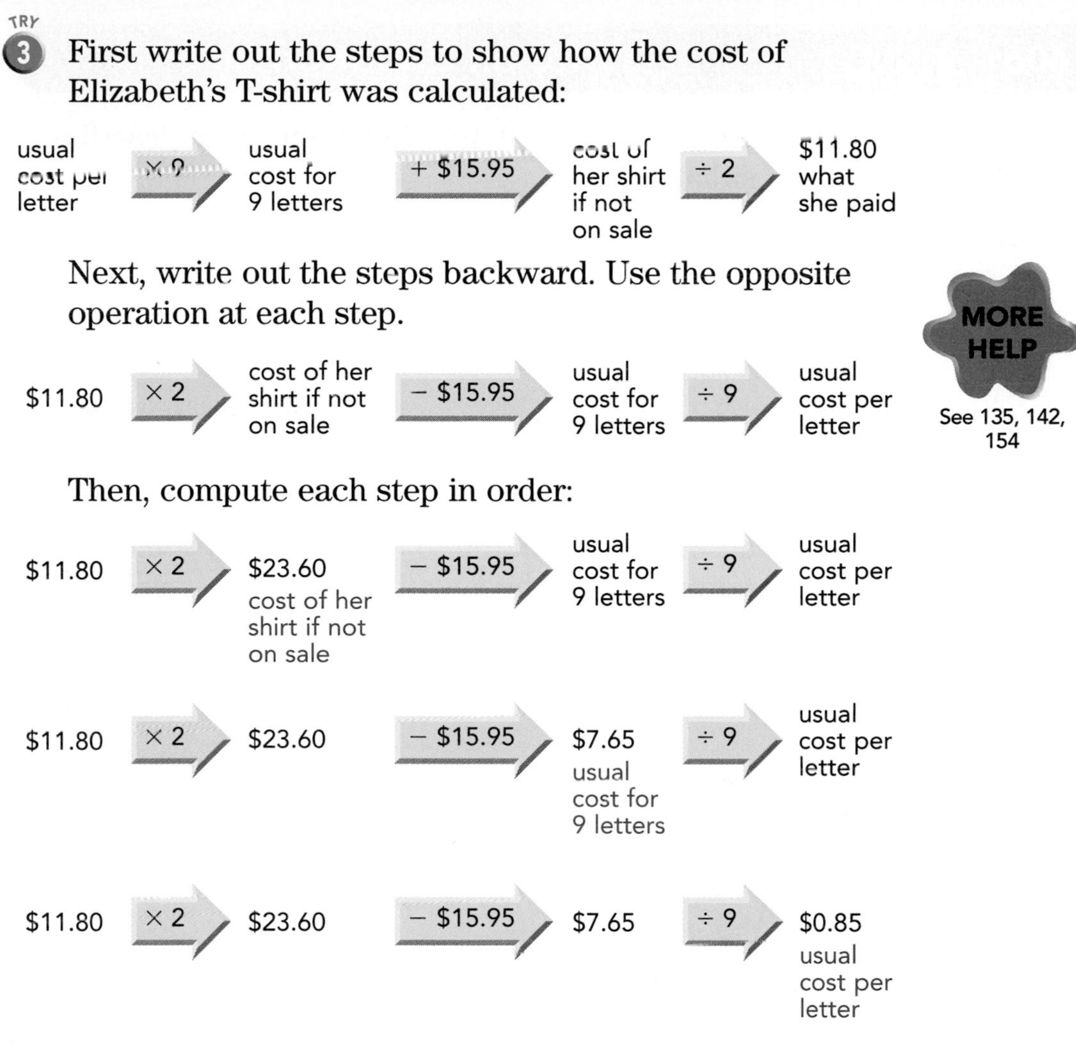

Next, write out the steps backward. Use the opposite operation at each step.

$11.80 → × 2 → cost of her shirt if not on sale → − $15.95 → usual cost for 9 letters → ÷ 9 → usual cost per letter

MORE HELP
See 135, 142, 154

Then, compute each step in order:

$11.80 → × 2 → $23.60 cost of her shirt if not on sale → − $15.95 → usual cost for 9 letters → ÷ 9 → usual cost per letter

$11.80 → × 2 → $23.60 → − $15.95 → $7.65 usual cost for 9 letters → ÷ 9 → usual cost per letter

$11.80 → × 2 → $23.60 → − $15.95 → $7.65 → ÷ 9 → $0.85 usual cost per letter

LOOK BACK

4 Check your answer by working forward.
It usually costs 9 × $0.85, or $7.65 for the 9 letters.
If not on sale, her shirt would cost $15.95 + $7.65, or $23.60. On sale, it should cost half of $23.60, or $11.80. That's what she paid. So the answer checks.

It usually costs $0.85 for each letter.

406 Write an Equation

See 125, 142, 237–243

You can use equations with missing numbers to describe problem situations.

EXAMPLE: A music store on the Internet charges \$14.95 for each CD plus a shipping charge of \$3.95 for your total order. How much will you pay in all for four CDs?

- Each CD costs \$14.95. You pay \$3.95 for shipping.
- How much will you pay in all for four CDs?

You can write an equation to describe the problem. Then you can solve the equation to find the missing number (the total amount you pay).

❶ Write an equation with words to show how things are related.	total = number of CDs × price per CD + shipping
❷ Write in the numbers you know.	4 × \$14.95 + \$3.95
❸ Solve the equation.	\$59.80 + \$3.95 = \$63.75

- Did you correctly write and solve the equation? Yes.
- Are your calculations correct? Estimate to check.
  \$14.95 ≈ \$15.00, and 4 × \$15.00 = \$60.00.
  \$3.95 ≈ \$4.00, and \$60.00 + \$4.00 = \$64.00.
  Your calculations are reasonable.

You will pay \$63.75 for four CDs.

Make a Graph

You have probably made graphs, to help people see data easily. You can also make graphs to solve math problems.

EXAMPLE: You want to save enough money to buy a bicycle speedometer that costs $22. You have only $2, but you earn $4 per hour mowing lawns. How many hours will you need to work to have enough to buy the speedometer?

See 185, 278

UNDERSTAND 1

- You have $2 now. You earn $4 per hour.
- After how many hours of work will you have at least $22?

PLAN 2

You can make a graph to show how your savings grow.

TRY 3

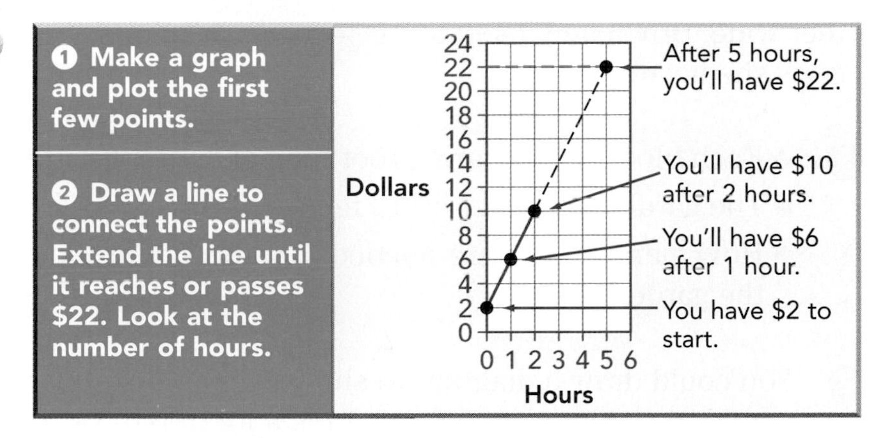

LOOK BACK 4

- Have you plotted your points accurately? Yes.
- Is it reasonable that you will have enough money after 5 hours of work? Yes.

After 5 hours of work, you'll have enough money.

408 Combine Strategies

Imagine your friend's mom is going to take the two of you to the movies. You walk to the garage to get your bicycle. You bike to your friend's house. Then your friend's mom drives you to the theater. You combined walking, biking, and riding in a car to get to the movies. In a similar way, you can combine different strategies to get to the solution of a math problem.

EXAMPLE: Mrs. Nu wants to put a border of paving blocks around the outside of her garden. The top of each block is a 1-foot square. The garden is 24 feet long and 13 feet wide. How many blocks does she need?

UNDERSTAND 1

- Each block is a square 1 foot long and 1 foot wide.
- The garden is 24 feet by 13 feet.
- How many blocks are needed for a border around the garden?

PLAN 2

You could draw a diagram to show every block, but that would mean drawing and counting a lot of blocks. You could start with a simpler problem, a very small garden.

TRY 3

Start with the simplest garden, 1 foot long and 1 foot wide. Then keep making it a little bigger. Look for a pattern. How does the number of blocks needed depend on the distance around the garden (the perimeter)?

MORE HELP

See 207, 398, 400–401, 404

Size	Diagram	Perimeter	Number of Blocks
1×1		4	8
2×1		6	10
2×2		8	12
3×1		8	12

A pattern is beginning to appear. The number of blocks seems to be 4 more than the perimeter. Try using the pattern to finish solving the problem.

The garden is 24 feet long and 13 feet wide.

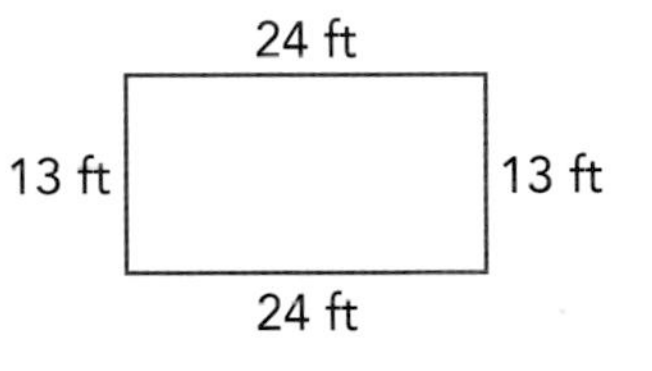

$24 + 24 + 13 + 13 = 74$
The perimeter is 74 feet.

$74 + 4 = 78$
So, it seems that 78 blocks are needed.

LOOK BACK 4 Does it make sense to use the pattern? Yes, because you always need 1 block for each foot of perimeter plus 4 more blocks for the corners.

You combined Make a Diagram, Use Simpler Numbers, Make a Table, and Look for a Pattern.

 Mrs. Nu needs 78 blocks.

409

Problem-Solving Skills

Playing basketball, soccer, football, baseball, or any sport involves more than one skill. In basketball, there's ball-handling, shooting, rebounding, and passing, to name a few. Problem solving, too, involves many skills. The more skills you have, the better all-around problem solver you'll be.

Take Notes

410

When you do research for a report, you take notes to help you keep track of what you read. This skill can also help you solve math problems.

See 168

EXAMPLE: Mr. Ortiz earns \$24.00 per hour fixing computers at an electronics store. If he works more than 40 hours in a week, he earns $1\frac{1}{2}$ times as much for each hour over 40. Last week he worked 48 hours. How much did he earn?

Taking notes can help you organize the important information in the problem.

Write down the different rates.

Write down how much he worked and what you are supposed to find out.

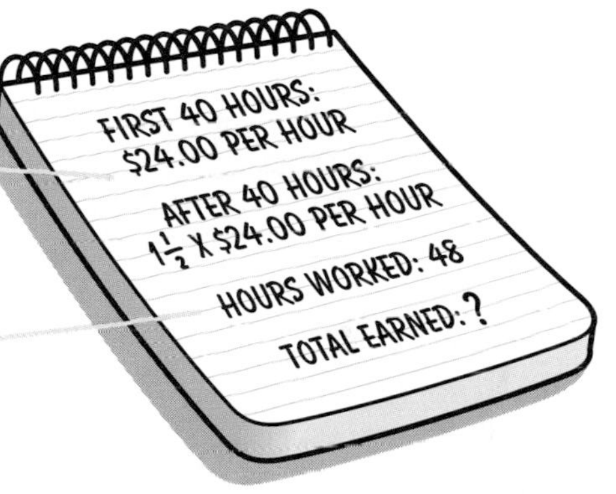

Your notes can help you think about how to solve the problem.

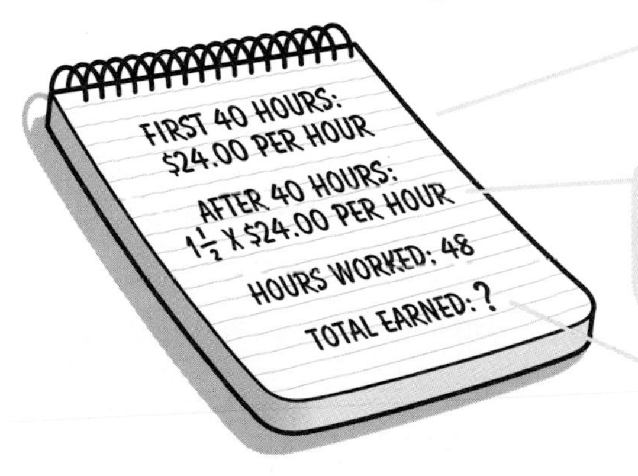

How much for first 40 hours? $40 \times 24 = 960$

How much for the other 8 hours?
$1\frac{1}{2} \times 24 = 36$ (\$36 for each one)
$8 \times 36 = 288$ (\$288 for all 8)

Total $= 960 + 288 = 1248$

Mr. Ortiz earned \$1248 last week.

411 Solve in More Than One Way

Suppose you look up the school lunch menu in the newspaper. Your friend checks the school's web site. If you each find a different lunch menu, then you know that something is wrong. You can check your answer to a math problem in a similar way. Solve the problem again with a different method and see if you get the same answer.

See 398, 403, 418

EXAMPLE: A sundae is made of 1 flavor of frozen yogurt, 1 kind of topping, and 1 flavor of syrup. How many different sundaes can be made from the choices shown?

Strategy: Make an organized list.
Solution:
List all the sundaes with each kind of yogurt.

Vanilla — Nuts — Hot fudge	Vanilla — Nuts — Butterscotch
Vanilla — Candy — Hot fudge	Vanilla — Candy — Butterscotch
Vanilla — Berries — Hot fudge	Vanilla — Berries — Butterscotch
Chocolate — Nuts — Hot fudge	Chocolate — Nuts — Butterscotch
Chocolate — Candy — Hot fudge	Chocolate — Candy — Butterscotch
Chocolate — Berries — Hot fudge	Chocolate — Berries — Butterscotch
Strawberry — Nuts — Hot fudge	Strawberry — Nuts — Butterscotch
Strawberry — Candy — Hot fudge	Strawberry — Candy — Butterscotch
Strawberry — Berries — Hot fudge	Strawberry — Berries — Butterscotch

Count the sundaes. There are 18 in the list.

Strategy: Use logical reasoning.
Solution:
There are 3 flavors of yogurt.

- With vanilla, you can choose from 3 toppings.
- With chocolate, you can choose from 3 toppings.
- With strawberry, you can choose from 3 toppings.

That's 9 ways to have yogurt and a topping.

For each yogurt and topping, there are 2 syrup flavors:
Since $9 \times 2 = 18$, you can make 18 different sundaes.

Strategy: Make a diagram.
Solution:

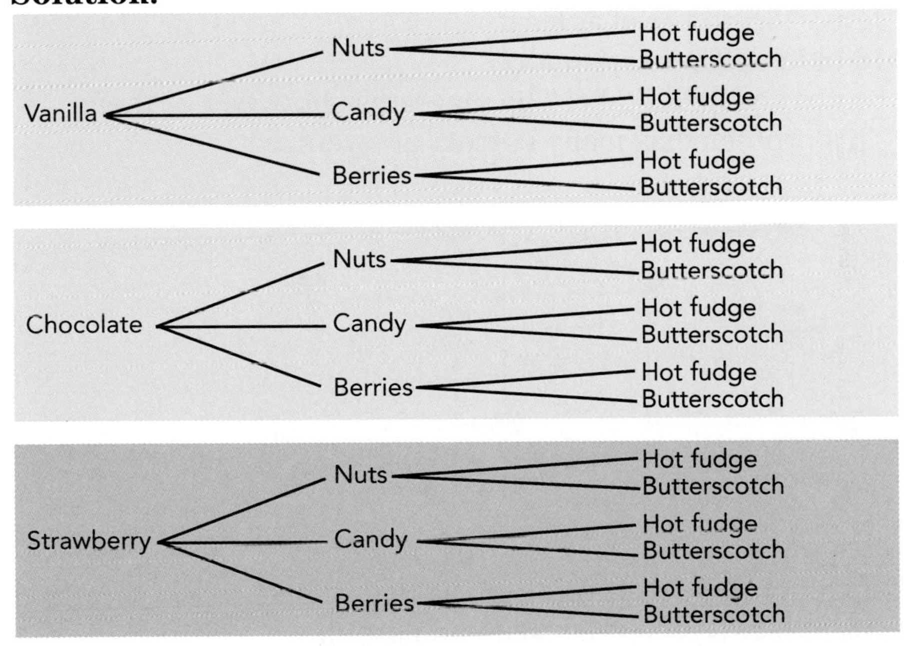

If you try at least two methods and get the same answer, you can be pretty sure the answer is correct.

No matter which strategy you use, 18 different sundaes can be made.

412 More Than One Answer

Have you ever bought a soft drink from a machine that required exact change? You can usually find exact change in more than one way. To make 75¢, you could use 3 quarters, or 7 dimes and a nickel, or other combinations. There are many math problems that may have more than one answer.

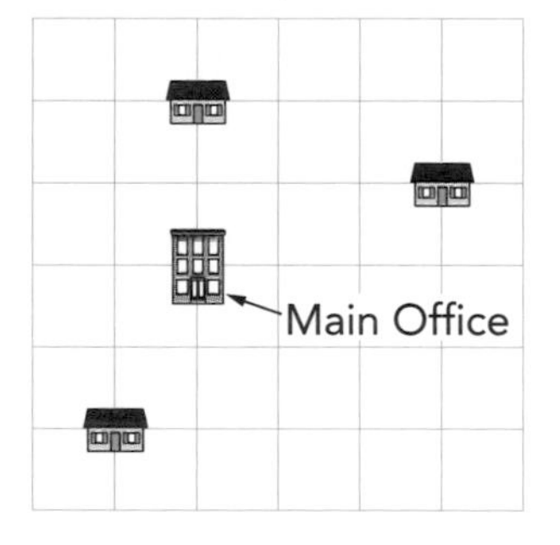

EXAMPLE 1: Jenny delivers packages. What is the shortest route from the main office to each house shown on the map and back to the office?

Jenny has to travel at least 16 blocks. There are many routes she could take. The diagrams show two of them. This problem has many correct answers.

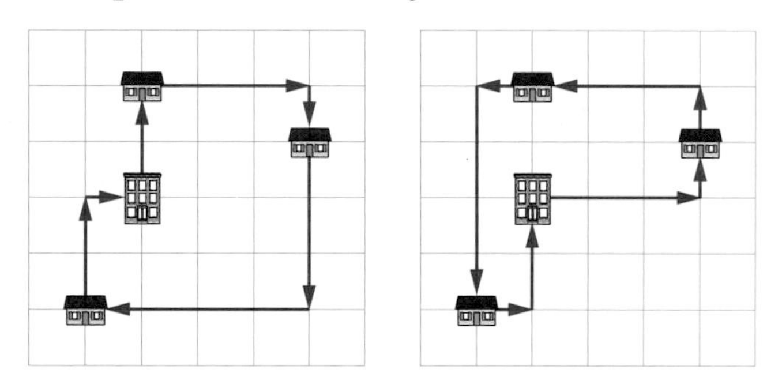

EXAMPLE 2: The list shows the weights of 6 packages. How can the packages be put into two groups so that each group weighs the same?

3 pounds
4 pounds
7 pounds
8 pounds
10 pounds
12 pounds

This problem has two correct answers. You could sort them like this:

3 lb, 7 lb, 12 lb and 4 lb, 8 lb, 10 lb

Or like this: 10 lb, 12 lb and 3 lb, 7 lb, 4 lb, 8 lb

Check for Reasonableness

413

When you're watching a cartoon, you know that most of the things you see cannot happen in real life. You have a sense of what's reasonable. You can use that sense with math problems. After you've found a solution, look it over and ask yourself about it.

- Does the answer fit the question that was asked?
- Is the size of the answer about right?
- Does the answer match what I know about the world?

EXAMPLE 1: Mr. Jenkins is planning a party for 26 people. Each table seats 6 people. How many empty places will there be?

To solve, you could divide. $26 \div 6 \longrightarrow$ 4R2

Suppose you answered that you need 5 tables (because 4 won't be enough). That's true, but it doesn't answer the question about how many empty places.

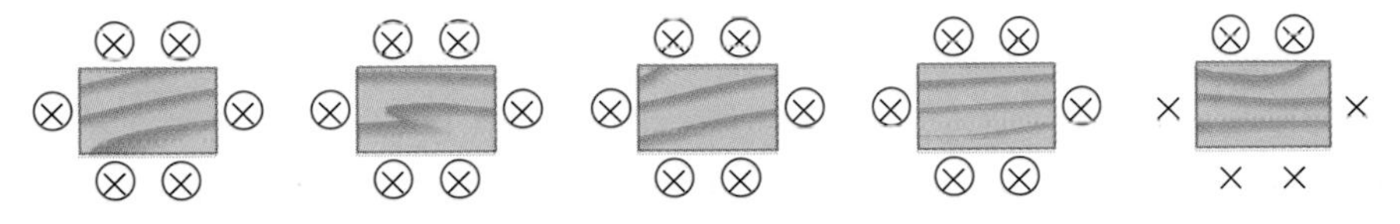

There will be 4 empty places.

EXAMPLE 2: Marisa bowled 3 games. Her scores were 97, 109, and 89. What was her total for the 3 games?

Suppose you added $97 + 109 + 89$ and got a sum of 195. Does that seem right?

No, it can't be right. Each score is around 100, so the total should be around 300, not 200.

Marisa's total is 295.

414 Choose an Estimate or Exact Amount

If you're supposed to meet your friend at 2:00 P.M. and you show up at 2:03 P.M., do you think you're late? What if you're just going to stand around and talk? What if you're planning to see a movie that starts at 2:00 P.M.? How exact you need to be depends on the situation. The same is true with math problems.

You can often estimate when you are comparing, and when you just need to know *about how many* or *about how much.*

EXAMPLE 1: You have \$20. Is that enough for 3 books that cost \$7.69 each?

Since you're comparing the amount you have with the total cost, you can try estimating. Even if each book cost only \$7, the total would be $3 \times \$7$, or \$21, which is more than you have.

No, \$20 is not enough.

See 185

EXAMPLE 2: You are riding in your family car on the way to go camping. A sign says the town you're going to is 146 miles away. If you continue to travel at about 50 miles per hour, how long will it take you to get there?

In this situation, you just need to know *about* how long it will take. Besides, you can't be exact because you won't be going *exactly* 50 miles per hour. So, you would estimate.

146 is about 150.
At 50 miles per hour, it would take 3 hours to go 150 miles.

It will take about 3 hours to get there.

Choose a Calculation Method

415

When you go to a friend's house, you probably take the easiest route. When you need to compute an exact answer to solve a problem, you can use whatever method is easiest.

CASE 1 With easy numbers, use mental math.

See 071

EXAMPLE 1: What is the cost of renting 3 videotapes at $0.98 each?

Since 98 cents is close to a dollar, you can use mental math to find 3 × $0.98. Think: Each tape costs 2¢ less than 1 dollar. So, 3 tapes cost 6¢ less than 3 dollars.

The 3 tapes cost $2.94 to rent.

CASE 2 With messy numbers, you might use a calculator.

EXAMPLE 2: What is the area of a rectangular playground that is 162 feet long and 118 feet wide?

These are messy numbers, so you might use a calculator.

See 301, 446–447

The area of the playground is 19,116 square feet.

CASE 3 Sometimes, using paper and pencil is best.

EXAMPLE 3: You have 415 baseball cards. You put 12 on each page of your album. If you put in all the cards, how many cards will be on the last page you use?

If you use a calculator, you may get an answer that's not easy to use for this problem. If you use paper and pencil, you can find the remainder more easily.

There will be 7 cards on the partly filled page.

416 Find Needed Information

You're setting your VCR to tape a special show about whales that's on tonight. To enter the correct start time, you would find out when the show is supposed to begin. Knowing what information you need and how to find it can come in handy when you solve math problems, too.

- What information could help solve the problem?
- Can I look up the information I need?
- Can I ask someone?
- Can I take a measurement?
- Can I use an estimate?

EXAMPLE: Your friend's birthday is May 29. On March 15, you want to know how many days away her birthday is.

You need to know how many days are in March and in April. You could look at a calendar.

March has 31 days and April has 30.

March 15 to March 31 ⟶	16 days
April 1 to April 30 ⟶	30 days
May 1 to May 29 ⟶	29 days
Total	75 days

Your friend's birthday is 75 days away.

Ignore Unneeded Information 417

Signs in a store give you lots of information: how much different items cost, when the store opens and closes, what types of things are in each aisle, what specials the store is having, and so on. But if you want to know whether you have enough money to buy a certain book, you look at the information you need and ignore the rest. Sometimes reading a math problem can be like walking through a store—there's a lot more information than you need.

See 324

EXAMPLE: How long is the film on dinosaurs?

Event	Location	Begins	Ends
Penguin Feeding	Aquarium	9:15 A.M.	9:45 A.M.
Star Search	Planetarium	10:30 A.M.	11:30 A.M.
Dinosaurs!	Theater	11:45 A.M.	12:15 P.M.
Wally the Woodchuck	Live Animal Stage	1:00 P.M.	1:30 P.M.

Since you need to find out the length of the film, you only need to know when the film starts and when it ends. You can ignore the rest of the information.

From 11:45 A.M. to 12:15 P.M. is 30 minutes.

The film is 30 minutes long.

418 Use Logical Reasoning

Have you ever used your reasoning abilities to figure out a mystery, like where you left your notebook or who ate the last cookie? You can use this skill to solve math problems as well.

EXAMPLE: Ana is thinking of a number from 65 to 85. The sum of the digits is less than 12. The product of the digits is greater than 20. If you divide the number by 3, the remainder is 1. What is the number?

To solve this problem, you can use logical reasoning to eliminate all the possible numbers except for one of them. Keep track of the numbers you eliminate by crossing them out.

1 Write all the numbers from 65 to 85.	65 66 67 68 69 70 71 72 73 74 75 76 77 78 79 80 81 82 83 84 85
2 Go through each number. If the sum of the digits is not less than 12, cross out the number.	65 ~~66~~ ~~67~~ ~~68~~ ~~69~~ 70 71 72 73 74 ~~75~~ ~~76~~ ~~77~~ ~~78~~ ~~79~~ 80 81 82 83 ~~84~~ ~~85~~
3 Go through the numbers that aren't crossed out. If the product of the digits is 20 or less, cross out the number.	65 ~~66~~ ~~67~~ ~~68~~ ~~69~~ ~~70~~ ~~71~~ ~~72~~ 73 74 ~~75~~ ~~76~~ ~~77~~ ~~78~~ ~~79~~ ~~80~~ ~~81~~ ~~82~~ 83 ~~84~~ ~~85~~
4 Go through the numbers that aren't crossed out. Find one that gives the right remainder.	21 R2 over 3)65 24 R1 over 3)73 24 R2 over 3)74 27 R2 over 3)83

Ana is thinking of the number 73.

Almanac

419

This almanac includes very useful tables and lists. It has tips on how to take notes in class and how to study for and take tests. It shows how to use a calculator and even has some computer math tips.

420

Prefixes

A prefix is added to the beginning of a word. It adds meaning to a word or suffix.

Prefix	Definition	Example
bi-	two	bicycle: two-wheeled
centi-, cent-	one hundredth	centimeter: one hundredth of a meter
circum-	around	circumference: the length around a circle
co-	joint, jointly, together	coplanar: lying in the same plane
dec-, deca-, deka-	ten	decahedron: polyhedron with ten faces
deci-	one tenth	deciliter: one tenth of a liter
di-	two, twice, double	dihedral: two sided
dodeca-	twelve	dodecagon: polygon with 12 sides
equi-	equal, equally	equiangular: having all angles equal (60° 60° 60°)
giga-	one billion	gigabyte: one billion bytes
hecto-	100	hectometer: 100 meters
hemi-	half	hemisphere: half of a sphere
in-	not or without	inequality: not equal $6 \neq 7$ or $6 < 7$
inter-	between, mutual	intersecting: crossing or meeting
iso-, is-	equal	isosceles triangle: a triangle that has two congruent angles (50° 50°)

Prefix	Definition	Example
kilo-	1000	kilogram: 1000 grams
mid-	middle	midpoint: point on a line segment that cuts it into two congruent segments midpoint 2 ft 2 ft
milli-	one thousandth	millimeter: one thousandth of a meter
nona-	ninth, nine	nonagon: polygon with nine sides and nine angles
octa-, octo-, oct-	eight	octagon: polygon with eight sides and eight angles
para-, par-	beside, alongside	parallel: an equal distance apart at every point
penta-, pent-	five	pentagon: polygon with five sides and five angles
per-	for each	percent: a ratio that compares a number to 100
poly-	many	polyhedron: a many-sided 3-dimensional figure
quad-	four	quadrilateral: a four-sided polygon
semi-	half	semiannually: happening once every half year
septi-, sept-	seven	septennial: occurring every seven years
sexa-, sex-	six	sexcentenary: referring to six centuries
tri-	three	triangle: polygon with three sides and three angles

421

Suffixes

A suffix is added to the end of a word. It adds meaning to a word or prefix.

Suffix	Definition	Example
-centenary	refers to a 100-year period	tercentenary: referring to three centuries
-gon	figure with a given number of interior angles	polygon: a figure with many angles hexagon: a figure with six angles
-hedral	surfaces or faces of a given number	dihedral: formed by two plane faces
-hedron	figure having a given number of faces or surfaces	polyhedron: a solid with faces that are polygons octahedron: polyhedron with eight faces
-lateral	of, at, or relating to sides	equilateral: having all sides equal
-metry	science or process of measuring	geometry: mathematics of properties, measurement, and relationships of points, lines, angles, surfaces, and solids
-sect	cut or divide	bisect: cut or divide into two equal parts

Study Tools

422

Creating and Using Graphic Organizers

423

You can use diagrams, charts, or graphs to help you organize your understanding of a topic. You can make a diagram to summarize data or even an idea. For example, time lines show data in a chronological order. Number lines can be used to show addition or subtraction. An order graph might be used to show different types of averages.

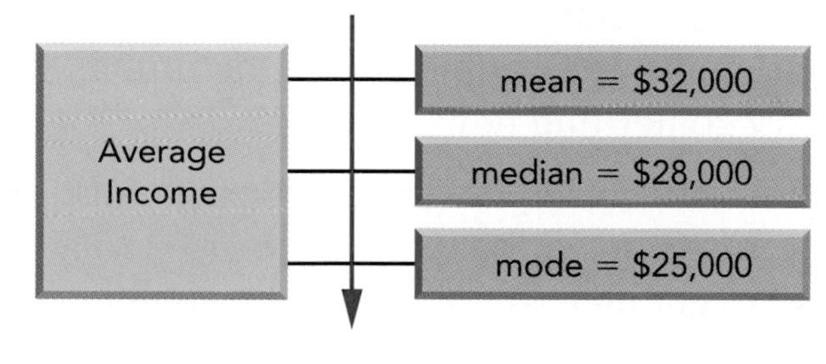

Taking Notes

424

Notes can help you remember what you learned in class. It's a good idea to keep all your math notes in one place—in a math journal. In your journal, you can write examples, facts, new math words, new symbols, diagrams, and descriptions of how math is used in the world around you. Wherever you take notes, have a good set of tools with you: a pencil or two with good erasers and plenty of lined paper.

425

How to Take Notes

Your math journal can take different forms. You can write math notes on cards and keep them in an envelope. To keep this card-journal handy, punch holes in the envelope and place it in your loose-leaf binder. You can also write your notes in a separate notebook. Date your notes to keep track of the order in which topics were introduced, studied, or reviewed in class.

Your notes can take different forms. You can use an outline form or write full sentences. You can even draw diagrams—you don't have to use the same form all the time.

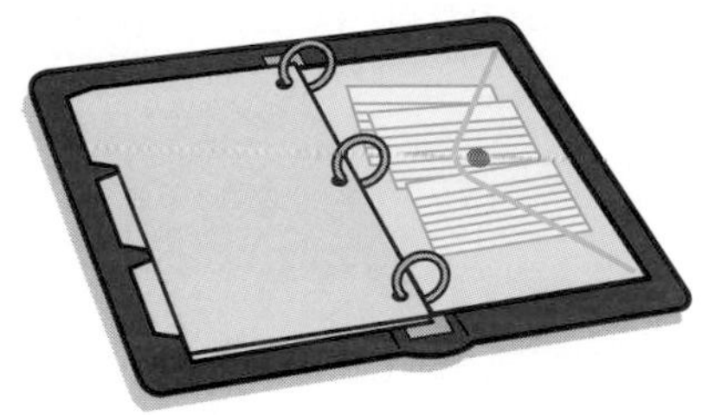

426

Evaluating Your Notes

If your notes are hard to read or they're not complete, they won't be much use when you study for tests. So, if one way of taking notes isn't as helpful as you'd like it to be, try something new.

- If writing in paragraphs doesn't work, write in outline form instead. Outlining the steps used to add mixed numbers might be easier than writing sentences describing them.

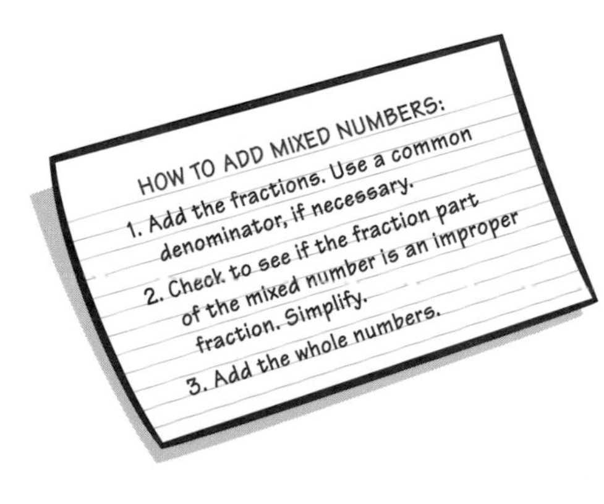

- If you can't find what you need in your notes because they are too brief, try to write more details next time.
- If you have too many details, try writing only the most important things instead of everything you can.

Guidelines for Improving Note-Taking Skills

- **Pay attention**. Listen carefully as your teacher explains how to solve a problem or an equation. Try to understand each step when a new topic is introduced. Carefully read through your textbook or this handbook. Write down the important ideas.

- **Keep it simple**. Remember, taking good notes does not mean writing down everything the teacher says. Write only things that help you understand. Use words, not just symbols, because words will help you make sense of your notes later. Use simple drawings to make things clearer.

- **Be organized**. When you make lists, number each item. Highlight or circle items that you do not understand. This way you can look them up or ask about them later. Review your notes every day or every other day. Try to read through all your notes once a week.

$\frac{3}{8} + \frac{2}{8} = \frac{5}{8}$

$\frac{7}{8} + \frac{5}{8} = \frac{12}{8} = 1\frac{4}{8}$

Can you simplify $\frac{4}{8}$? Yes!

$\frac{7}{8} + \frac{5}{8} = 1\frac{1}{2}$

Now try:

$\frac{5}{6} + \frac{1}{3}$

Different denominators

LCD = 6 so

$\frac{5}{6} + \frac{1}{3} = \frac{5}{6} + \frac{2}{6} = \frac{7}{6} = 1\frac{1}{6}$

Reminders:

*Ask Mr. Carson to explain how to tell when a fraction is in simplest form.

*Review common denominators

*Get help on writing mixed numbers as fractions and fractions as mixed numbers.

428

Working Together

In school, community, business, or family, people usually need to work together to solve problems. That's why learning to work with a group is so important. Here are good work habits that help groups to be successful.

Cooperation

The only way a group can succeed is by cooperating. Cooperating means using common courtesy and sharing a common goal. Everyone in the group should take part in the project and no one should be left out. Compliment and support other group members. You can disagree with someone without being mean. Try it. For example, if you think someone's computation is wrong, you could say something like "Let's check that on a calculator," instead of "No way, you don't know how to add."

Responsibility

A group does its job best when everyone chips in. Do what you are supposed to do carefully and help others do their tasks as well.

Listening

No one knows everything. There's always something to learn by listening to what others have to say. Maybe someone in your group has a different way to solve the problem, or a different way to look at it. These ideas, even if they don't work, may give you another great idea to try.

Encouragement

Offer and ask for help. Also, when someone does a good job or has a good idea, say so. People work better when they feel appreciated.

Decision Making

Groups have lots of decisions to make. For starters, they need to decide what the project will be. Then they must decide when, where, and how to do it. Help the group make decisions, but don't feel you have to have your way. State your points clearly, but be willing to follow a different group plan that is just as good.

Share the Work Fairly

Try to make sure that one person does not have more work than others in the group. If one part of a topic is longer or more complicated than another part, break it up so that the work load is the same for all group members. Make sure that you do your part of the project thoroughly. Encourage others to keep up with their parts.

Listening Skills 429

When you watch the teacher work through a problem, it all seems easy. But, do you really understand the problem? It is important to try to listen as actively as possible. Try to work through similar problems on your own as soon as you can.

Listen carefully. That means think about what the teacher or group leader is saying and how the problem is being solved. If you are uncertain about a particular step or part of a solution and have a question, ask it. Don't be afraid to ask; others probably have the same question in mind. By listening carefully, you will also be ready to answer a question or help to solve a part of a problem.

430

Managing Your Time

If you can't seem to get things done on time (or at all), try organizing your time. By planning study times and free time, you may find you actually have more free time!

Keep a Weekly Schedule

A weekly planner helps you organize your homework and set aside time to complete it. The planner makes it easier for you to prepare for tests and get projects and other homework done well and on time. You can even set a time to ask your teacher to explain things that you do not understand. Keep this schedule in your notebook or in your math journal.

Make a Daily List

Write down things you need to do today and tomorrow. Number them in order of importance. Hang the list someplace where you won't forget it and can check it. Another place to keep it is in a special section of your math journal. Check off items as you complete them.

- **Have a Homework Schedule**

Do your homework at the same time each day. Let yourself have short breaks between assignments. To finish your homework correctly and quickly, don't make the breaks longer than 5 minutes.

- **Set Goals**

Be realistic. Make sure the tasks you set for yourself and the time you allow are reasonable. Do not be too hard on yourself. When you don't quite achieve a goal, try to figure out why so you can do better next time. Reward yourself for reaching key goals.

- **Get It Done and Turn It in on Time**

Go over any instructions your teacher has given for an assignment. Read directions carefully. Check your notes and your textbook or math handbook to make sure that you know how to do your homework as accurately as possible. Keep a list of things you don't understand to ask your teacher about. After you're finished, check the directions again to make sure your homework is really complete.

Pick an easy thing to do first, just to get started.

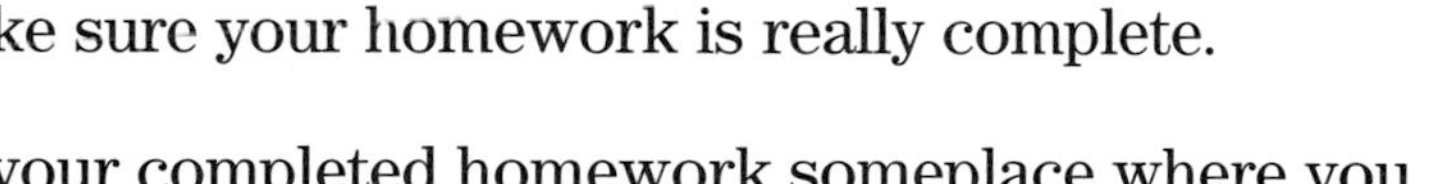

Keep your completed homework someplace where you won't forget to take it to school:

- on the floor by the door
- in the front pocket of your book bag
- in a pouch on your bike
- with your lunch or lunch money

431 Writing in Mathematics

When you write in mathematics, as in other subjects, you want your work to be clear. You should be able to show your understanding. This can seem hard to do because symbols are used so often. However, symbols can help make things clear because they have meanings, just like words. You should also use words, diagrams, or anything else that will help someone know what you're trying to say.

You can use the outline below to help you organize your thoughts to write clearly in mathematics.

Statement of Problem	
Solution	Detail: Step 1
	Detail: Step 2
	. . .
	. . .
	Detail: Last Step
Conclusion/Answer	

You can use the following outline, or **rubric**, to help you when you are writing in mathematics.

See 438

Rubric Level:

- **Well done** The writing shows a full understanding of the topic. It shows how you think through a problem. Each step is explained with both examples and words. The examples are worked out correctly and in detail. Diagrams may also be used. The problem is organized so that steps are in order. It is more than just calculations. An example of a problem that is worked out at this level can be found in 438.

- **Adequate** The writing shows a good understanding of the topic. You include examples, calculations, and maybe even diagrams. Steps are followed in order with each example or diagram explained. There may be some errors in computation but your descriptions show that you can think through the details of the problem.
- **Weak** The writing does not really explain what is going on in the problem. There are errors in both understanding and in computation. There may be examples, but there is nothing to explain what the examples mean or why the calculations were made. Steps may not be in order.
- **Inadequate** The writing is unclear and it is not well-organized. Examples and computation are not explained. There are many errors in both understanding and computation.
- **Too brief to evaluate**
- **Blank**

You can improve your ability to write in mathematics by using the rubric to evaluate your own work. Check to see how much you understand from your own writing. Decide which level you think a solution you have worked through would be. Then ask a parent or teacher to evaluate the solution. Discuss any differences. Keep in mind that it takes a lot of effort to reach the top level. But if you keep trying, you'll get there.

432

Test-Taking Skills

Success in school depends on you! Be prepared and be involved, especially before a test. Don't wait until the last minute to study for a test.

433

Organizing and Preparing Test Material

- **What will be on the test?** Ask the teacher to be as specific as possible.
- **When is the test?** If you think it is about time for a test, ask your teacher if there will be a test soon.
- **What will the test questions be like?** Will there be short-answer computation, word problems, multiple-choice, or true/false questions?
- **Be sure you understand the topics.** Make sure that everything makes sense to you. If not, get help.
- **Review your notes carefully.** Compare class notes and examples with those in your math book.
- **Get any notes or assignments that you missed.** Your teacher or classmates will help if you ask ahead of time.

- **Make a list of questions.** If you are uncertain about anything, talk to your teacher several days before the test.
- **Make an outline of topics.** Find an example for each type of problem that will be on the test.
- **Talk about the topics with other students.** By talking about how to work through problems, you will test your own understanding.

Reviewing Test Material 434

- **Start reviewing early.** You will remember more by studying a little over several days than you will by studying everything the night before the test.
- **Find a good place to study.** Find a quiet place with room to spread out your papers.
- **Make sure you get enough rest.** If you get tired, take a break or get some sleep. If you're tired on test day, you'll have trouble concentrating.
- **Set up a specific time to study.** Keep to your study time. Review your notes, look through your textbook, and try sample problems.
- **Practice, practice, practice.** Redo old homework problems or quiz problems.
- **Use study aids.** Silly sayings, called mnemonic devices, can help you remember complicated rules or procedures. Use the ones in this handbook or make some up yourself.
- **Think about how things fit together.** The more you understand, the less you'll have to memorize. For example, if you understand why a formula works, you'll have an easier time remembering it.
- **Study with other students.** As often as possible, talk out loud about how to work through a problem.
- **Study by yourself.** You will be taking the test alone, so practice doing problems on your own.

435 Taking the Test

- **Check that you have all the materials you need for the test.** You may need a ruler, sharp pencils, paper, and so on.
- **Know the rules of the test.**

How much time do I have to complete the test? Do all the questions count equally? Can I use my textbook, my handbook, my calculator, or notes? Does the teacher want explanations for short-answer test items?

- **Look over the entire test quickly.** Then begin the test. Check the time once in a while.
- **Read the directions carefully.** If you don't understand something, ask your teacher. Follow all instructions.
- **Answer the questions you are sure of first.** If you are stuck, move on to the next question.
- **If time is running out, try not to panic.** Do what you can do. Panicking will only waste time that you could be using to finish a problem or move on to another problem.
- **Double-check.** Make sure that you have answered all the questions that you can. Recheck your work if you have time.
- **Check that all your answers are readable.**
- **If you have no idea how to solve a problem, move on.** After you have completed the rest of the problems, return to the ones you don't know.
- **Only change your answers if you are sure they are wrong.** Do not rush through a problem a second time and quickly change your answer. You are often right the first time!

Tips for Taking Multiple-Choice Tests 436

In a multiple-choice problem, usually only one answer is correct. You can often eliminate at least one of the choices.

See 095, 107, 148, 230

- **Use number sense to save time and work.**

EXAMPLE 1: Compute $5769 + (1921 \times 2 \times 0)$.

A. 3842 **B.** 5769 **C.** 9611 **D.** 15,380

Notice that the product $(1921 \times 2 \times 0)$ is zero. Now it's easy: $5769 + 0 = 5769$. The answer is **B**.

- **Estimate when you can.**

EXAMPLE 2: Which statement is true?

A. $28 \times 416 > 54 \times 284$ **B.** $538 + 2805 + 117 = 3459$

C. $400 - 382 < 250 - 229$ **D.** $8 \times 712 < 5600$

You can use estimation in this problem.
In **A**, round to the nearest ten: $30 \times 400 = 12{,}000$ and $50 \times 300 = 15{,}000$. **A** is not true.

In **B**, add the ones digits: $8 + 5 + 7 = 20$. The ones digit must be a 0, not 9, so **B** is not true.

Use rounding again for **D**: $8 \times 700 = 5600$, and 8×712 is greater than 8×700, so **D** is not true.

The only possible true statement is **C**.

- **Eliminate obviously wrong choices.**

EXAMPLE 3: Complete the number sentence.
$971 \div 9 = 107$ R■

A. 1 **B.** 8 **C.** 10 **D.** 61

Both **C** and **D** are obviously wrong since the remainder cannot be greater than the divisor (9). So the answer is either **A** or **B**. Since **A** is an easy number, try it first. If it works, great. If not, the answer is **B**. (The answer is **B**.)

437 Problem Solving on Multiple-Choice Tests

- **Read the problem carefully.** Underline or circle words in the problem that help you understand what you are asked to do.
- **Decide how hard the problem will be to solve.** If it looks easy, do it now. If not, come back to it later.
- **Look over the answers to see if you can rule out any choices.** Use clues to help you solve the problem. For example, think about the size of the units—should the answer be in the same units, square units, cubic units?

EXAMPLE: Find the area of a rectangular floor that is 8 feet long by 6 feet wide.

A. 28 ft **B.** 28 ft^2 **C.** 48 ft **D.** 48 ft^2

Since this is an area problem, you know that the answer must be in square feet. So choices **A** and **C** cannot be correct. The answer is **D**.

438 Tips for "Explain-Your-Thinking" Tests

See 431

As you work through a problem that asks you to explain your answer, include details you think the teacher should see to understand how you think. Grading usually combines how you do and how you explain each step of the problem.

SAMPLE PROBLEM: Some people think of summer as lasting from Memorial Day through Labor Day. In 1997, Memorial Day was on May 26 and Labor Day was on September 1. In 1998, Memorial Day was on May 25 and Labor Day was on September 7. Which "summer" was longer, 1997 or 1998? How much longer? Explain your thinking.

1. First you need to decide which "summer" was longer. If you stop at this point, you will only receive partial credit.

Since Memorial Day was earlier in 1998 than it was in 1997 and Labor Day was later in 1998 than it was in 1997, "summer" was longer in 1998.

2. Now you must figure out how much longer the "summer" was. First, find how many days are in each month from June through August.

May: 31 June: 30 July: 31 August: 31
Add the total number of days
June–August: 92

3. Now find a way to add on the extra days in May and September. A diagram or a table might make your thinking clearer.

	Number of Summer Days 1997	Number of Summer Days 1998
May	May 26–May 31: 6 days	May 25–May 31 7 days
June–August	92 days	92 days
September	September 1: 1 day	Sept. 1–Sept. 7 8 days
Total	99 days	107 days

4. After solving the problem, use the original question to form your answer.

"Summer" was 8 days longer in 1998 than in 1997.

439 Tips for Short-Answer or Fill-In Problems

- Do exactly what the directions tell you to do.
- These questions are often either right or wrong. So you want to be very careful with your computation.
- Know and use the order of operations.
- If the directions ask for a fraction in simplest form, make sure that the answer really is in simplest form.
- If you are asked for a diagram, make sure you draw and clearly label the diagram.

Problem	Solution	Answer
A. $36 \times 42 =$ ____	36×42: $72 + 1440 = 1512$	1512
B. $85 \div ■ = 7$ R1	■ $= 12$ because $12 \times 7 = 84$ and $84 + 1 = 85$	12
C. Complete the figure so that it is symmetrical. 	The other half of the figure has to fold over the line of symmetry and exactly match the first half.	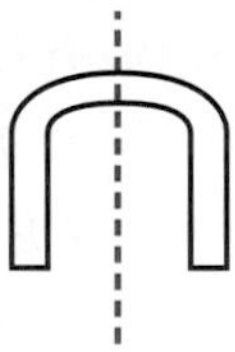
D. Compare. Write $>$ or $<$. $\frac{2}{3}$ ____ $\frac{3}{4}$	$\frac{2}{3} = \frac{8}{12}$ and $\frac{3}{4} = \frac{9}{12}$, $\frac{8}{12} < \frac{9}{12}$, so $\frac{2}{3} < \frac{3}{4}$	$<$
E. $3 \times (4 + 2) - 5 =$ ____	$3 \times (4 + 2) - 5$ ↓ $3 \times 6 - 5$ ↓ $18 - 5 = 13$	13

After a Test

- Keep a record in your math journal of how you got ready for the test and how you did. Use it to help you plan for future tests.
- Describe how you studied for the test. Think about how you could improve your study habits, the way you study for a test, and how you actually take a test.
- Be an editor and correct your errors. Go through each item on the test. For the ones you got right, think about how your preparation helped you. For the questions you missed, write down what you will do differently when you prepare for the next test.

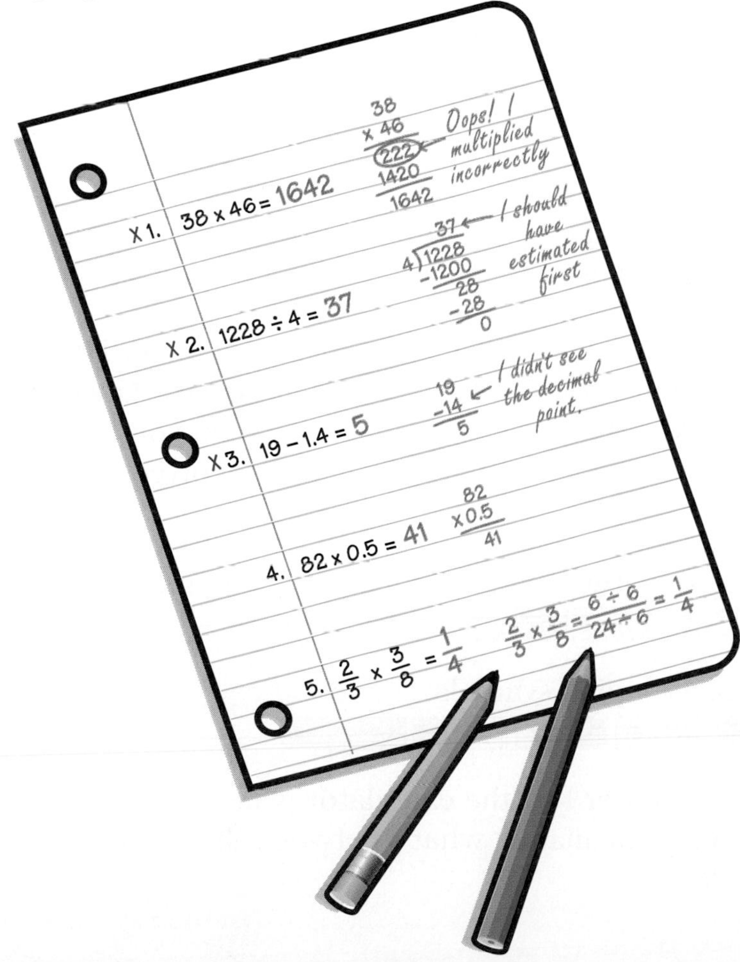

441

Using a Calculator

Order of Operations on the Calculator

Some calculators automatically follow the rules for order of operations. This is known as the **algebraic ordering system**. Other calculators carry out operations in the order that you enter them, even if that order is different from the rules.

EXAMPLE: Check your calculator for order of operations with this expression: $8 + 4 \div 2$.

[8] [+] [4] [÷] [2] [=]

If your display shows [10.], your calculator has order of operations. If it shows [6.], it doesn't. Once you know how your calculator computes, then you can enter the calculations so they are done in the correct order.

For example, to find $8 + 4 \div 2$ on a calculator that does not use the order of operations, you can press:

[4] [÷] [2] [+] [8] [=] [10.].

OR, use grouping symbols:

[8] [+] [(] [4] [÷] [2] [)] [=] [10.].

When you enter [(], the calculator waits until you enter [)] before calculating what's between those grouping symbols.

MATH ALERT **Your Calculator May Be Different**

Your calculator may have different keys that do the same thing as the keys we show you in this handbook. Check the directions for your calculator.

MATH ALERT **What Happens When a Number Won't Fit on the Display?**

Sometimes your answer will be larger (or smaller) than the calculator display will hold. There are many ways your calculator might handle this problem. Check the calculator instruction book to find out.

- Does the calculator cut off (truncate) or round the answer? (Try pressing 2 ÷ 3. If the display ends in a 6, it cut off the answer. If the display ends in 7, it rounded the answer.)
- Some calculators show large numbers with exponents.

The Constant Function

The **constant key** stores an operation and a number. For example, if you press + 6 Cons, the calculator will remember the + 6.

EXAMPLE: Use the constant key to show 15 + 6 + 6 + 6 + 6.
Clear the memory.
Then press: + 6 Cons 1 5 Cons Cons Cons Cons 39..

Some calculators have displays that tell you how many times you've pressed Cons.

446 Using a Calculator to Compute

You can use a calculator to compute with whole numbers, decimals, fractions, or integers.

447 Using a Calculator to Compute with Whole Numbers

You can use a calculator to add, subtract, multiply, or divide whole numbers.

EXAMPLE 1: Use a calculator to add. 598 + 654 = ■

5 9 8 + 6 5 4 = 1252.

EXAMPLE 2: Use a calculator to subtract. 4009 − 978 = ■

4 0 0 9 − 9 7 8 = 3031.

EXAMPLE 3: Use a calculator to multiply. 36 × 104 = ■

3 6 × 1 0 4 = 3744.

EXAMPLE 4: Use a calculator to divide. 27 ÷ 5 = ■

2 7 ÷ 5 = 5.4

The 0.4 of the 5.4 can also be shown as a fraction ($\frac{4}{10}$) or a remainder. ($\frac{4}{10} = \frac{2}{5}$, so the remainder is 2.) Your answer will depend on the problem.

Some calculators have a key that shows the remainder as a whole number instead of as a decimal. The key may look like INT÷.

EXAMPLE 5: Divide. 26 ÷ 3 = ■

2 6 INT÷ 3 = Q 8 R 2

The Q shows that the whole number part of the quotient is 8. The R shows that the remainder is 2. The answer is 8 R2.

Using a Calculator to Compute with Decimals

You can use a calculator to add, subtract, multiply, or divide decimals.

EXAMPLE 1: Use a calculator to add. 3.73 + 6.298 = ■

3 . 7 3 + 6 . 2 9 8 = 10.028

EXAMPLE 2: Use a calculator to subtract. 12.06 − 4.39 = ■

1 2 . 0 6 − 4 . 3 9 = 7.67

EXAMPLE 3: Use a calculator to multiply. 7.2 × 31.6 = ■

7 . 2 × 3 1 . 6 = 227.52

EXAMPLE 4: Use a calculator to divide. 8.8128 ÷ 2.04 = ■

8 . 8 1 2 8 ÷ 2 . 0 4 = 4.32

Using a Calculator to Compute with Percents

Calculators are very useful when you need to compute with percents. Some calculators have a percent key % that you can use when you work with percents.

EXAMPLE: Find 15% of 700.

7 0 0 × 1 5 % = 105.

Remember to check how your calculator works. On some calculators, the percent must be the second factor in the multiplication. Some of these calculators do not require you to press = after using the % key.

450 Using a Calculator to Compute with Fractions

Special Keys

Some calculators have keys for working with fractions.

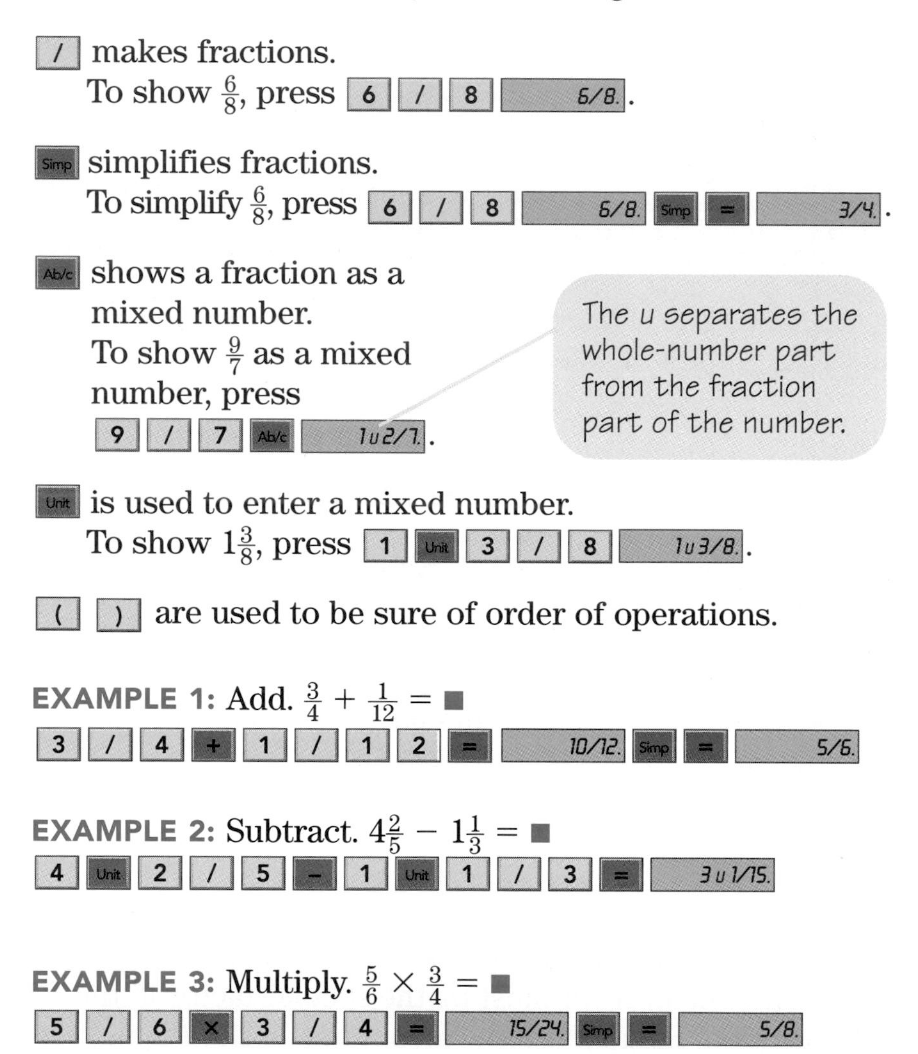

EXAMPLE 1: Add. $\frac{3}{4} + \frac{1}{12} = \blacksquare$

3 / 4 + 1 / 1 2 = 10/12. Simp = 5/6.

EXAMPLE 2: Subtract. $4\frac{2}{5} - 1\frac{1}{3} = \blacksquare$

4 Unit 2 / 5 − 1 Unit 1 / 3 = 3 u 1/15.

EXAMPLE 3: Multiply. $\frac{5}{6} \times \frac{3}{4} = \blacksquare$

5 / 6 × 3 / 4 = 15/24. Simp = 5/8.

EXAMPLE 4: Divide. $3\frac{2}{3} \div \frac{4}{5} = ■$

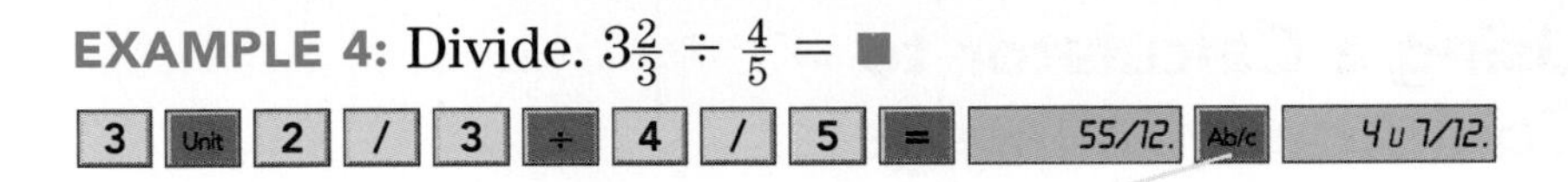

Use Ab/c when the display shows a fraction greater than 1.

Computing with Fractions Without Using Fraction Keys

Even if your calculator doesn't have special keys for fractions, you can still compute with fractions. Just enter each fraction as the numerator divided by the denominator and then compute. Your answer will be a decimal. If your answer needs to be a fraction, you may not want to use your calculator in this way.

See 043, 212

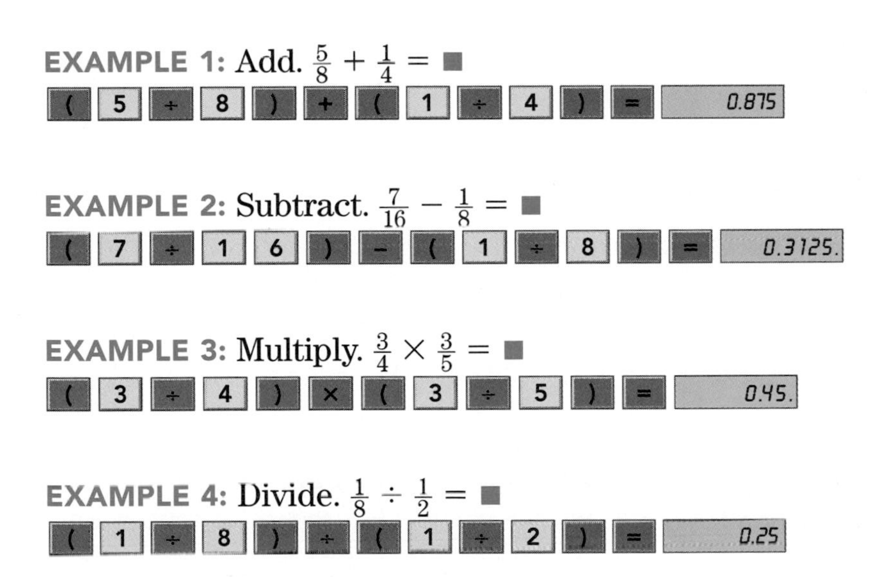

EXAMPLE 1: Add. $\frac{5}{8} + \frac{1}{4} = ■$

EXAMPLE 2: Subtract. $\frac{7}{16} - \frac{1}{8} = ■$

EXAMPLE 3: Multiply. $\frac{3}{4} \times \frac{3}{5} = ■$

EXAMPLE 4: Divide. $\frac{1}{8} \div \frac{1}{2} = ■$

452

Using a Calculator to Compute with Integers

See 201–211

You can use some calculators to add, subtract, multiply, or divide signed numbers. Find this key on your calculator. [+/-]

EXAMPLE: Use a calculator to add. $^-5 + {}^-4 = ■$

[5] [+/-] [+] [4] [+/-] [=] -9.

Using a Calculator with Exponents and Roots

MORE HELP

See 065–067

You can use a calculator to evaluate numbers with exponents and to find square roots of numbers.

You can use the exponent key. [y^x]

EXAMPLE 1: Evaluate. $15^2 = ■$

[1] [5] [y^x] [2] [=] 225.

EXAMPLE 2: Find $\sqrt{25}$.

You can use the square root key: [√]

Press:

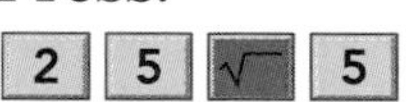

454

MATH ALERT Do Your Calculator Answers Make Sense?

Always check the answer on your calculator display to see that it makes sense. It's easy to press a wrong key. If you estimate first, you will know when the answer is not reasonable.

Using a Computer for Math

455

The Internet

456

A home computer must have a **modem** to get onto the **Internet**. The modem and special software use telephone or cable lines to connect to a **server** and its services. Servers are computers that are only used to provide **electronic mail** (e-mail), and access to the Internet.

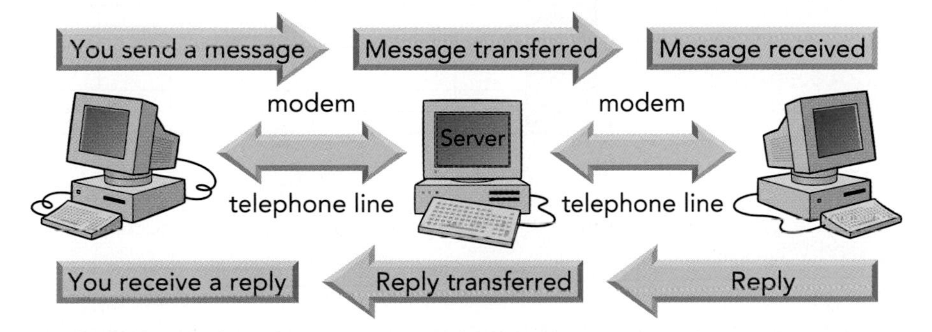

The **World Wide Web** (WWW) allows Internet users to interact with each other. They can look up information, including newspaper and magazine articles, government reports, and weather forecasts on **web pages**. A user can point to text or an item on the screen and select it to immediately be linked to another Web page.

457 Spreadsheets

Computer spreadsheets can help you organize and analyze data quickly and efficiently. If some data change, the spreadsheet can automatically recalculate the rest of the data. Computer spreadsheets are faster and more accurate than paper worksheets.

458 Terminology

A document that is created with a spreadsheet program may be called either a **worksheet** or a **spreadsheet**.

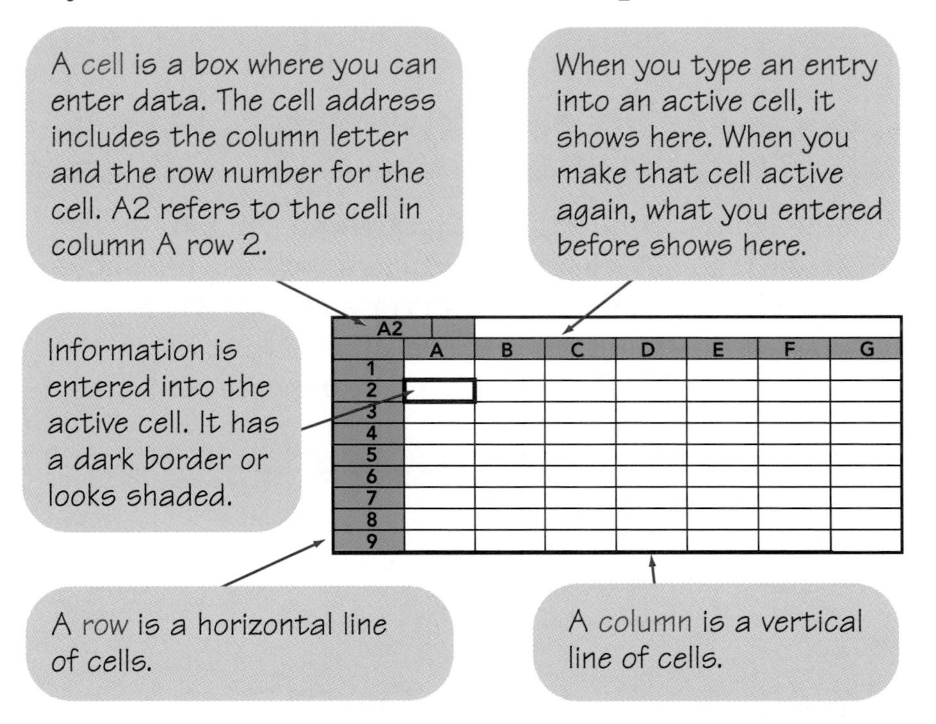

- **Labels** are words or abbreviations. They may be entered into cells to identify information used in the spreadsheet.
- **Values** are numbers.
- **Formulas** or **functions** tell the computer what calculations to do. Functions are certain formulas that are built into spreadsheets for calculations that need to be done often.

B8		= Sum (B1:B6)					
	A	B	C	D	E	F	G
1		89					
2		76					
3		92					
4		85					
5		90					
6		81					
7							
8	Sum:	513					
9							
10							

The entry window at the top of the spreadsheet shows the function used for B8.

Sum is a function that adds a list of numbers.

Cell B8 contains this function: = Sum(B1:B6). This means it will add the numbers in cells B1 through B6. When you press **Enter** after typing the function, the computer adds the numbers and puts the sum in cell B8.

What-if Analysis

A wonderful feature of spreadsheets is that you can see what happens if the data change. For example, if you double the length of the side of a rectangle, you can immediately see how the area will change. If you want to see how doubling the width changes the area, you only have to change the width. The spreadsheet will calculate the new area for you. That makes it easy to get answers to "what if?" questions. "What if I change this?" "What if I change that?" Of course, you have to be sure you entered the correct formulas. The spreadsheet will calculate whatever formula you enter, even a wrong one.

On a computer, * means multiply.

B4		= B1 * B2					
	A	B	C	D	E	F	G
1	Length	8	16	8	16		
2	Width	6	6	12	12		
3							
4	Area	48	96	96	192		
5							
6							
7							
8							

460 Graphs

You know graphs are useful, but they can take a lot of time to make. Spreadsheets can create graphs automatically. Spreadsheets usually let you make bar graphs, scatter plots, pie charts, line graphs, and 3-dimensional graphs. You choose which kind of graph. The spreadsheet makes the graph and even creates a legend that explains the colors or symbols used in the graph.

The spreadsheet leads you through steps to create the graph. It does not check to see if your graph makes sense. You have to do that. You must be careful to choose the kind of graph that applies to the data you have. Suppose you want to graph the data about temperatures below.

See 279

B2

	A	B	C	D	E	F	G	H	I
1	Week	1	2	3	4	5	6	7	8
2	Water Only	2	3	6	7	10	11	13	15
3	Plant Food	2	4	6	8	11	12	15	18
4									
5									

You may decide to make a double-line graph from the data in the spreadsheet. This choice of a graph makes sense because a line graph is a good way to show change over time.

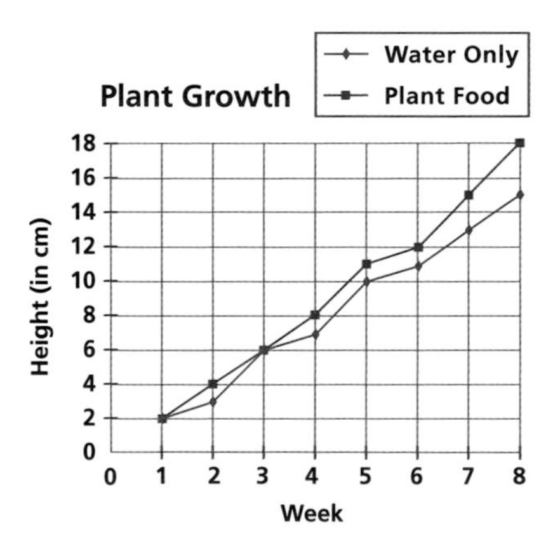

Databases

A database is a collection of information, like a telephone book, a dictionary, an address book, or a list of CDs you own. The nice thing about computer databases is that they're fast, neat, and easy to change. Your electronic address book won't be full of crossouts like your paper one probably is.

You can use a database to collect and organize information. A **file** is made up of a collection of records. You may think of each **record** as an index card containing information about a single subject.

FIELD Each piece of data is stored in a field. Each field has its own name to keep it separate from other fields. In an address book, there are different fields for the name, the address, and so on.

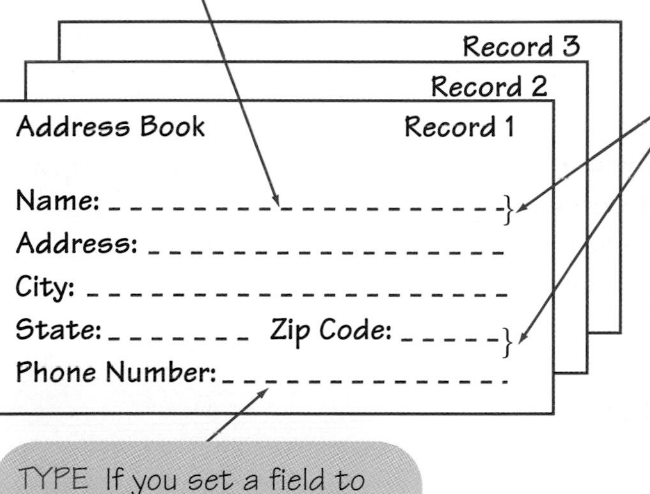

SIZE Each field might be limited in size. You can type in only as many letters or numerals as the size of the field allows. So, if you have a very long name, and the field is not long enough, some of the letters may get cut off!

TYPE If you set a field to **numeric**, you may only enter digits from 0–9. This will guarantee that you do not accidentally write names in number-only fields.

462 Draw Tools

Drawing programs are fun! They're great for designing floor plans, creating diagrams, or even experimenting with patterns.

A drawing program has its own tool palette. It also has a fill pattern palette. Examples are shown below.

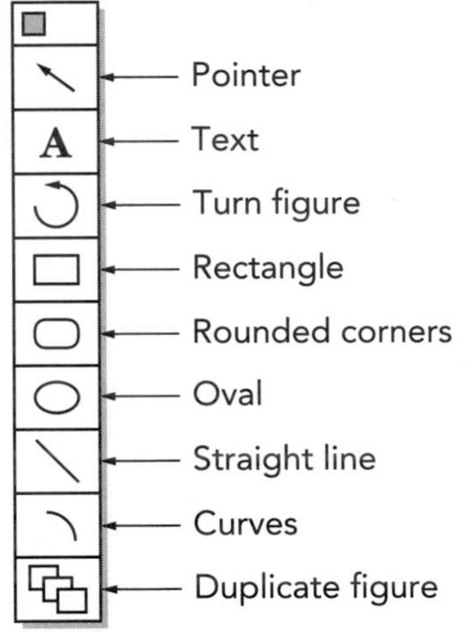

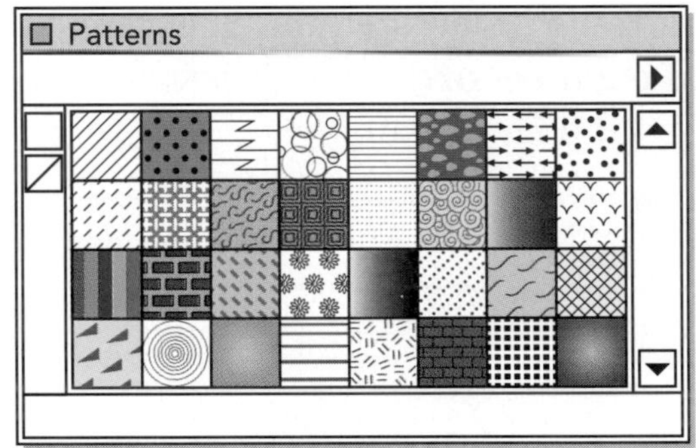

You select a tool from the palette by clicking on it. Then you move the cursor into your drawing and use the tool. For example, if you want to draw an oval, click on the oval tool. Then click in the drawing-box while holding down the mouse button. Drag down and right until the oval is as large as you want it. Then let go of the mouse button. You can even use a fill pattern from the pattern palette to fill the oval.

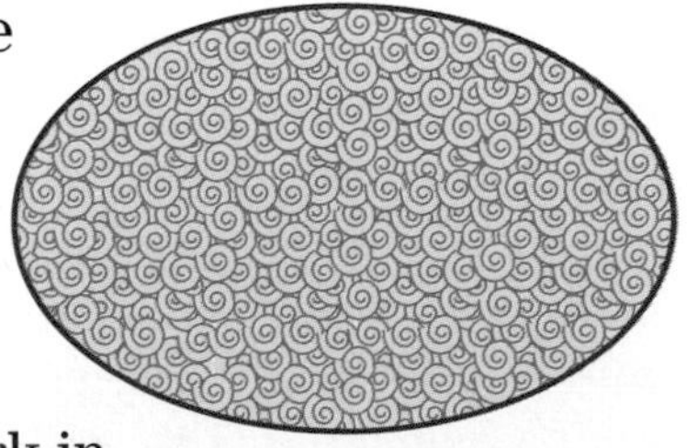

Other types of programs that work in a manner similar to drawing programs include paint programs and combination draw/paint programs.

Using a Computer Keyboard

463

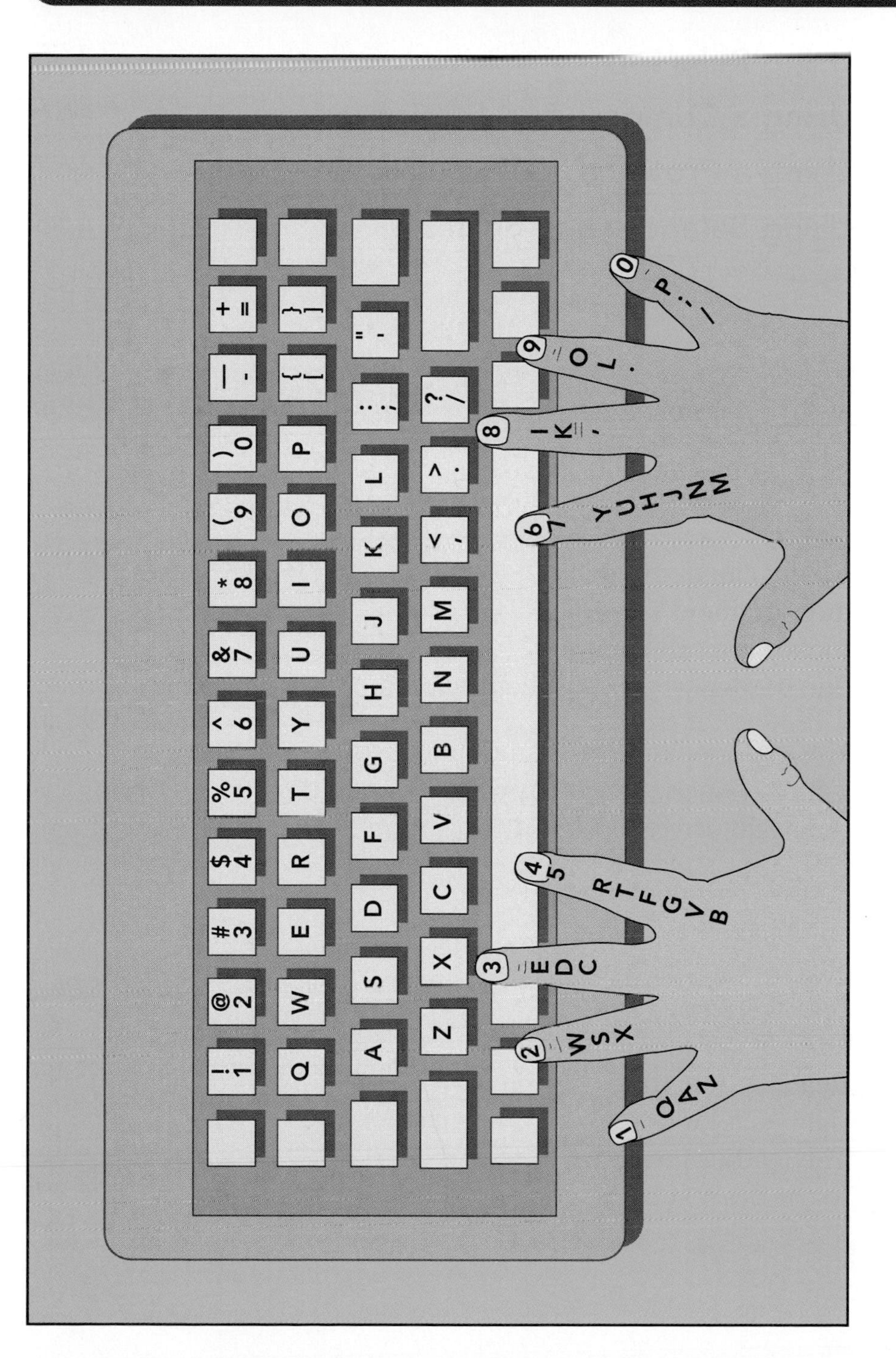

464

Math Tools

465

Using Rulers

A **ruler** is a straight-edge (usually wood, metal, or plastic) used for measuring lengths and for drawing straight lines.

Rulers can show two measurement systems. Inches are often along one border while centimeters are along the opposite border.

You can measure the same object to the nearest inch, $\frac{1}{2}$-inch, $\frac{1}{4}$-inch, $\frac{1}{8}$-inch, or even $\frac{1}{16}$-inch, depending upon the scale of your ruler.

To measure the length of an object, line up the left edge of the ruler (where zero would be) with the left-edge of the object. Read the length at the right-edge of the object. The feather measures 3 in. to the nearest inch.

If the object you want to measure is longer than the ruler,

- use a longer ruler or tape measure, or
- mark the object at the end of the ruler; move the zero point of the ruler to that point and continue measuring. Write down each measurement and then add them all together.

Using a Compass and Straightedge

A **straightedge** is just that—anything that will help you draw a straight line segment. A ruler is a straightedge, but you don't need the measurement marks of a ruler to draw a straight line.

See 367, 469

A **compass** is a tool for making circles or arcs. You can use the fact that a compass can be set for a specific radius to help you copy segments and angles. These segments and angles can help you make other geometric constructions.

You can set a compass at the length you want your radius.

You can also set a compass by putting the point on one endpoint of a segment and moving the pencil until it is on the other end of the segment.

To make a circle or arc, hold the point steady where you want the center of your circle to be. Spin the pencil-end of the compass around that point.

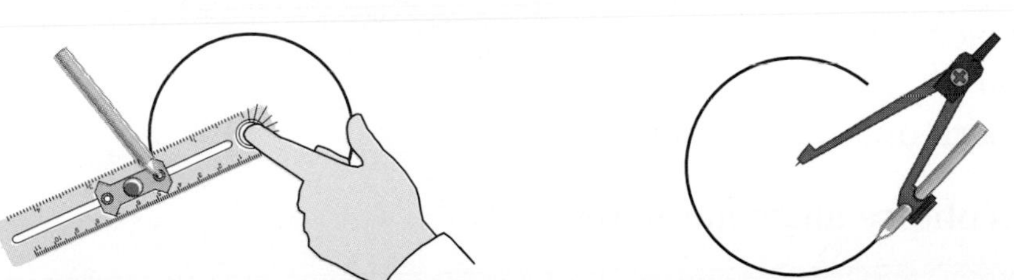

467 Using a Protractor

A **protractor** is used to measure angles. Some protractors are semicircular and some are circular, but both kinds work the same way. Almost all protractors have two scales. To decide which scale to read, think about whether you're measuring an acute or obtuse angle.

See 344–346

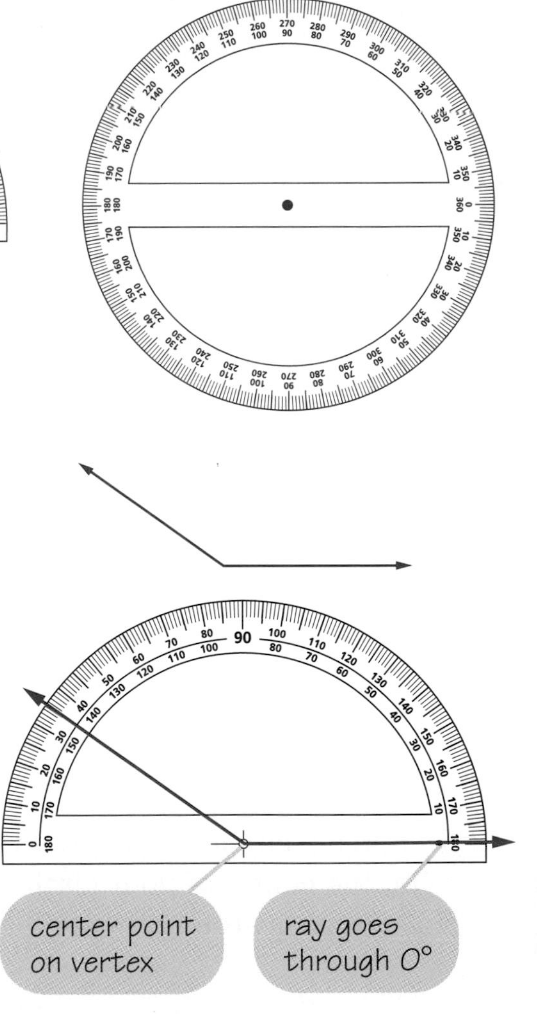

EXAMPLE 1:

1. To measure this obtuse angle, choose one ray for the bottom.
2. Place the protractor on the angle. If the rays aren't long enough, extend them with your straightedge. If you can't write in your book, trace the angle on paper and extend the rays on your copy. Place the center point on the vertex. Place the 0° mark on the bottom ray.

This obtuse angle measures 145°.

EXAMPLE 2:

1. To measure this acute angle, choose one ray for the bottom.
2. Place the protractor on the angle. If the rays aren't long enough, extend them with your straightedge or trace the angle on paper and extend the rays on your copy. Place the center point on the vertex. Place the 0° mark on the bottom ray.

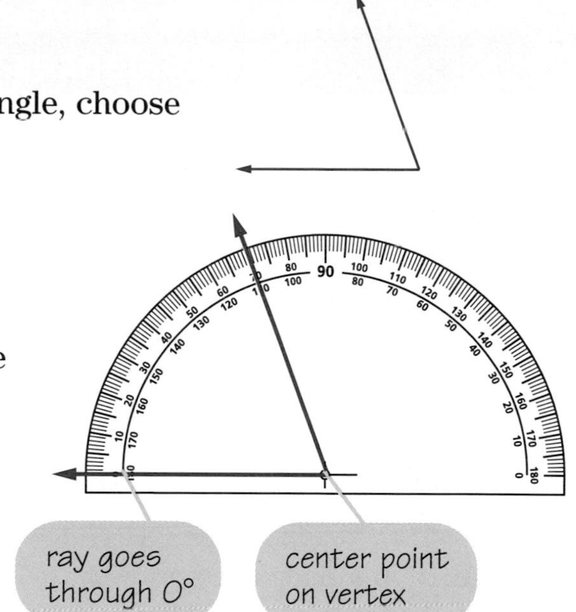

This acute angle measures 70°.

EXAMPLE 3:

1. You can measure the same acute angle from the other ray as well.
2. Place the protractor on the angle. If the rays aren't long enough, extend them with your straightedge or trace the angle on paper and extend the rays on your copy. Place the center point on the vertex. Place the 0° mark on the bottom ray.

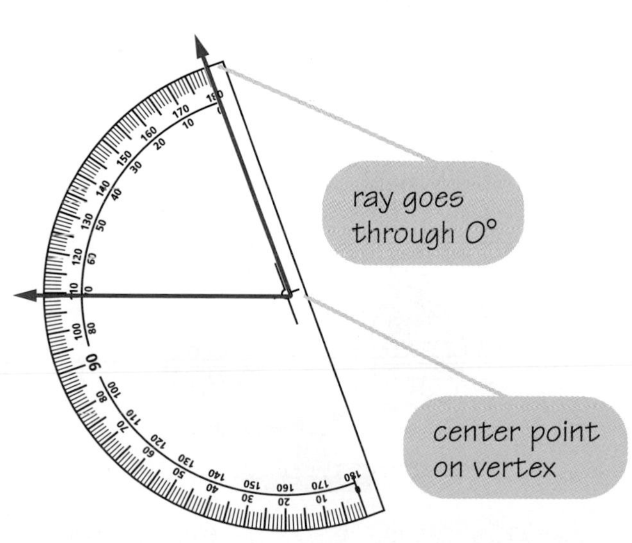

The acute angle still measures 70°.

468 Using Balances and Scales

Balances and scales are instruments used for weighing.

A balance usually has a beam that is supported in the center. On each side of the beam are two identical trays. When the trays hold equal weights, the beam is level. It is *balanced.*

You can use a balance to find the weight of an object. Suppose you have a bag of apples and a balance with some weights.

1. If you place the apples on one side of the balance, the side with the apples will tip down. You try to balance the apples using weights on the other tray.

2. If the weights are not heavy enough, the balance will remain tipped down toward the apples.

3. If the weights are too heavy, the balance will tip down toward the weights.

4. The beam will balance when the weights match the weight of the apples.

The bag of apples weighs 1 pound 7 ounces.

A **scale** is another tool used for weighing.

You can read the weight of an object directly from the scale. The display can be digital, or it can be graduated like a ruler.

The scale shows that the person weighs 85 pounds.

This person weighs 103 pounds.

People often have smaller scales to weigh items, such as food, at home.

The apple weighs 6 ounces.

Did You Know...

that there are some loose measurements in the form of words that are used to describe groups of animals?

A *flock* is a group of animals that live, travel, or feed together. A *gaggle* is a flock of geese. A *pack* is a group of animals, usually dogs or wolves, that run and hunt together. A *school* is a group of water animals, especially fish, swimming together.

469

Geometric Constructions

470 Copying a Line Segment

1. Draw a segment longer than $\overline{AB}$. Mark point A' on that segment. 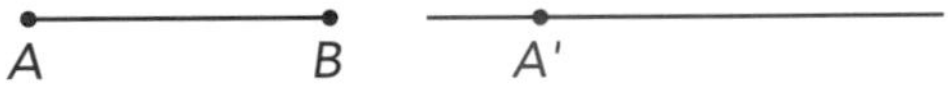
2. To copy $\overline{AB}$, set your compass to the length of $\overline{AB}$.
3. Place the point of the compass on A' and draw an arc that intersects the segment. Label the point of intersection B'.

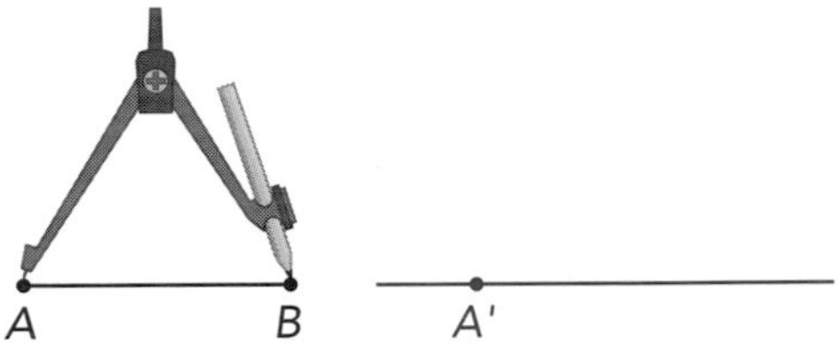

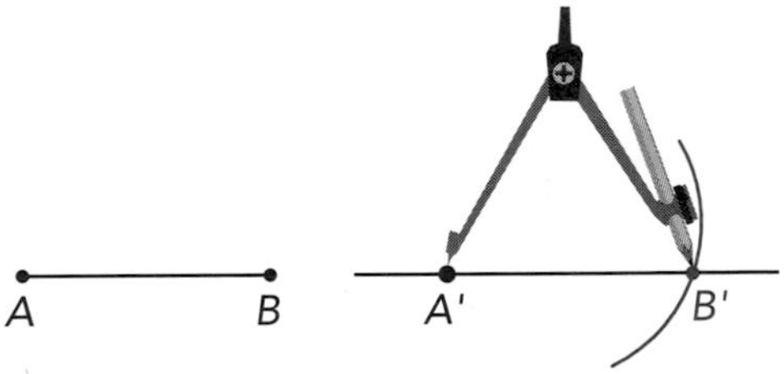

A B A'

A B A'

A B A' B'

MORE HELP

See 515

Since you set your compass to the length of $\overline{AB}$, the length of $\overline{A'B'}$ is the same as the length of $\overline{AB}$. So, $\overline{AB} \cong \overline{A'B'}$.

471 Bisecting a Line Segment

1. Place the point of the compass on A. Open the compass to a length that is greater than half the length of $\overline{AB}$. Use this compass opening to draw an arc above $\overline{AB}$ and below $\overline{AB}$.

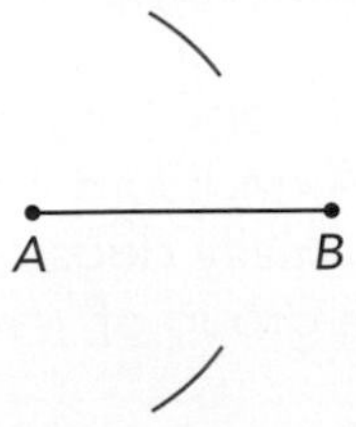

A B

2. Place the point of the compass on B and use the same compass opening from Step 1. Draw two more arcs, one above $\overline{AB}$ and one below $\overline{AB}$.

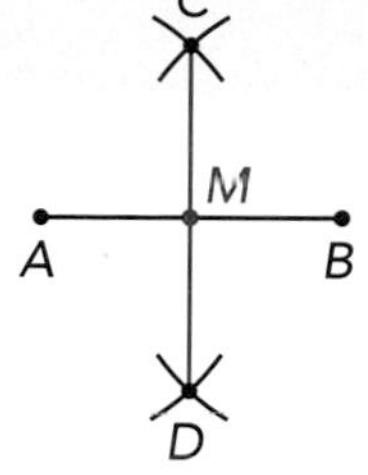

3. Use a straightedge and draw $\overline{CD}$.

Because both ends of segment $\overline{CD}$ are the same distance from A as from B, point M bisects segment $\overline{AB}$.

Copying an Angle

1. To construct an angle congruent to $\angle MNO$, draw a ray. Label the endpoint Z.

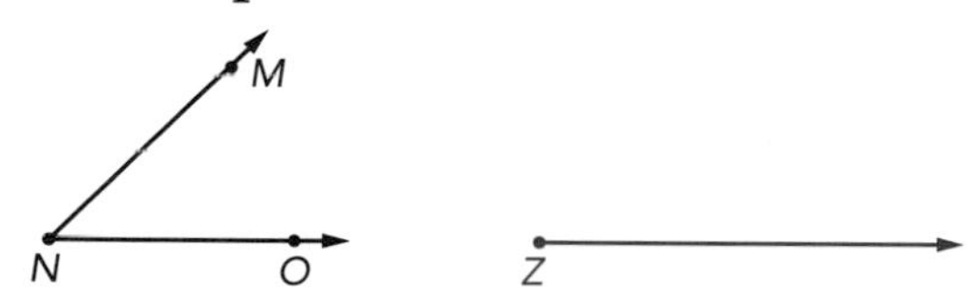

2. Place the point of your compass at N and draw an arc that intersects both rays of the angle. Keep this setting. Place the point of your compass at Z and draw a long arc that intersects ray Z.

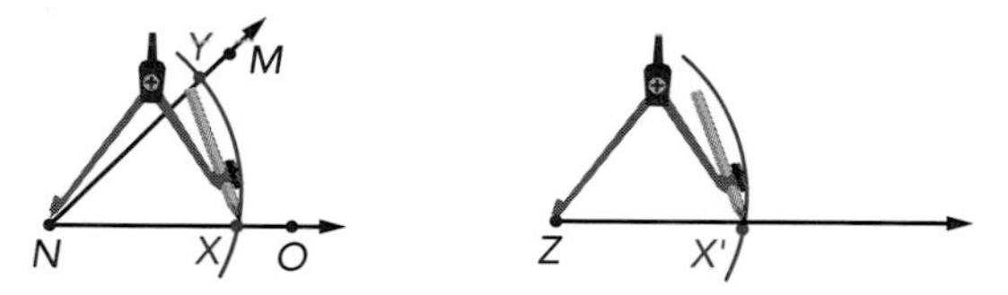

3. Place the point at X and the pencil at Y and set the compass. Place the point at X' and draw an arc. Use your straightedge to draw a ray from Z through Y'.

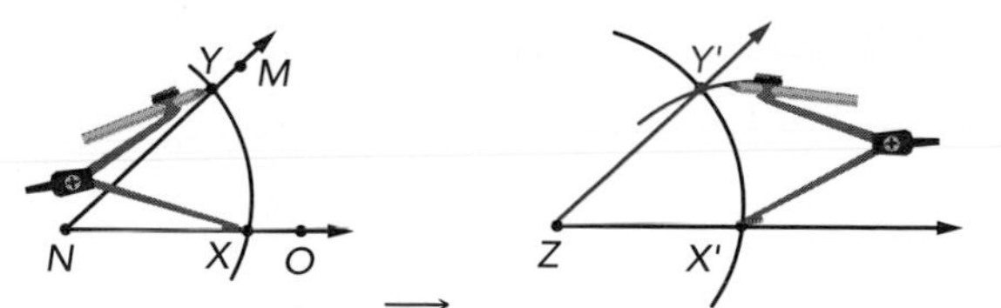

Point Y' is related to $\overrightarrow{ZY'}$ in exactly the same way as point Y is related to $\overrightarrow{NY}$. That makes $\angle Y'ZX'$ congruent to $\angle MNO$.

473 Bisecting an Angle

1. Place the point of the compass on C. Open the compass and draw an arc.

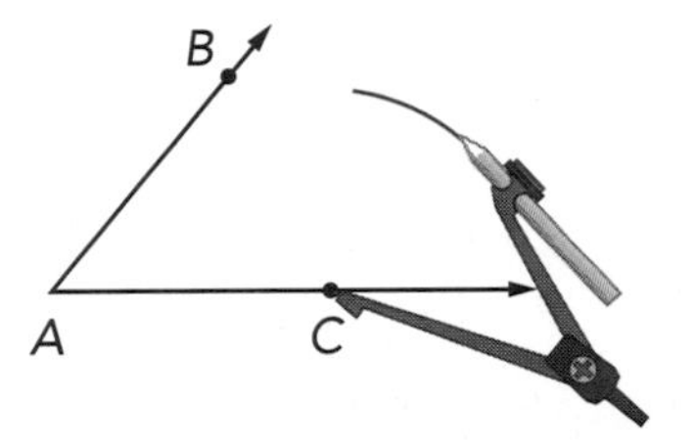

2. Place the point of the compass on B and use the same compass opening from Step 1. Draw another arc that intersects your first arc.

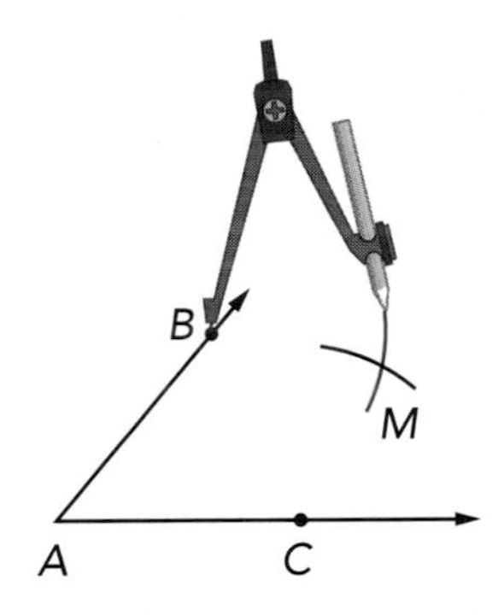

3. Use a straightedge and draw $\overrightarrow{AM}$.

Because point M is the same distance from point C as from point B, $\overrightarrow{AM}$ bisects $\angle BAC$.

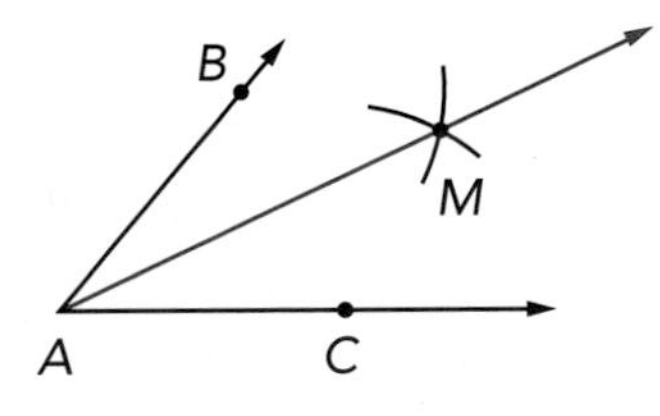

Drawing Solid Figures

474

Drawing Cubes and Rectangular Prisms

475

① Draw two overlapping rectangles with the corresponding sides parallel and congruent.	② Connect corresponding vertices. Use dashed lines for the edges that would be hidden from view.

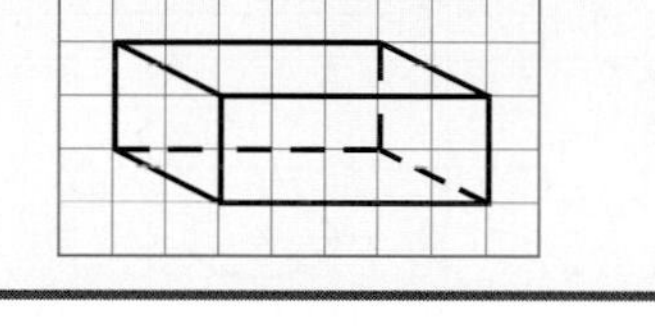

Drawing Pyramids

476

① Draw a square standing on one corner. Draw a point that is above the square.

② Connect each vertex of the square to the point. Use dashed lines for the edges that would be hidden from view.

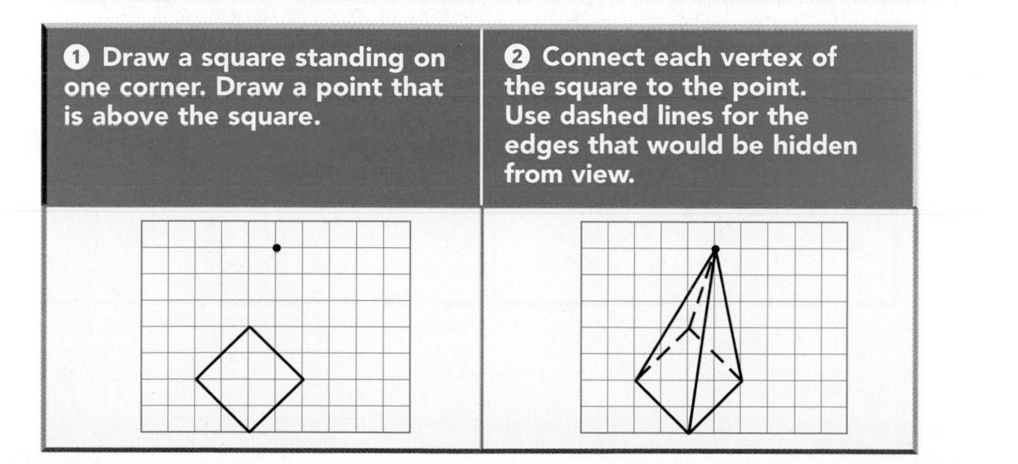

Drawing Cylinders

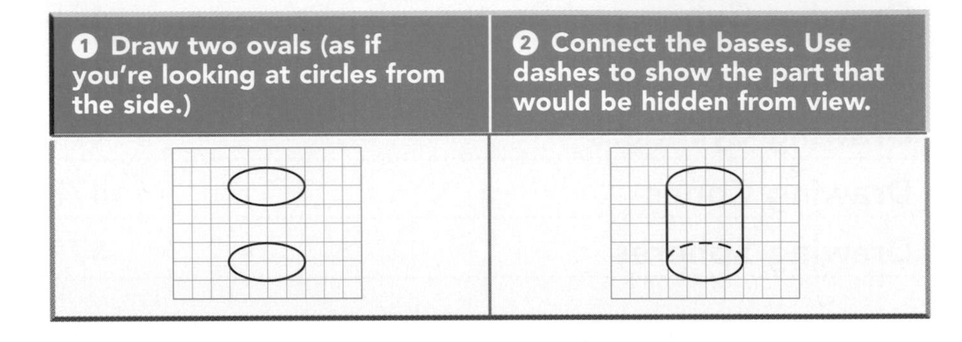

478 Drawing Cones

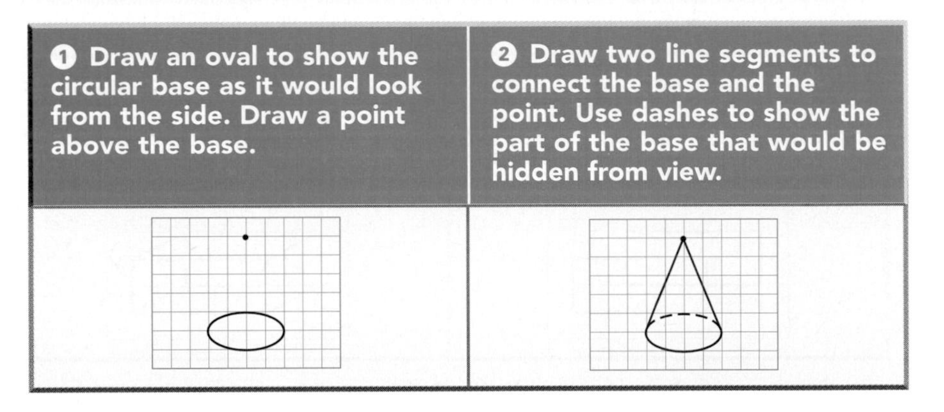

479 Drawing Spheres

Handy Tables

Addition/Subtraction Table

+	0	1	2	3	4	5	6	7	8	9
0	0	1	2	3	4	5	6	7	8	9
1	1	2	3	4	5	6	7	8	9	10
2	2	3	4	5	6	7	8	9	10	11
3	3	4	5	6	7	8	9	10	11	12
4	4	5	6	7	8	9	10	11	12	13
5	5	6	7	8	9	10	11	12	13	14
6	6	7	8	9	10	11	12	13	14	15
7	7	8	9	10	11	12	13	14	15	16
8	8	9	10	11	12	13	14	15	16	17
9	9	10	11	12	13	14	15	16	17	18

- You can use the table to add.

  Here is how you can use the table to add 4 and 7. Look along the top row and find 4. The sum will be in this column. Now look down the first column and find 7. Move right across the row for 7.
  The square where the column for 4 and the row for 7 meet is the sum: $4 + 7 = 11$.

- You can use the table to subtact.

  To subtract $11 - 4$, look along the top row and find 4. Now look down this column and find the box with 11. Follow the row with 11 to the left-most column. The number there is the difference: $11 - 4 = 7$.

Multiplication/Division Table

×	1	2	3	4	5	6	7	8	9	10
1	1	2	3	4	5	6	7	8	9	10
2	2	4	6	8	10	12	14	16	18	20
3	3	6	9	12	15	18	21	24	27	30
4	4	8	12	16	20	24	28	32	36	40
5	5	10	15	20	25	30	35	40	45	50
6	6	12	18	24	30	36	42	48	54	60
7	7	14	21	28	35	42	49	56	63	70
8	8	16	24	32	40	48	56	64	72	80
9	9	18	27	36	45	54	63	72	81	90
10	10	20	30	40	50	60	70	80	90	100

- You can use the table to multiply.

  Here is how you can use the table to multiply 6 by 8. Look along the top row and find 6. The product will be in this column. Now look down the first column and find 8. Move right across the row for 8. The square where the column for 6 and the row for 8 meet is the product: $6 \times 8 = 48$.

- The multiplication table can also be used for division.

  Here is how you can use the table to find $48 \div 6$. Look along the top row and find 6. Now look down this column and find the box with 48. Follow the row with 48 to the left-most column. The number there is the quotient: $48 \div 6 = 8$.

Fraction/Decimal Equivalents

Fraction	Decimal	Fraction	Decimal
$\frac{1}{2}$	0.5	$\frac{1}{10}$	0.1
$\frac{1}{3}$	$0.\overline{3}$	$\frac{2}{10}$	0.2
$\frac{2}{3}$	$0.\overline{6}$	$\frac{3}{10}$	0.3
$\frac{1}{4}$	0.25	$\frac{4}{10}$	0.4
$\frac{2}{4}$	0.5	$\frac{5}{10}$	0.5
$\frac{3}{4}$	0.75	$\frac{6}{10}$	0.6
$\frac{1}{5}$	0.2	$\frac{7}{10}$	0.7
$\frac{2}{5}$	0.4	$\frac{8}{10}$	0.8
$\frac{3}{5}$	0.6	$\frac{9}{10}$	0.9
$\frac{4}{5}$	0.8	$\frac{1}{16}$	0.0625
$\frac{1}{8}$	0.125	$\frac{2}{16}$	0.125
$\frac{2}{8}$	0.25	$\frac{3}{16}$	0.1875
$\frac{3}{8}$	0.375	$\frac{4}{16}$	0.25
$\frac{4}{8}$	0.5	$\frac{5}{16}$	0.3125
$\frac{5}{8}$	0.625	$\frac{6}{16}$	0.375
$\frac{6}{8}$	0.75	$\frac{7}{16}$	0.4375
$\frac{7}{8}$	0.875	$\frac{8}{16}$	0.5
		$\frac{9}{16}$	0.5625
		$\frac{10}{16}$	0.625
		$\frac{11}{16}$	0.6875
		$\frac{12}{16}$	0.75
		$\frac{13}{16}$	0.8125
		$\frac{14}{16}$	0.875
		$\frac{15}{16}$	0.9375

EQUIVALENT FORMS

1

$\frac{1}{2}$ or 0.50 | $\frac{1}{2}$ or 0.50

$\frac{1}{3}$ or $0.\overline{3}$ | $\frac{1}{3}$ or $0.\overline{3}$ | $\frac{1}{3}$ or $0.\overline{3}$

$\frac{1}{4}$ or 0.25 | $\frac{1}{4}$ or 0.25 | $\frac{1}{4}$ or 0.25 | $\frac{1}{4}$ or 0.25

$\frac{1}{5}$ or 0.2 | $\frac{1}{5}$ or 0.2 | $\frac{1}{5}$ or 0.2 | $\frac{1}{5}$ or 0.2 | $\frac{1}{5}$ or 0.2

$\frac{1}{8}$ or 0.125 | $\frac{1}{8}$ or 0.125 | $\frac{1}{8}$ or 0.125 | $\frac{1}{8}$ or 0.125 | $\frac{1}{8}$ or 0.125 | $\frac{1}{8}$ or 0.125 | $\frac{1}{8}$ or 0.125 | $\frac{1}{8}$ or 0.125

$\frac{1}{10}$ or 0.1 | $\frac{1}{10}$ or 0.1 | $\frac{1}{10}$ or 0.1 | $\frac{1}{10}$ or 0.1 | $\frac{1}{10}$ or 0.1 | $\frac{1}{10}$ or 0.1 | $\frac{1}{10}$ or 0.1 | $\frac{1}{10}$ or 0.1 | $\frac{1}{10}$ or 0.1 | $\frac{1}{10}$ or 0.1

$\frac{1}{16}$ or 0.0625 | $\frac{1}{16}$ or 0.0625 | $\frac{1}{16}$ or 0.0625 | $\frac{1}{16}$ or 0.0625 | $\frac{1}{16}$ or 0.0625 | $\frac{1}{16}$ or 0.0625 | $\frac{1}{16}$ or 0.0625 | $\frac{1}{16}$ or 0.0625 | $\frac{1}{16}$ or 0.0625 | $\frac{1}{16}$ or 0.0625 | $\frac{1}{16}$ or 0.0625 | $\frac{1}{16}$ or 0.0625 | $\frac{1}{16}$ or 0.0625 | $\frac{1}{16}$ or 0.0625 | $\frac{1}{16}$ or 0.0625 | $\frac{1}{16}$ or 0.0625

484

Squares and Roots

n	n^2	$\sqrt{n}$	n	n^2	$\sqrt{n}$
1	1	1.000	51	2601	7.141
2	4	1.414	52	2704	7.211
3	9	1.732	53	2809	7.280
4	16	2.000	54	2916	7.348
5	25	2.236	55	3025	7.416
6	36	2.449	56	3136	7.483
7	49	2.646	57	3249	7.550
8	64	2.828	58	3364	7.616
9	81	3.000	59	3481	7.681
10	100	3.162	60	3600	7.746
11	121	3.317	61	3721	7.810
12	144	3.464	62	3844	7.874
13	169	3.606	63	3969	7.937
14	196	3.742	64	4096	8.000
15	225	3.873	65	4225	8.062
16	256	4.000	66	4356	8.124
17	289	4.123	67	4489	8.185
18	324	4.243	68	4624	8.246
19	361	4.359	69	4761	8.307
20	400	4.472	70	4900	8.367
21	441	4.583	71	5041	8.426
22	484	4.690	72	5184	8.485
23	529	4.796	73	5329	8.544
24	576	4.899	74	5476	8.602
25	625	5.000	75	5625	8.660
26	676	5.099	76	5776	8.718
27	729	5.196	77	5929	8.775
28	784	5.292	78	6084	8.832
29	841	5.385	79	6241	8.888
30	900	5.477	80	6400	8.944
31	961	5.568	81	6561	9.000
32	1024	5.657	82	6724	9.055
33	1089	5.745	83	6889	9.110
34	1156	5.831	84	7056	9.165
35	1225	5.916	85	7225	9.220
36	1296	6.000	86	7396	9.274
37	1369	6.083	87	7569	9.327
38	1444	6.164	88	7744	9.381
39	1521	6.245	89	7921	9.434
40	1600	6.325	90	8100	9.487
41	1681	6.403	91	8281	9.539
42	1764	6.481	92	8464	9.592
43	1849	6.557	93	8649	9.644
44	1936	6.633	94	8836	9.695
45	2025	6.708	95	9025	9.747
46	2116	6.782	96	9216	9.798
47	2209	6.856	97	9409	9.849
48	2304	6.928	98	9604	9.899
49	2401	7.000	99	9801	9.950
50	2500	7.071	100	10,000	10.000

(Roots are rounded to the nearest thousandth.)

The Metric System

LINEAR MEASURE		
1 centimeter	0.01 meter	0.3937 inch
1 decimeter	0.1 meter	3.937 inches
1 meter		39.37 inches
1 dekameter	10 meters	32.8 feet
1 hectometer	100 meters	328 feet
1 kilometer	1000 meters	0.621 mile

CAPACITY MEASURE		
1 centiliter	0.01 liter	0.338 fluid ounce
1 deciliter	0.1 liter	3.38 fluid ounces
1 liter		1.056 quarts
1 dekaliter	10 liters	2.642 gallons
1 hectoliter	100 liters	26.42 gallons
1 kiloliter	1000 liters	264.2 gallons

VOLUME MEASURE		
1 cubic centimeter	1000 cubic millimeters	0.06102 cubic inch
1 cubic decimeter	1000 cubic centimeters	61.02 cubic inches
1 cubic meter	1000 cubic decimeters	35.315 cubic feet

MASS		
1 centigram	0.01 gram	0.0003527 ounce
1 decigram	0.1 gram	0.003527 ounce
1 gram	10 decigrams	0.03527 ounce
1 dekagram	10 grams	0.3527 ounce
1 hectogram	100 grams	3.527 ounces
1 kilogram	1000 grams	2.2046 pounds
1 metric ton	1,000,000 grams	2204.6 pounds

LAND MEASURE		
1 centare	1 square meter	1.196 square yards
1 are	100 square meters	119.6 square yards
1 hectare	10,000 square meters	2.471 acres
1 square kilometer	1,000,000 square meters	0.386 square mile

486

The Customary System

LINEAR MEASURE		
1 inch		2.54 centimeters
1 foot	12 inches	0.3048 meter
1 yard	3 feet	0.9144 meter
1 mile	5280 feet	1609.3 meters

SQUARE MEASURE		
1 square foot	144 square inches	929.0304 square centimeters
1 square yard	9 square feet	0.83761 square meter
1 acre	43,560 square feet	4046.86 square meters

DRY MEASURE			
1 pint		33.60 cubic inches	0.5505 liter
1 quart	2 pints	67.20 cubic inches	1.1012 liters
1 peck	16 pints	537.61 cubic inches	8.8096 liters
1 bushel	64 pints	2150.42 cubic inches	35.2383 liters

LIQUID MEASURE			
1 cup	8 fluid ounces	14.438 cubic inches	0.2366 liter
1 pint	16 fluid ounces	28.875 cubic inches	0.4732 liter
1 quart	32 fluid ounces	57.75 cubic inches	0.9463 liter
1 gallon	128 fluid ounces	231 cubic inches	3.7853 liters

WEIGHT (AVOIRDUPOIS)		
1 ounce	0.0625 pound	28.3495 grams
1 pound	16 ounces	453.59 grams
1 ton	2000 pounds	907.18 kilograms

TIME			
60 seconds	1 minute	168 hours	1 week
60 minutes	1 hour	12 months	1 year
24 hours	1 day	52 weeks	1 year
7 days	1 week	365.25 days	1 year

General Measurement

487

BENCHMARK MEASURES

1 inch	≈	tip of your thumb
1 centimeter	≈	width of the tip of your index finger
1 foot	≈	length of your notebook
1 kilogram	≈	mass of your math textbook
1 minute	≈	time it takes to count to 60 saying *one thousand* between each number
1 pound	≈	weight of a loaf of bread
1 ounce	≈	weight of a slice of bread
1 gram	≈	mass of a shoelace

COMMON ABBREVIATIONS

c	cup	mg	milligram
cm	centimeter	mi	mile
d	day	min	minute
dm	decimeter	mL	milliliter
fl oz	fluid ounce	mm	millimeter
ft	foot	mo	month
gal	gallon	oz	ounce
g	gram	pt	pint
h	hour	qt	quart
in.	inch	s	second
kg	kilogram	t	ton
L	liter	wk	week
lb	pound	yd	yard
m	meter	y	year

CONVERSION FACTORS

To change	to	multiply by	To change	to	multiply by
centimeters	inches	0.3937	kilometers	miles	0.6214
centimeters	feet	0.03281	liters	gallons	0.2642
feet	meters	0.3048	meters	feet	3.2808
feet	miles	0.0001894	meters	yards	1.0936
gallons	liters	3.7853	miles	kilometers	1.6093
grams	ounces	0.0353	ounces	grams	28.3495
grams	pounds	0.002205	ounces	pounds	0.0625
hours	days	0.04167	pounds	kilograms	0.3782
inches	centimeters	2.54	pounds	ounces	16
kilograms	pounds	2.2046	yards	meters	0.9144

488

Maps

489

How to Read a Map

On most maps, north is at the top. But you should always check the **compass rose** or **directional finder** to make sure. If there is no symbol, you can assume that north is at the top.

compass rose

Most important marks and symbols used in a map are explained in a legend or key. The **legend** is usually enclosed in a box. It can include symbols for capital cities, for boundary lines, for rivers, or for different types of roads.

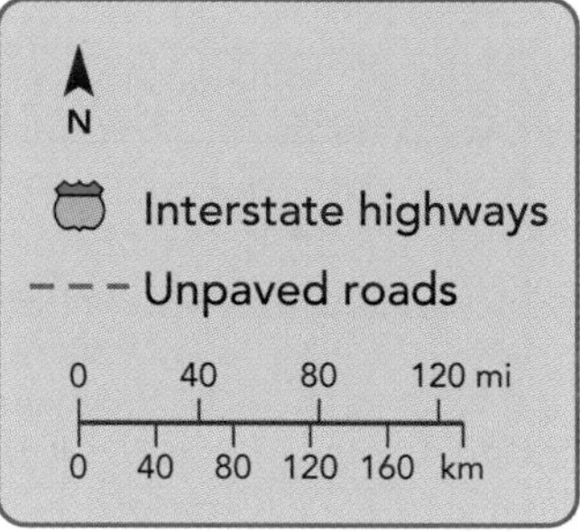

The scale helps you figure out how far it really is between places. For example, a scale might show that one inch on the map equals fifty miles. Map scales differ from map to map. Always check the scale on the map you are using, if you want to know the real distances.

490

Latitude and Longitude

Latitude and longitude refer to imaginary lines that mapmakers use. Both are measured in degrees. The imaginary lines that go from east to west around the earth are called lines of **latitude**. Latitude numbers may be printed along the left- and right-hand sides of a map.

The imaginary lines that run from the North Pole to the South Pole are lines of **longitude**. Longitude numbers may be printed at the top and bottom of a map.

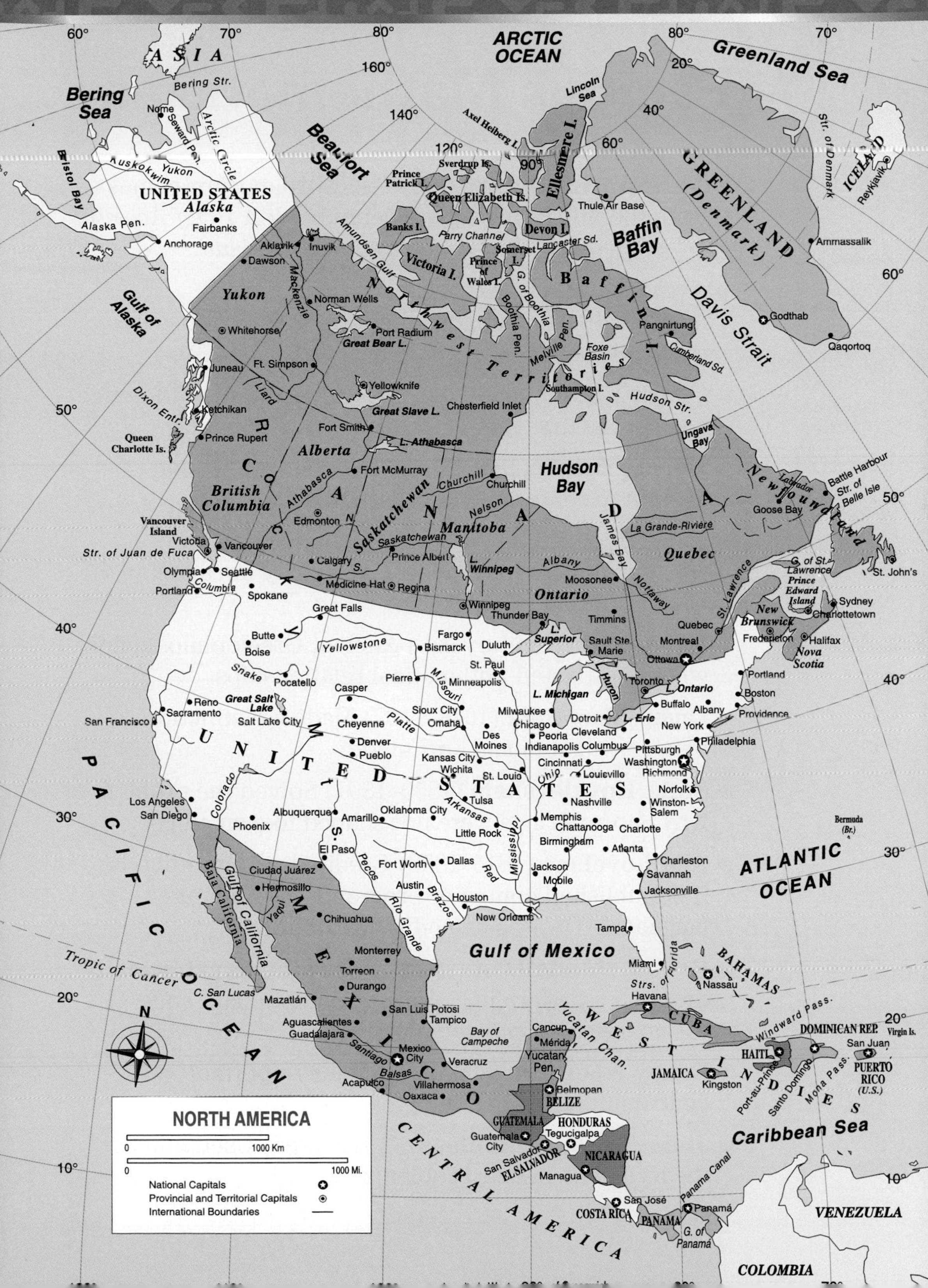
NORTH AMERICA
0 1000 Km
0 1000 Mi.
National Capitals
Provincial and Territorial Capitals
International Boundaries
ARCTIC OCEAN
Greenland Sea
Lincoln Sea
Bering Sea
Bering Str.
ASIA
Beaufort Sea
Baffin Bay
Davis Strait
GREENLAND (Denmark)
ICELAND
Reykjavik
Str. of Denmark
Ammassalik
Godthab
Qaqortoq
Thule Air Base
Axel Heiberg I.
Ellesmere I.
Sverdrup Is.
Prince Patrick I.
Queen Elizabeth Is.
Devon I.
Banks I.
Parry Channel
Lancaster Sd.
Victoria I.
Prince of Wales I.
Somerset I.
Amundsen Gulf
G. of Boothia
Boothia Pen.
Melville Pen.
Foxe Basin
Baffin I.
Pangnirtung
Cumberland Sd.
Southampton I.
Hudson Str.
Ungava Bay
Hudson Bay
James Bay
UNITED STATES
Alaska
Nome
Seward Pen.
Arctic Circle
Yukon
Kuskokwim
Bristol Bay
Alaska Pen.
Fairbanks
Anchorage
Gulf of Alaska
Juneau
Ketchikan
Dixon Entr.
Queen Charlotte Is.
Prince Rupert
Yukon
Aklavik
Inuvik
Dawson
Mackenzie
Norman Wells
Whitehorse
Port Radium
Great Bear L.
Northwest Territories
Ft. Simpson
Liard
Yellowknife
Great Slave L.
Chesterfield Inlet
Fort Smith
L. Athabasca
Alberta
Fort McMurray
Athabasca
Churchill
British Columbia
Edmonton
Saskatchewan
Manitoba
Nelson
Vancouver Island
Victoria
Vancouver
Str. of Juan de Fuca
Calgary
Prince Albert
L. Winnipeg
Albany
Medicine Hat
Regina
Winnipeg
Ontario
Moosonee
Nottaway
La Grande-Rivière
Quebec
St. Lawrence
Labrador
Newfoundland
Battle Harbour
Str. of Belle Isle
Goose Bay
G. of St. Lawrence
St. John's
Prince Edward Island
Sydney
Charlottetown
New Brunswick
Fredericton
Halifax
Nova Scotia
CANADA
ROCKY MTS.
Thunder Bay
Timmins
L. Superior
Sault Ste. Marie
Quebec
Montreal
Ottawa
Olympia
Seattle
Portland
Columbia
Spokane
Great Falls
Butte
Boise
Yellowstone
Fargo
Bismarck
Duluth
St. Paul
Minneapolis
Snake
Pocatello
Pierre
Missouri
Casper
L. Huron
Toronto
L. Ontario
L. Michigan
Portland
Boston
Reno
Great Salt Lake
Sacramento
San Francisco
Salt Lake City
Cheyenne
Platte
Sioux City
Omaha
Milwaukee
Chicago
Detroit
L. Erie
Buffalo
Albany
Providence
New York
Philadelphia
Des Moines
Peoria
Cleveland
Denver
Pueblo
Indianapolis
Columbus
Pittsburgh
Kansas City
Wichita
Cincinnati
Washington
Louisville
Richmond
St. Louis
Ohio
Norfolk
UNITED STATES
Colorado
Los Angeles
San Diego
Phoenix
Albuquerque
Amarillo
Oklahoma City
Tulsa
Arkansas
Nashville
Winston-Salem
Memphis
Chattanooga
Charlotte
Little Rock
Birmingham
Atlanta
Charleston
Savannah
Jacksonville
El Paso
Pecos
Fort Worth
Dallas
Red
Mississippi
Jackson
Mobile
Austin
Brazos
Houston
New Orleans
Rio Grande
Tampa
Miami
Bermuda (Br.)
ATLANTIC OCEAN
PACIFIC OCEAN
Ciudad Juárez
Hermosillo
Baja California
Gulf of California
Yaqui
Chihuahua
Monterrey
Torreon
Durango
Tropic of Cancer
C. San Lucas
Mazatlán
San Luis Potosi
Tampico
Aguascalientes
Guadalajara
Santiago
Mexico City
Veracruz
Balsas
Acapulco
Villahermosa
Oaxaca
MEXICO
Gulf of Mexico
Bay of Campeche
Cancun
Mérida
Yucatan Pen.
Yucatan Chan.
Strs. of Florida
BAHAMAS
Nassau
Havana
CUBA
WEST INDIES
Windward Pass.
DOMINICAN REP.
Virgin Is.
HAITI
Port-au-Prince
Santo Domingo
Mona Pass.
San Juan
PUERTO RICO (U.S.)
JAMAICA
Kingston
Caribbean Sea
Belmopan
BELIZE
GUATEMALA
Guatemala City
HONDURAS
Tegucigalpa
San Salvador
EL SALVADOR
NICARAGUA
Managua
San José
COSTA RICA
PANAMA
Panamá
Panama Canal
G. of Panama
CENTRAL AMERICA
VENEZUELA
COLOMBIA
N
60°
70°
80°
160°
140°
120°
90°
60°
40°
20°
80°
70°
60°
50°
40°
30°
20°
10°

491 Famous Mathematicians

- **Al-Khwarizmi** (c.680–750) The word *algebra* comes from the title of his early algebra book *A Brief Account of the Methods of al-Jabar and al-Muqabala.*
- **Archimedes** (287?–212 B.C.) Ancient Greek mathematician who made important discoveries in mechanics.
- **Charles Babbage** (1792–1871) Invented the difference engine: the first general-purpose computer.
- **George Boole** (1815–1864) Invented a new kind of algebra to represent logical statements.
- **Rene Descartes** (1596–1650) Remade geometry. He made modern geometry possible by introducing the coordinate plane.
- **Euclid** (third century B.C.) Ancient Greek philosopher who investigated arithmetical relationships.
- **Omar Khay Yám** (1048–1122) Made a distinction between algebra and arithmetic.
- **Sofia Kovalevskaia** (1850–1891) Solved the problem of why Saturn's rings were not shaped like an ellipse.
- **Ada Lovelace** (1815–1852) Credited with writing the first published computer program. The computer language ADA was named in her honor.
- **Benoit Mandelbrot** (1924–) Used the term fractal to describe a pattern in clumps of random data and "discovered" fractal geometry.
- **Emmy Noether** (1882–1935) Made important contributions to ring theory.
- **Julia Bowman Robinson** (1919–1985) Used number theory to solve logic problems.

Math Topics for Science Fairs

Research Topics

- Fibonacci numbers in nature
- Number systems for computers
- Different cultures' approximations of pi
- The golden ratio
- Patterns in finger prints
- Carbon dating and half-life
- Mathematics in simple machines
- The mathematics of soap bubbles
- Really big numbers (such as a googol) or really small numbers and their applications
- Symmetry in nature

Things to Make or Do

- Make a Moebius strip and investigate applications in everyday life
- Make a straw balance scale; devise weights and discuss accuracy
- Make a galvanometer with a lemon and investigate how to calibrate your galvanometer
- Make a light meter; use distance to measure brightness
- Make a clock with a pendulum
- Prove that air has weight
- Measure the size of raindrops
- Make a weather station and explain how to read the instruments
- Use toothpicks and modeling clay to build and investigate 3-dimensional figures

493

Number Patterns

494 Triangular Numbers

The pattern of dots below shows how to make the first five triangular numbers. They are called **triangular numbers** because they name triangular-shaped pictures of dots.

Add the number of dots in the new row to the total number of dots in the previous triangle to get the next trianglar number.

495 Square Numbers

The pattern of dots below shows how to make the first five square numbers. They are called **square numbers** because they name square-shaped pictures of dots.

To find other square numbers, multiply the number of the square by itself.

Pascal's Triangle

Blaise Pascal was a French mathematician, philosopher, and scientist who lived from 1623 to 1662. Pascal, together with another French mathematician, Pierre de Fermat (1601–1665), studied the theory of probability and explored its applications.

This pattern is called Pascal's Triangle because he studied it. It may have been invented at least 500 years before that by a Chinese mathematician and by Omar Khay Yám. Over the years, many patterns have been found within the triangle. The pattern itself can be used to find probabilities and to solve algebra problems.

Row 0 1
Row 1 1 1
Row 2 1 2 1
Row 3 1 3 3 1
Row 4 1 4 6 4 1
Row 5 1 5 10 10 5 1
Row 6 1 6 15 20 15 6 1
Row 7 1 7 21 35 35 21 7 1
Row 8 1 8 28 56 70 56 28 8 1
Row 9 1 9 36 84 126 126 84 36 9 1

To find the numbers in a row, look at the row above. The sum of two side-by-side numbers goes between them in the row below.

497 Fibonacci Numbers

Leonardo Fibonacci, an Italian mathematician who lived from about 1180 to about 1250, found this pattern. Mathematicians are still finding things in nature that can be described with this series of numbers.

0, 1, 1, 2, 3, 5, 8, 13, 21, 34, 55, 89, 144, 233, 377, 610, . . .

Each number in the series is the sum of the two numbers before it.

498 The Golden Ratio

Artists often use the golden ratio because it produces shapes that are pleasing to the eye. One of the most famous buildings of ancient Greece, the Parthenon, was designed using the golden ratio.

An average person's total height compared to waist height is the golden ratio.

Golden ratio ≈ 1.618

The golden ratio appears in surprising places, even in the Fibonacci sequence. If you divide the third number in the Fibonacci sequence by the second, then the fourth by the third, then the fifth by the fourth, and so on, the quotient gets closer and closer to the golden ratio.

$2 \div 1 = 2.0$
$3 \div 2 = 1.5$
$5 \div 3 = 1.\overline{6}$
$8 \div 5 = 1.6$
$13 \div 8 = 1.625$
$21 \div 13 \approx 1.61538$

Number Systems

Egyptian Numerals

In ancient Egypt, over 5000 years ago, the Egyptians used these numerals:

|	1
||	2
|||	3
||||	4
|||||	5
||||||	6
|||||||	7
||||||||	8
|||||||||	9
𓎆	10
𓎆|	11
𓍢	100
𓆼	1000
𓂭	10,000

The Egyptians did not use place value. So, the number we write as 1999 would have been written like this:

𓆼 𓍢𓍢𓍢𓍢𓍢𓍢𓍢𓍢𓍢 𓎆𓎆𓎆𓎆𓎆𓎆𓎆𓎆𓎆 |||||||||

501 Babylonian Numerals

At about the same time as the ancient Egyptians, the ancient peoples of Babylonia used different numerals.

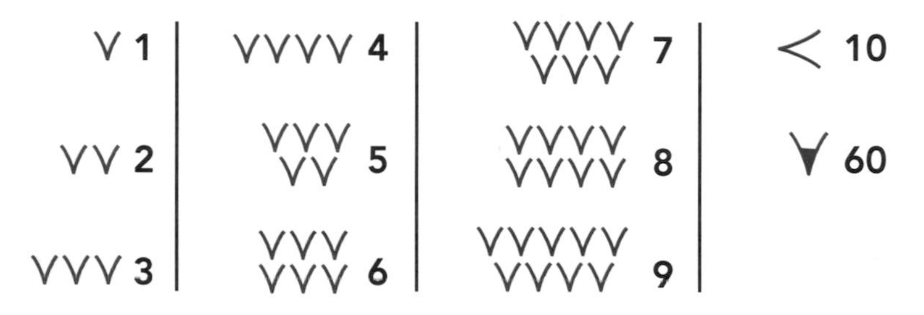

Like the Egyptians, the Babylonians did not use place value. So, the number we write as 75 would have been written like this:

502 Roman Numerals

The Romans used numerals that are still used today.

I	1	VIII	8	XXV	25	D	500
II	2	IX	9	XL	40	M	1,000
III	3	X	10	XLV	45	$\overline{\text{V}}$	5,000
IV	4	XI	11	L	50	$\overline{\text{X}}$	10,000
V	5	XII	12	LX	60	$\overline{\text{L}}$	50,000
VI	6	XIX	19	XC	90	$\overline{\text{C}}$	100,000
VII	7	XX	20	C	100	$\overline{\text{D}}$	500,000

The Romans didn't use place value. But they did use addition and subtraction in their number system. The symbol for 12 is XII $(10 + 2)$. When a symbol for a smaller number is to the left of another symbol, it represents a difference. So, IV is $5 - 1 = 4$.

Here is how you would write our number **1999** using Roman numerals: **MCMXCIX**

Mayan Numerals 503

The Maya Indians of Central America had a number system based on 20. They used concepts of place value and of zero.

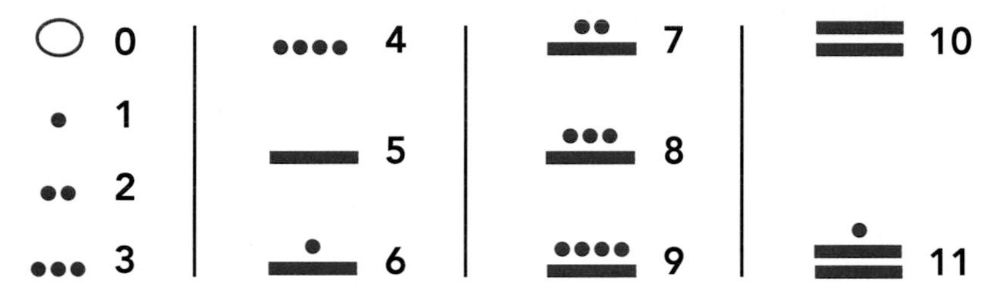

So, the number we write as 19 would look like this:

Indian Numerals 504

Before the Romans used letters for numerals, the Hindus in India used nine number symbols with which they could write any number. These numerals have a place value like our numerals.

1 | 4 | 7
2 | 5 | 8
3 | 6 | 9

Early in their history the Hindus would write ३ ३ for 303.
Later, they invented a zero and wrote ३ • ३.

505 Arabic Numerals

The Arabs adapted the ideas of the Hindus and made some changes. Although they have changed several times, the numerals we use today are based on these.

0	1	2	3	4
5	6	7	8	9

506 Binary Numbers

Computers are electronic devices. They can only read whether a microcircuit is ON or OFF. ON can be represented by the digit 1 and OFF by 0. This is a **binary**, or **base-two**, number system.

In the binary system, each time you move one place to the left, the place value doubles.

Decimal	Binary	Computer
1	1	ON
2	10	ON-OFF
3	11	ON-ON
4	100	ON-OFF-OFF
5	101	ON-OFF-ON
6	110	ON-ON-OFF
7	111	ON-ON-ON
8	1000	ON-OFF-OFF-OFF
9	1001	ON-OFF-OFF-ON
10	1010	ON-OFF-ON-OFF

128	64	32	16	8	4	2	1
2^7	2^6	2^5	2^4	2^3	2^2	2^1	2^0

So: $20_{ten} = 10100_{two}$
$40_{ten} = 101000_{two}$
$75_{ten} = 1001011_{two}$
$100_{ten} = 1100100_{two}$

The "$_{two}$" means the number is written in the binary system (base two). The "$_{ten}$" means the number is in base ten. If you see a number with no base listed, assume it's base ten.

Yellow Pages

The numbers in black at the end of most entries refer you back to topic numbers, not page numbers. You will find topic numbers at the top of each page and next to each new piece of information in *Math at Hand*.

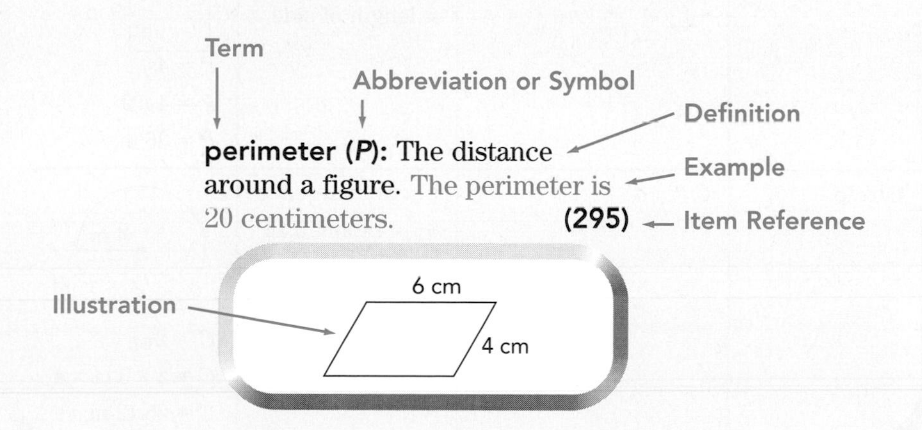

Glossary of Mathematical Formulas

Perimeter: The distance around a plane figure.

MORE HELP
See 295–298

Figure	Formula	Variables	Example
Polygon	$P = s_1 + s_2 + s_3 + \ldots$ Triangle: $P = s_1 + s_2 + s_3$ Quadrilateral: $P = s_1 + s_2 + s_3 + s_4$ Parallelogram: $P = s_1 + s_2 + s_3 + s_4$ Pentagon: $P = s_1 + s_2 + s_3 + s_4 + s_5$	P: Perimeter s_1: length of one side s_2: length of another side s_3: length of third side … and so on	6 cm, 3cm, 3 cm, 5 cm $P = s_1 + s_2 + s_3 + s_4$ $P = 5 + 3 + 3 + 6$ $P = 17$ cm
Rectangle or Parallelogram	$P = 2l + 2w$	P: Perimeter l: length w: width	3 ft, 5 ft $P = 2l + 2w$ $P = (2 \times 5) + (2 \times 3)$ $P = 16$ ft
Square	$P = 4s$	P: Perimeter s: length of side	9 in. $P = 4s$ $P = 4 \times 9$ $P = 36$ in.
Circle	$C = \pi d$ or $C = 2\pi r$	C: circumference π: pi (about 3.14 or $\frac{22}{7}$) d: diameter r: radius	4 m $C = 2\pi r$ $C \approx 2 \times 3.14 \times 4$ $C \approx 25.12$ m

Area: The number of square units a figure contains.

Figure	Formula	Variables	Example
Triangle	$A = \frac{1}{2}bh$	A: Area b: length of base h: height **The height is the perpendicular line segment from the base to the opposite vertex. You can choose any side as the base as long as the height is perpendicular to that side.**	h = 4 mm, b = 6 mm, 5 mm, 5 mm $A = \frac{1}{2}bh$ $A = \frac{1}{2} \times 6 \times 4$ $A = 12 \text{ mm}^2$
Parallelogram	$A = bh$	A: Area b: length of base h: height **You can choose any side as the base, as long as you use the height that is perpendicular to that side.**	4 cm, 5 cm, 11 cm $A = bh$ $A = 11 \times 4$ $A = 44 \text{ cm}^2$

MORE HELP

See 299–305

Figure	Formula	Variables	Example
Rectangle	$A = lw$	A: Area l: length w: width	3 m 5 m $A = lw$ $A = 5 \times 3$ $A = 15\ m^2$
Square	$A = s^2$	A: Area s: length of a side	7 m $A = s^2$ $A = 7 \times 7$ $A = 49\ m^2$
Trapezoid	$A = \frac{1}{2} \times h \times (b_1 + b_2)$	A: Area h: height b_1: length of one base b_2: length of other base **The bases are the 2 parallel sides and the height is perpendicular to both.**	8 ft 7 ft 6 ft 8 ft 16 ft $A = \frac{1}{2} \times h \times (b_1 + b_2)$ $A = \frac{1}{2} \times 6 \times (8 + 16)$ $A = 72\ ft^2$
Circle	$A = \pi r^2$	A: Area π: pi (about 3.14 or $\frac{22}{7}$) r: radius	7 in. $A = \pi r^2$ $A \approx \frac{22}{7} \times (7)^2$ $A \approx \frac{22}{7} \times 49$ $A \approx 154\ in.^2$

Surface Area: The total area of the faces (including the bases) of a solid figure.

Figure	Formula	Variables	Example
Rectangular prism	$SA = 2lw + 2lh + 2wh$ or $2 \times (lw + lh + wh)$	SA: Surface Area l: length w: width h: height	3 cm, 8 cm, 12 cm $SA = (2 \times 12 \times 8) + (2 \times 12 \times 3) + (2 \times 8 \times 3)$ $SA = (192 + 72 + 48)$ $SA = 312 \text{ cm}^2$
Cube	$SA = 6s^2$	SA: Surface Area s: length of side	3 m, 3 m, 3 m $SA = 6 \times 3^2$ $SA = 6 \times 9$ $SA = 54 \text{ m}^2$
Cylinder	$SA = 2\pi r^2 + 2\pi rh$ **$2\pi r^2$ is the area of the 2 circular bases.** **$2\pi r$ is the circumference of the circular base.**	SA: Surface Area π: pi (about 3.14 or $\frac{22}{7}$) r: radius h: height	3 cm, 8 cm $SA \approx (2 \times 3.14 \times 3^2) + (2 \times 3.14 \times 3 \times 8)$ $SA \approx 56.52 + 150.72$ $SA \approx 207.24 \text{ cm}^2$
Sphere	$A = 4\pi r^2$	SA: Surface Area π: pi (about 3.14 or $\frac{22}{7}$) r: radius	5 ft $SA \approx 4 \times 3.14 \times 5^2$ $SA \approx 4 \times 3.14 \times 25$ $SA \approx 314 \text{ ft}^2$

MORE HELP

See 306–308

Volume: The amount of space inside a solid figure. Volume is measured in cubic units.

Figure	Formula	Variables	Example
Rectangular prism	$V = lwh$	V: Volume l: length w: width h: height	4 cm, 3 cm, 8 cm $V = lwh$ $V = 8 \times 3 \times 4$ $V = 96 \text{ cm}^3$
Cube	$V = s^3$	V: volume s: length of side	5 m, 5 m, 5 m $V = s^3$ $V = 5^3$ $V = 125 \text{ m}^3$

MORE HELP

See 309–312

Statistics: Collecting, representing, summarizing, comparing, and interpreting data.

Statistic	Formula	Example
Mean	$\frac{\text{sum of numbers}}{\text{number of numbers}}$	Find the mean test score: 85, 92, 84, 89, 95 $(85 + 92 + 84 + 89 + 95) \div 5$ Mean = 89

MORE HELP

See 255–263

Statistics (continued)

Statistic	Formula	Example
Median	**(a)** middle piece of data of a set of data when numbers are arranged from least to greatest	Find the median daily high temperature: **(a)** 38°C, 24°C, 34°C, 29°C, 26°C Order the data: 24, 26, <u>29</u>, 34, 38 Median is the middle piece of data: 29°C
	(b) When a set has two middle pieces of data, the median is halfway between them.	**(b)** 41°C, 35°C, 46°C, 41°C, 38°C, 39°C Order the data: 35, 38, <u>39</u>, <u>41</u>, 41, 46 Median is halfway between middle two: $(39 + 41) \div 2 = 80 \div 2$ Median = 40°C
Mode	piece of data that appears most frequently in a set of data	Find the mode of the temperatures: **(a)** 56°F, 64°F, 48°F, 56°F, 63°F Mode: 56°F **(b)** 72°F, 68°F, 80°F, 68°F, 75°F, 80°F Mode: 68°F and 80°F (called bimodal) **(c)** 92°F, 89°F, 85°F, 94°F Mode: none
Range	difference between the greatest and least values in a set of data	Find the range of the temperatures: 86°F, 72°F, 88°F, 80°F Range = high − low $88 - 72 = 16$ Range = 16°F

Glossary of Mathematical Symbols

Symbol	Meaning	Example
$+$	plus (addition)	$6 + 7 = 13$
$^{+}$	positive	$^{+}3$: the integer 3 units to the right of zero on a number line
$-$	minus (subtraction)	$15 - 7 = 8$
$^{-}$	negative	$^{-}6$: the integer 6 units to the left of zero on a number line
$\times$, $\cdot$, $a(b)$	multiplied by or times	$4 \times 5 = 20$; $8 \cdot 3 = 24$; $3(4) = 12$
$\div$ or $\overline{)}$	divided by	$4 \div 2 = 2$ $\quad \begin{array}{r} 2 \\ 2\overline{)4} \end{array}$
$=$	is equal to	$3 + 2 = 5$
$\neq$	is not equal to	$8 - 5 \neq 8$
$\cong$	is congruent to	In an equilateral triangle, all of the sides are congruent: side $AB \cong$ side $BC \cong$ side CA 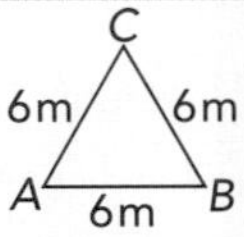
$\sim$	is similar to	$\Delta ABC \sim \Delta DEF$ 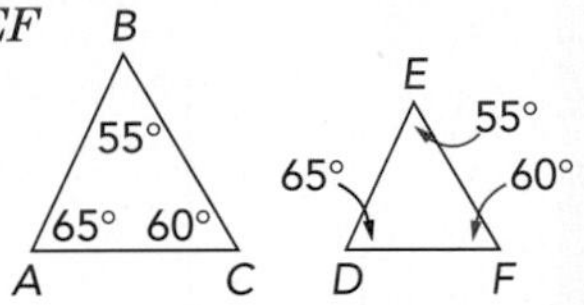
$\approx$	is approximately equal to	$\pi \approx 3.14$
$<$	is less than	$7 + 6 < 15$
$\leq$	is less than or equal to	$3 \leq 4$
$>$	is greater than	$8 > 2$
$\geq$	is greater than or equal to	$6 \geq 6$
( )	parentheses: used as grouping symbols	$(3 + 4) - (3 - 1) \rightarrow$ $7 - 2 = 5$
%	percent	50%: 50 percent
¢	cents	35¢: 35 cents
$	dollars	$5.25; say: *5 dollars and 25 cents*
°	degree(s)	360°
°F	degrees Fahrenheit	60°F
°C	degrees Celsius	36°C

Symbol	Meaning	Example
'	minute(s)	4': 4 minutes
'	foot (or feet)	8': 8 feet
"	second(s)	35": 35 seconds
"	inch(es)	9": 9 inches
:	is to (ratio)	4:3; *4 is to 3*
π	the irrational number pi	usually use $\pi \approx 3.14$ or $\pi \approx \frac{22}{7}$
$\lvert \ \rvert$	absolute value	$\lvert ^{-}3 \rvert = 3$ and $\lvert 3 \rvert = 3$
$\sqrt{\ }$	square root	$\sqrt{16} = 4$
$2.\overline{3}$	repeating decimal	$2.\overline{3} = 2.333333\ldots$
$\angle$	angle	$\angle S$ 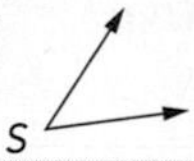 (S)
Δ	triangle	ΔQRS (S, Q, R) 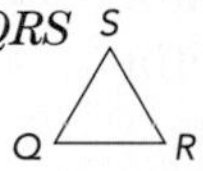
$\overleftrightarrow{AB}$	line *AB*	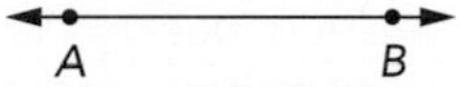 (A, B)
$\overline{JK}$	line segment *JK*	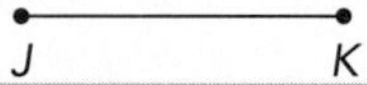 (J, K)
$\overrightarrow{LM}$	ray *LM*	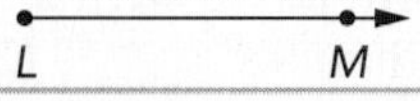 (L, M)
$\overset{\frown}{QR}$	arc *QR*	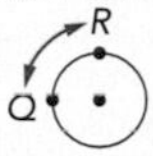 (R, Q)
∟	right angle	$\angle ABC$ is a right angle. 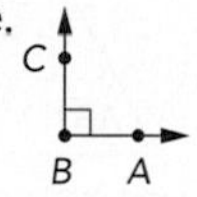 (C, B, A)
$\perp$	is perpendicular to	$\overline{AB} \perp \overline{CD}$ 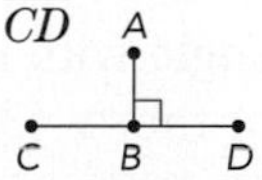 (A, C, B, D)
$\parallel$	is parallel to	$\overline{AB} \parallel \overline{CD}$ 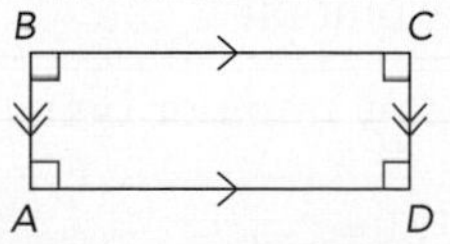 (B, C, A, D)
!	factorial	$5! = 5 \times 4 \times 3 \times 2 \times 1 = 120$
a^0	1	$3^0 = 17^0 = 1$
a^n	n is the number of times a is a factor	$10^3 = 10 \times 10 \times 10 = 1000$
∞	infinity	

Glossary of Mathematical Terms

A

abacus: A device for computation, usually a frame that allows beads to slide along rods that have different place values.

abscissa: *See x-coordinate*

absolute value (||): The distance of a number from zero on the number line. Always positive. $|^{-}4| = 4$; $|4| = 4$

abundant number: A number whose factors (except for the number itself) have a sum greater than the number. 20 is an abundant number because $1 + 2 + 4 + 5 + 10 > 20$.

acute angle: An angle with a measure less than 90°. **(347)**

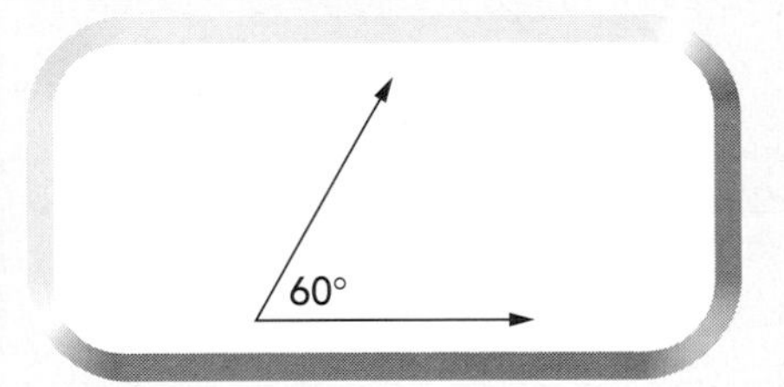

acute triangle: A triangle with no angle measuring 90° or more. **(361)**

add (+): Combine. **(118)**

addend: Any number being added. **(118)**

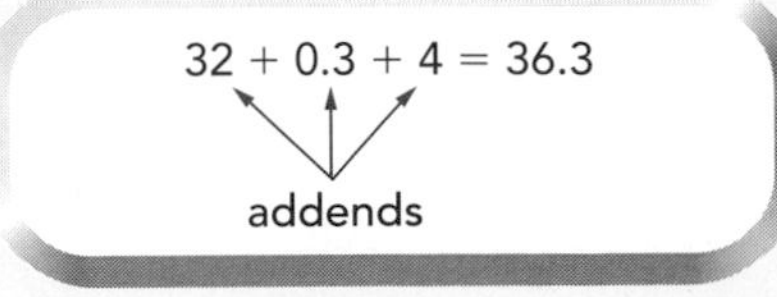

adjacent: Next to. $\angle A$ and $\angle B$ are adjacent angles. $\angle A$ and $\angle C$ are not adjacent angles. **(353)**

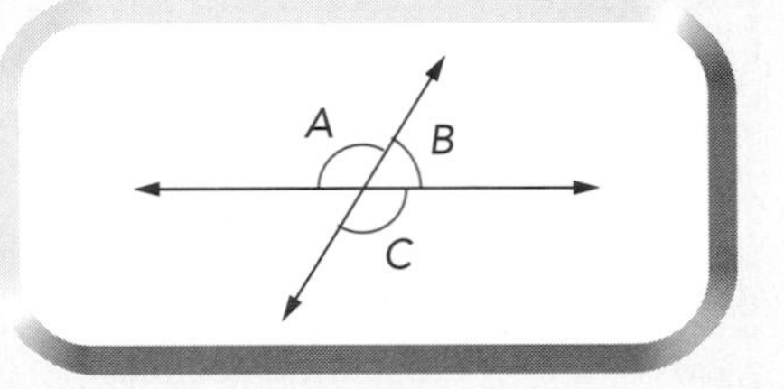

algebra: Algebra uses variables to express general rules about numbers, number relationships, and operations. **(198)**

algebraic expression: A group of numbers, symbols, and variables that express an operation or a series of operations. 3×2 is an expression for three times two. **(235)**

algorithm: A step-by-step method for computing.

alternate angles: Some of the pairs of angles formed when a line crosses two other lines. Angles 1 and 7 are alternate exterior angles (so are angles 2 and 8). Angles 3 and 5 are alternate interior angles (so are angles 4 and 6).

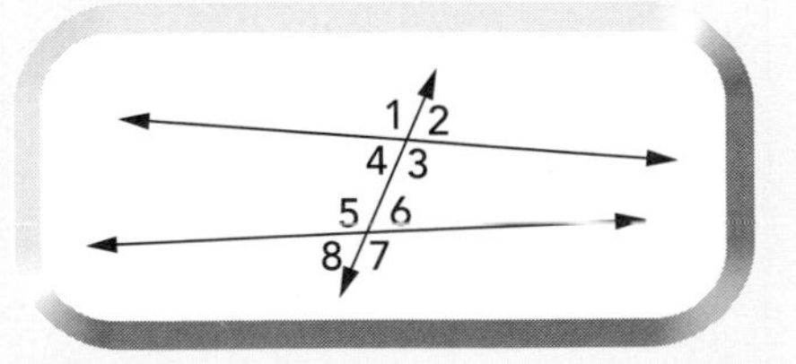

altitude: *See height*

angle (∠): Two rays that share an endpoint. **(344)**

apex: The point on a geometric figure that is farthest from the base. **(385)**

approximate number: A number that describes another number without specifying it exactly. $3.14 \approx \pi$.

Arabic numerals: The digits used in our base ten number system: 0, 1, 2, 3, 4, 5, 6, 7, 8, 9. **(505)**

arc ($\frown$): Part of a curve between any two of its points.

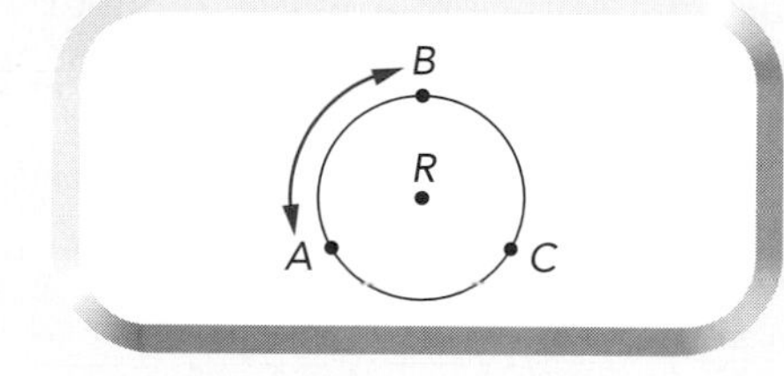

area (*A*): The measure, in square units, of the interior region of a 2-dimensional figure or the surface of a 3-dimensional figure. The area of the rectangle is 8 square units. **(299)**

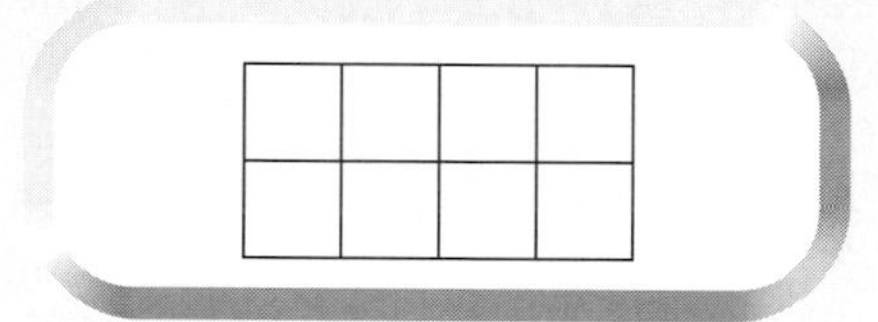

arithmetic: Calculation using addition, subtraction, multiplication, and division.

array: An arrangement of objects in equal rows.

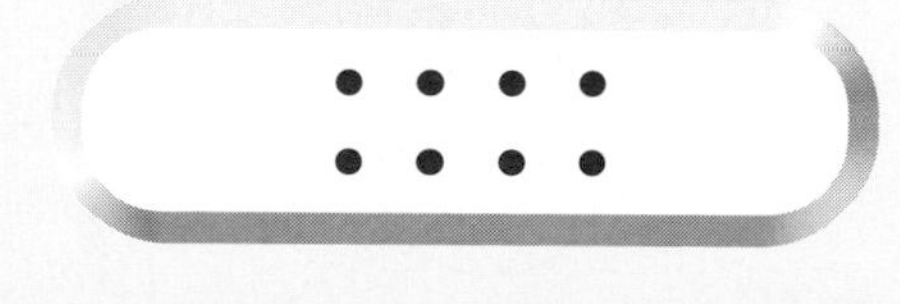

Associative Law: *See Associative Property*

Associative Property of Addition: The sum stays the same when the **grouping** of addends is changed.
$(6 + 4) + 2 \longrightarrow 10 + 2 = 12$
$6 + (4 + 2) \longrightarrow 6 + 6 = 12$ **(221)**

Associative Property of Multiplication: The product stays the same when the **grouping** of factors is changed.
$(6 \times 4) \times 2 \longrightarrow 24 \times 2 = 48$
$6 \times (4 \times 2) \longrightarrow 6 \times 8 = 48$ **(222)**

average: A single number that describes all the numbers in a set. Usually, the average is the mean, but sometimes it is the median or the mode. *See also mean, median, mode* **(259)**

axes: Plural of *axis*.

axis: A reference line from which distances or angles are measured on a coordinate grid. **(270)**

base of an exponent: The number used as the factor in exponential form. In 3^2, the base is 3 and the exponent is 2. **(065)**

base of a percent: The number for which the percent is found. In 10% of 55, the base is 55.

base of a polygon (*b*): The side of a polygon that contains one end of the altitude. **(359)**

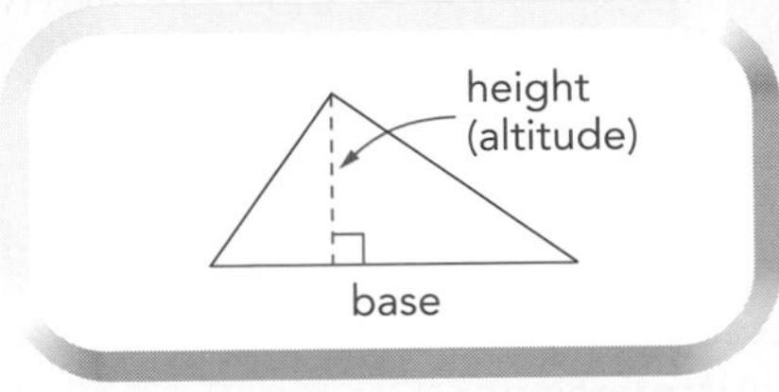

base of a solid figure (*B*): A special face of a solid figure. If the solid is a cylinder or prism, there are two bases that are parallel and congruent. **(382)**

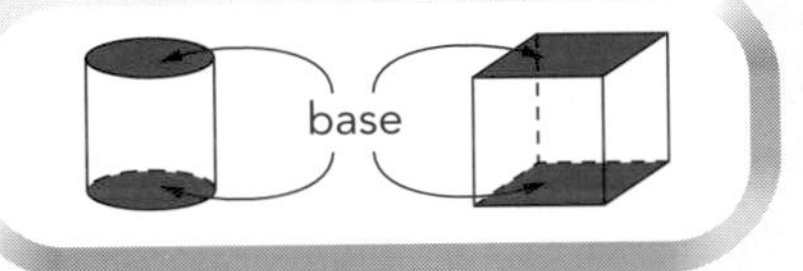

base ten: A number system in which each digit has ten times the value of the same digit one place to its right. $77 = 70 + 7$ **(002)**

binary number system: A number system with place values that are powers of 2. The only digits this system uses are zero and 1. 110 in binary is 6 (one 4, one 2, and no 1s) in base ten. **(506)**

bisect: To cut or divide into two equal parts. The midpoint of a line segment bisects it. **(470)**

bit: Abbreviation for binary digit—a zero or a 1 in the binary number system. 110 in binary contains three bits (two ones and a zero).

borrow: Regroup from one place value to a lower place value in order to subtract. *See regroup*

C

cancel: To remove equal factors from both sides of an equation or from the numerator and denominator of a fraction. $\frac{6}{8} = \frac{\cancel{2} \times 3}{\cancel{2} \times 4}$ Since both numerator and denominator have 2 as a factor, the 2s can be canceled to simplify the fraction to $\frac{3}{4}$. **(169)**

capacity: The maximum amount that can be contained by an object. **(313)**

cardinal number: A whole number that names *how many* objects are in a group.

carry: Move an extra digit from one place-value column to the next. *See regroup*

Cartesian coordinate system: *See coordinate grid*

Celsius (C): The metric-system scale for measuring temperature. **(321)**

center: (1) A point that is the same distance from all points on a circle or a sphere. (2) A point that is the same distance from all the vertices of a regular polygon. (3) Occupying a middle position. **(367)**

centigrade: *See Celsius*

central angle: An angle that has the center of a circle as its vertex. $\angle CDE$ is a central angle. **(355)**

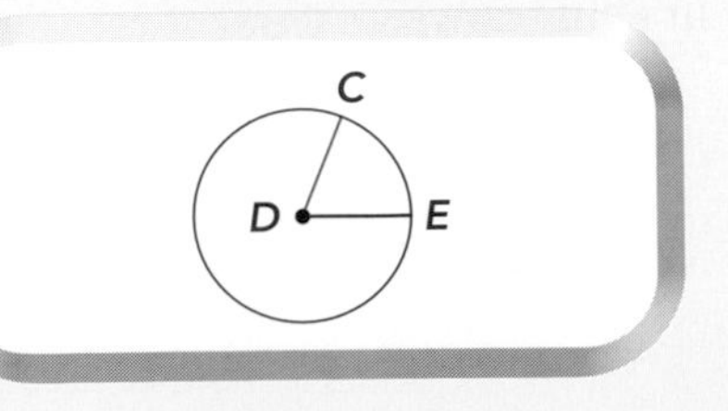

certain event: An event that will *definitely* happen. A certain event has a probability of 1.

chord: A line segment joining any two points on a circle. **(367)**

circle: A closed curve with all its points in one plane and the same distance from a fixed point (the center). **(367)**

circumference (*C*): The perimeter of a circle. **(298)**

clockwise: In the same direction that the hands of a clock rotate. **(377)**

collecting terms: *See combining like terms*

collinear: On the same line. Points A, B, and C are collinear. Points A, B, and D are not collinear.

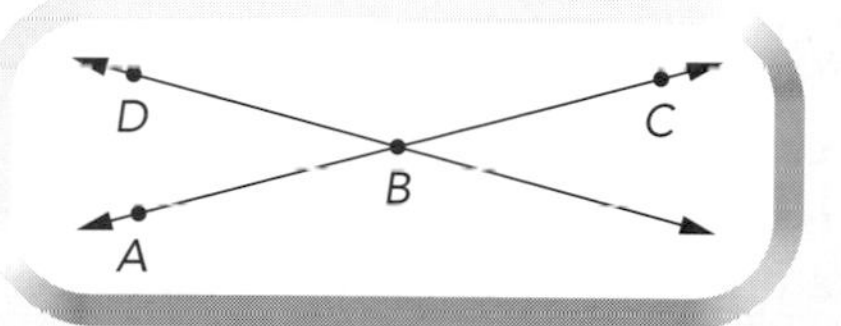

combination: A group of items or events. Placing these items or events in a different order does not create a new combination. A nickel, a dime, and a penny are a combination of coins. A dime, a nickel, and a penny are the same combination.

combining like terms: When working with an expression, combining terms that have the same variable with the same exponent. $4x + 3y + 8x^2 - x + y$ is the same as $8x^2 + 3x + 4y$.

common: Shared. These two triangles have a common vertex.

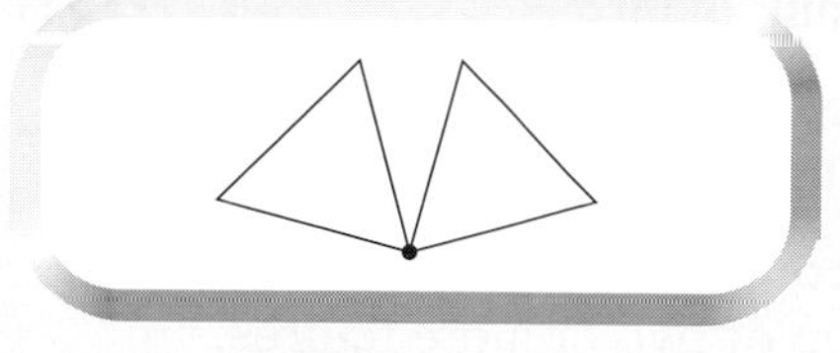

common factor: A number that is a factor of two or more numbers. The factors of 18 are 1, 2, 3, 6, 9, 18. The factors of 24 are 1, 2, 3, 4, 6, 8, 12, 24. The numbers 1, 2, 3, and 6 are the common factors of 18 and 24. **(058)**

common fraction: Any fraction whose numerator and denominator are whole numbers. $\frac{4}{5}$ and $\frac{8}{3}$ are both common fractions.

common multiple: A number that is a multiple of two or more numbers. The numbers 6, 12, 24, and 30 are some of the common multiples of 2 and 3. **(061)**

Commutative Property of Addition: The sum stays the same when the **order** of the addends is changed. $6 + 4 = 4 + 6$ **(217)**

Commutative Property of Multiplication: The product stays the same when the **order** of the factors is changed.
$8 \times 5 = 5 \times 8$ **(218)**

commutativity: *See Commutative Property*

compatible numbers: A pair of numbers that are easy to work with mentally. The numbers 25 and 70 are compatible numbers for estimating 22 + 73. **(105)**

complementary angles: Two angles that have measures with a sum of 90°. **(351)**

composite: Made up of several different things.

composite figure: A figure made up of two or more figures.

composite number: A number that has more than two factors. 8 is a composite number because it has four factors. **(055)**

computation: The process of computing. **(117)**

compute: To find a numerical result, usually by adding, subtracting, multiplying, or dividing. **(117)**

concave polygon: A polygon with one or more diagonals that have points outside the polygon. *See also convex polygon*

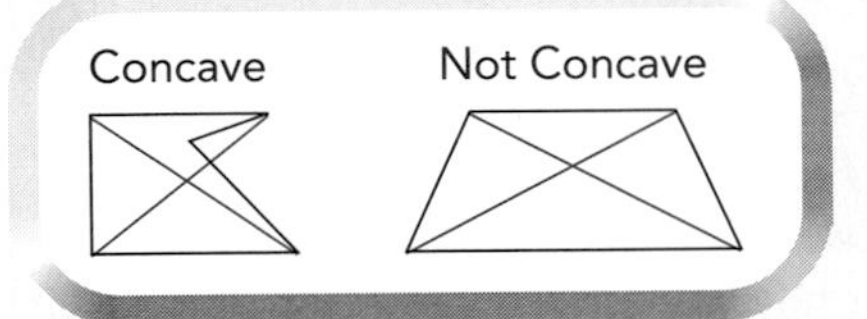

cone: A 3-dimensional figure with one curved surface, one flat surface (usually circular), one curved edge, and one vertex. **(390)**

congruent (≅): Having exactly the same size and shape. ΔABC is congruent to ΔQRS. **(372)**

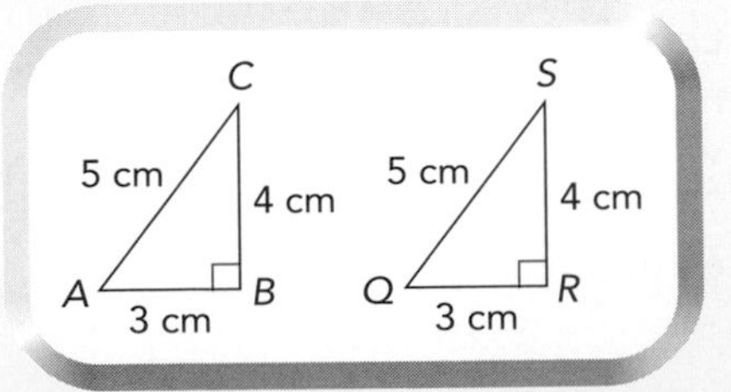

consecutive: In order, with none missing. 8, 9, 10 are consecutive whole numbers. 2, 4, 6 are consecutive even numbers.

convex polygon: A polygon with all interior angles measuring less than 180°. All diagonals of a convex polygon are inside the figure. *See also concave polygon*

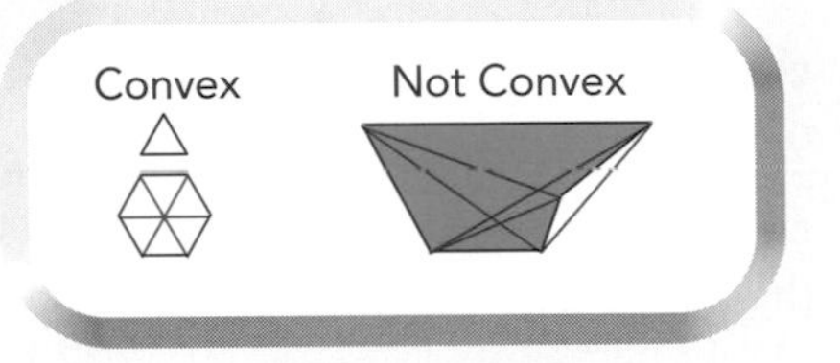

coordinate grid: A 2-dimensional system in which a location is described by its distances from two intersecting, usually perpendicular, straight lines called axes. **(266)**

coordinate plane: *See coordinate grid*

coordinates: An ordered pair of numbers that give the location of a point in a coordinate grid. The coordinates of point A in the coordinate grid are (2, 3). The x-coordinate 2 tells how many units to move horizontally starting at the origin. The y-coordinate 3 tells how many units to move in the vertical direction. **(265)**

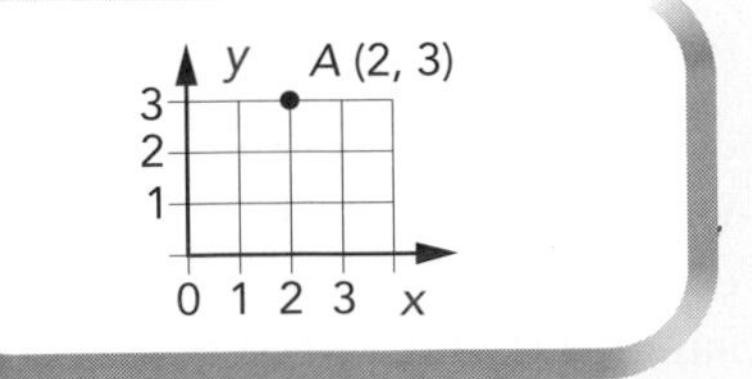

coplanar: In the same plane.

counterclockwise: In a direction opposite to the direction that the hands of a clock rotate. **(377)**

counting principle: A method of multiplying to find out how many different ways two events can happen together. If one event can happen in 3 different ways and a second event can happen in 5 different ways, the two can occur together in 3×5, or 15 different ways. **(292)**

cross multiplication: A method for finding a missing numerator or denominator in equivalent fractions or ratios by making the cross products equal. If $\frac{2}{3} = \frac{■}{9}$, then using *cross multiplication*:
$3 \times ■ = 2 \times 9$
$3 \times ■ = 18$
Since $3 \times 6 = 18$, $■ = 6$.

cross product: The product of one numerator and the opposite denominator in a pair of equivalent fractions. $\frac{3}{6} = \frac{4}{8}$; $3 \times 8 = 24$ and $6 \times 4 = 24$. **(184)**

cross section: A shape formed when a plane cuts through a 3-dimensional figure.

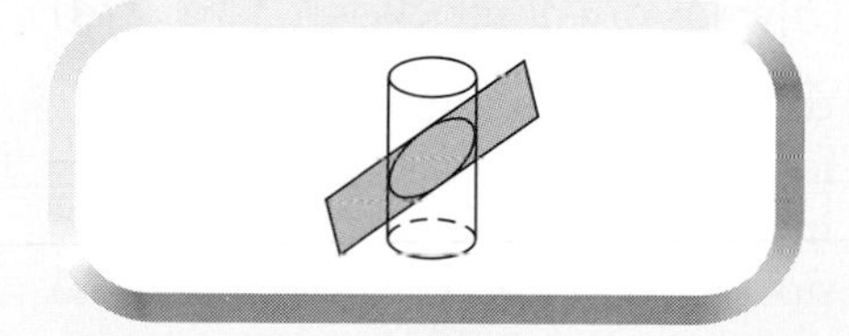

cube: (1) A regular solid with six congruent square faces. (2) The third power of a number. $10^3 = 10 \times 10 \times 10 = 1000$ **(383)**

cubic measure: *See cubic unit*

cubic unit: A unit such as a cubic meter used to measure volume or capacity. **(309)**

curve: A smooth line that is continuously not straight. **(343)**

customary system: A system of measurement used in the U.S. The system includes units for measuring length, capacity, weight, and temperature. Inches, teaspoons, and pounds are examples of customary units of measure. **(486)**

cylinder: A 3-dimensional figure with two parallel and congruent circles as bases, one curved surface, two curved edges, and no vertices. **(388)**

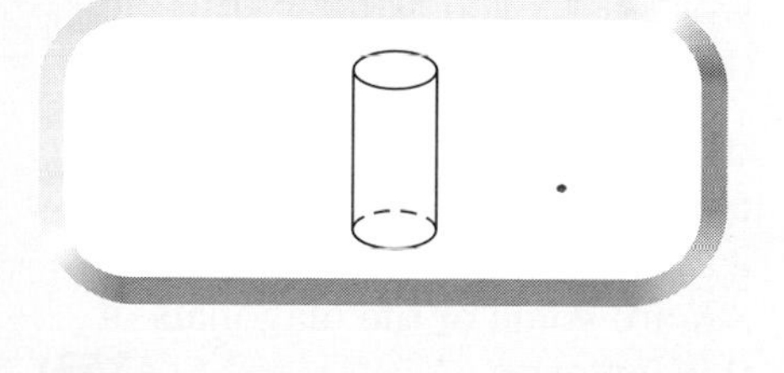

D

data: Information, especially numerical information. Usually organized. **(248)**

decimal number: (1) A number written using base ten. (2) A number containing a decimal point. **(011)**

decimal point: A dot separating the ones and tenths places in a decimal number. **(011)**

deficient number: A number whose factors (except for the number itself) have a sum less than the number. 10 is a deficient number because $1 + 2 + 5 < 10$.

degree (angle measure): A unit for measuring angles. **(346)**

degree Celsius (°C): The metric unit of measurement for temperature. *See also Celsius* **(321)**

degree Centigrade: *See degree Celsius*

degree Fahrenheit (°F): The customary unit of measurement for temperature. *See also Fahrenheit* **(320)**

denominate number: A number used with a unit. 24 inches or 95°

denominator: The quantity below the line in a fraction. It tells the number of equal parts into which a whole is divided. In the fraction $\frac{5}{8}$, the denominator is 8. **(028)**

diagonal: A line segment that joins two vertices of a polygon but is not a side of the polygon. $\overline{AC}$, $\overline{AD}$, and $\overline{AE}$ are some of the diagonals of this hexagon. **(357)**

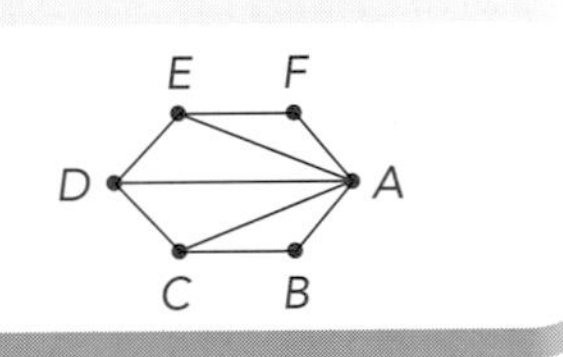

diagram: A drawing that represents a mathematical situation. **(398)**

diameter: A chord that goes through the center of a circle. **(367)**

difference: The amount that remains after one quantity is subtracted from another. **(100)**

digit: Any one of the ten symbols 0, 1, 2, 3, 4, 5, 6, 7, 8, or 9. **(002)**

dimensions: (1) The lengths of sides of a geometric figure. (2) The number of ways a figure can be measured. A line segment is 1-dimensional; polygons, circles, and other 2-dimensional figures lie in a single plane; cones, prisms, and other 3-dimensional figures occupy more than one plane.

directed number: A number with a positive or negative sign to show its direction from zero. Temperatures and altitudes are directed numbers. Area and volume are always positive so they are not directed numbers.

Distributive Property: When one of the factors of a product is written as a sum, multiplying each addend before adding does not change the product.

$3 \times (4 + 5) = (3 \times 4) + (3 \times 5)$
$3 \times 9 = 12 + 15$
$27 = 27$ **(224)**

divide (÷): To separate into equal groups. **(144)**

dividend: A quantity to be divided.
dividend ÷ divisor = quotient
$\text{divisor}\overline{)\text{dividend}}$ with quotient above **(144)**

divisible: One number is divisible by another if their quotient is an integer. 16 is divisible by 2 but is not divisible by 3. **(062)**

division: The operation of making equal groups. **(144)**

divisor: The quantity by which another quantity is to be divided.
dividend ÷ divisor = quotient
$\text{divisor}\overline{)\text{dividend}}$ with quotient above **(144)**

E

edge: The line segment where two faces of a solid figure meet. **(302)**

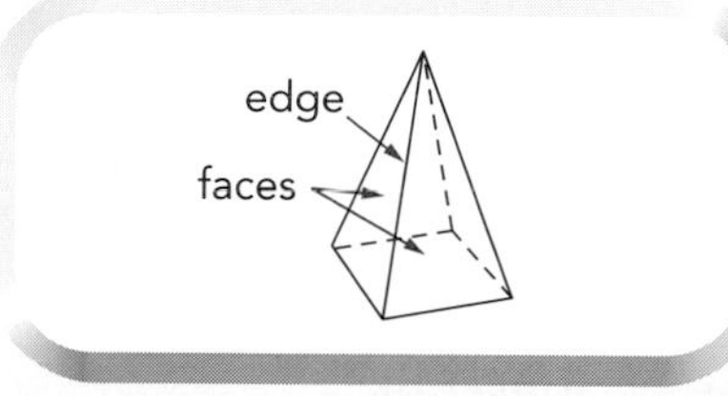

endpoint: A point marking either end of a line segment. **(338)**

equal (=): Having the same value.

equal ratios: *See proportion*

equally likely: Having the same chance, or probability. If you flip a coin, the two outcomes—the coin lands heads up and the coin lands tails up—are equally likely to occur. **(286)**

equation: A statement that two mathematical expressions are equal. $n + 50 = 75$ means that $n + 50$ must have the same value as 75. **(235)**

equiangular: All angles have the same measure. In an equiangular triangle, each angle measures 60°. **(361)**

equidistant: Equally distant.

equilateral triangle: A triangle with all sides the same length. **(362)**

equivalent: Having the same value. 4.6 and 4.60 are equivalent decimals. $\frac{2}{3}$ and $\frac{4}{6}$ are equivalent fractions. 2:6 and 1:3 are equivalent ratios. **(015, 035, 181)**

estimate (es' ti mate): To find a number close to an exact amount. **(393)**

estimate (es' ti mit): A number close to an exact amount; an estimate tells *about* how much or *about* how many.

evaluate: To find the value of a mathematical expression.

even number: A whole number that is divisible by 2. Even numbers have 0, 2, 4, 6, or 8 in the ones place. 138 is an even number. **(063)**

event: A possible outcome in probability. If you toss a fair coin, an event is *the coin lands on "tails."* **(286)**

expanded form: A way to write numbers that shows the place value of each digit. $789 = (7 \times 100) + (8 \times 10) + (9 \times 1)$ **(006)**

experimental probability: A statement of probability based on trials.

exponent: The number that tells how many equal factors there are. $10 \times 10 \times 10 \times 10 = 10^4$; the exponent is 4. **(007)**

exponential form: A way of writing a number using exponents. $425 = (4 \times 10^2) + (2 \times 10^1) + (5 \times 10^0)$ **(007)**

expression: A variable or combination of variables, numbers, and symbols that represents a mathematical relationship. $4r^2$; $3x + 2y$; $\sqrt{25}$ **(235)**

F

face: A plane figure that serves as one side of a solid figure. The faces of a cube are squares. **(282)**

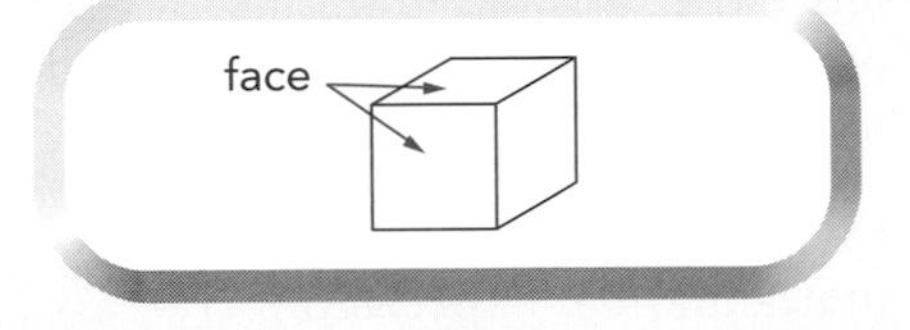

factor: (1) An integer that divides evenly into another. $2 \times 6 = 12$; 2 and 6 are factors of 12. (2) *See factorize* **(050)**

factorial (!): The product of a whole number and every positive whole number less than itself.
4 factorial = 4!
$4! = 4 \times 3 \times 2 \times 1 = 24$ **(052)**

factorize: (1) To find the factors of a number or expression. (2) To write a number or expression as a product of its factors.

Fahrenheit (F): Temperature scale. 32°F is the freezing point of water at sea level. 212°F is the boiling point of water at sea level. **(320)**

favorable outcome: In probability, the outcome you are interested in measuring. Suppose the possible outcomes of picking a marble out of a bag are red, blue, and yellow. If you want to know the probability of picking blue, then blue is the favorable outcome. **(287)**

Fibonacci sequence: A special series of numbers in which each number is the sum of the two numbers before it. 1, 1, 2, 3, 5, 8, 13, . . . **(497)**

figure: A closed shape in 2 or 3 dimensions. **(382)**

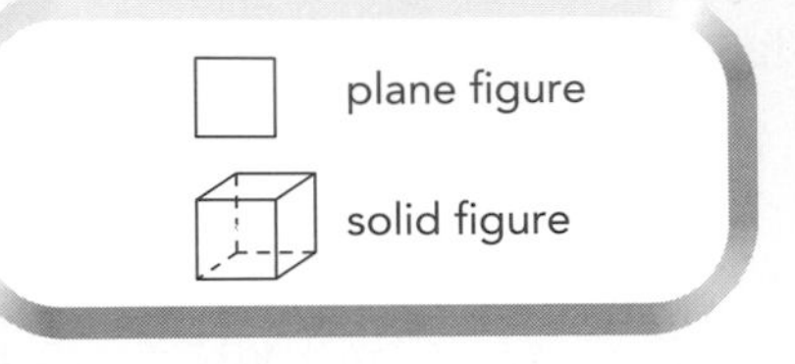

finite: Countable.

flip: *See reflection*

formula: A general equation or rule. **(246)**

fraction: A way of representing part of a whole or part of a group by telling the number of equal parts in the whole and the number of those parts you are describing.
$\frac{4}{5}$ 4 ← numerator (4 parts)
5 ← denominator (5 equal parts in the whole) **(028)**

frequency: The number of times something occurs in an interval or set of data. **(268)**

front-end estimation: Estimating by computing with the front digits. **(103)**

G

geometric: Having to do with geometry.

geometry: The mathematics of the properties and relationships of points, lines, angles, surfaces, and solids. **(334)**

graph: A drawing that shows a relationship between sets of data. **(269)**

greatest common divisor (GCD): *See greatest common factor*

greatest common factor (GCF): The largest number that divides evenly into two or more numbers. The greatest common factor of 12, 18, and 30 is 6. **(058)**

grid: A pattern of horizontal and vertical lines, usually forming squares. **(265)**

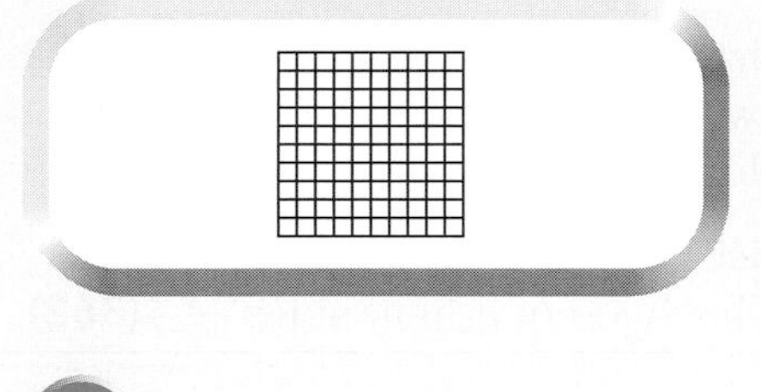

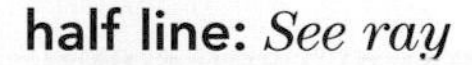

half line: *See ray*

half turn: A rotation of 180° (or half of one revolution) about a point.

height (altitude, *h*): (1) The length of a perpendicular from a vertex to the opposite side of a plane figure. (2) The length of a perpendicular from the vertex to the base of a pyramid or cone. (3) The length of a perpendicular between the bases of a prism or cylinder. If there is more than one side or edge that can be used as a base in a figure, then the figure may have more than one possible height. **(359)**

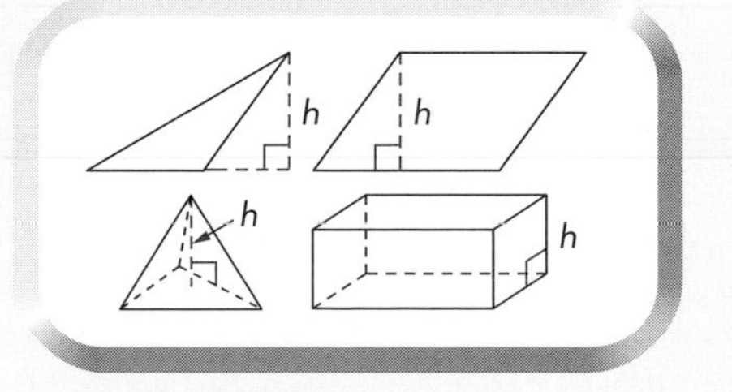

hexagon: A polygon with six sides.

hexagonal prism: A prism with six-sided bases.

highest common factor: *See greatest common factor*

histogram: A bar graph in which the labels for the bars are consecutive groups of numbers.

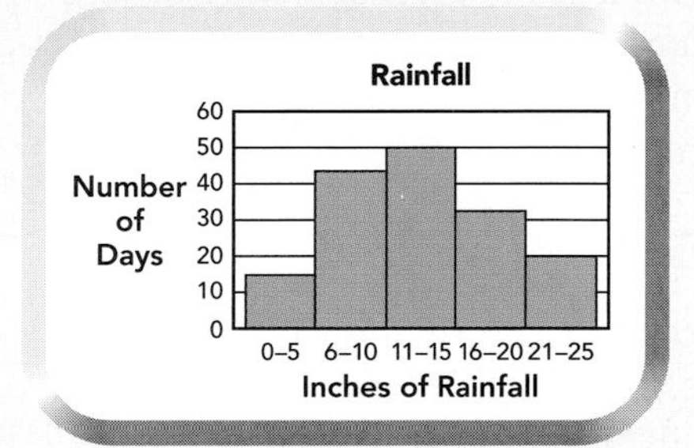

horizontal: Parallel to or in the plane of the horizon. In a coordinate grid, the x-axis is a horizontal line.

hypotenuse: The longest side of a right triangle. This side is opposite the right angle.

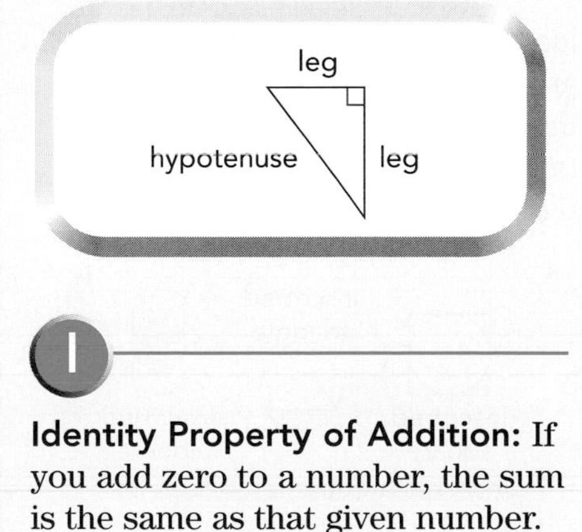

I

Identity Property of Addition: If you add zero to a number, the sum is the same as that given number. $8 + 0 = 8$ and $0 + 8 = 8$ **(227)**

Identity Property of Multiplication: If you multiply a number by 1, the product is the same as the given number. **(227)**

impossible event: An event with a probability of zero. If you roll a 1–6 number cube, rolling a 7 is an impossible event. **(286)**

improper fraction: A fraction with a value greater than 1 or less than $^{-}1$ which is not written as a mixed number. $\frac{5}{3}$ is an improper fraction. **(034)**

indirect measurement: Finding a measurement by measuring something else and then using relationships to find the measurement you need.

inequality: A mathematical sentence that compares two unequal expressions using one of the symbols $<$, $>$, $\leq$, $\geq$, or $\neq$.

infinite: Having no boundaries or limits. Lines are infinitely long.

inscribed: A figure whose vertices are part of another figure is inscribed in that figure. A figure tangent to all surfaces of another figure is inscribed in that figure.

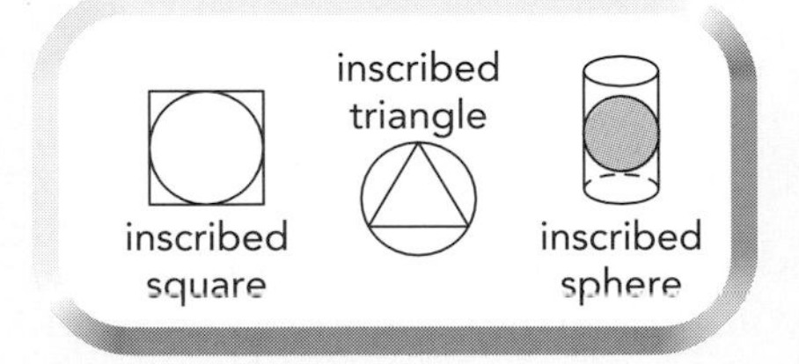

integers: Whole numbers and their opposites.
$\ldots {}^{-}2, {}^{-}1, 0, 1, 2, \ldots$ **(046)**

integral: Refers to an integer.

intersect: To meet or cross. **(340)**

irrational numbers: Numbers that cannot be written as a ratio of two integers. The digits in an irrational number never terminate and never repeat. 0.10110111011110 . . . or 3.14159 . . . **(022)**

irregular polygon: A polygon whose sides are not all the same length.

isosceles triangle: A triangle that has two congruent sides. **(362)**

L

latitude: Distance north and south of the Equator. Measured in degrees from 0° at the Equator to 90° at each pole. **(049)**

least common denominator (LCD): The smallest common multiple of the denominators of two or more fractions. The LCD of $\frac{1}{4}$ and $\frac{5}{6}$ is 12. **(036)**

least common multiple (LCM): The smallest common multiple of a set of two or more numbers. The LCM of 4 and 6 is 12. **(061)**

leg: In a right triangle, one of the two sides that form the right angle.

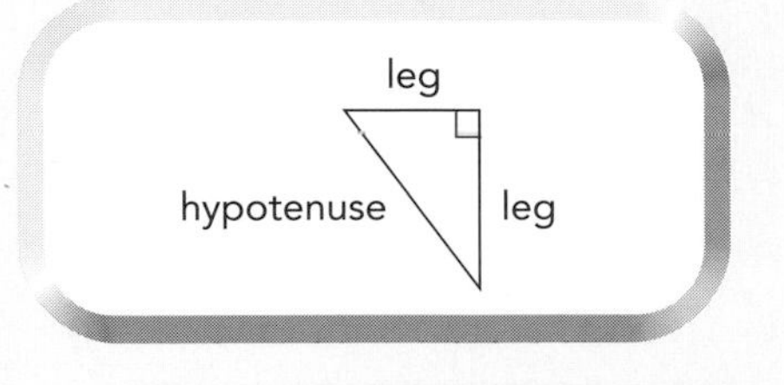

length (*l*): (1) The distance along a line or figure from one point to another. (2) One dimension of a 2- or 3-dimensional figure. **(362)**

like terms: Terms that have the same variables with the same exponents. In $3x + 2x + 5y + 6$, $3x$ and $2x$ are like terms.

line (⟷): An infinite set of points forming a straight path extending in two directions. **(337)**

line of symmetry: A line that divides a figure into two congruent halves that are mirror images of each other. **(380)**

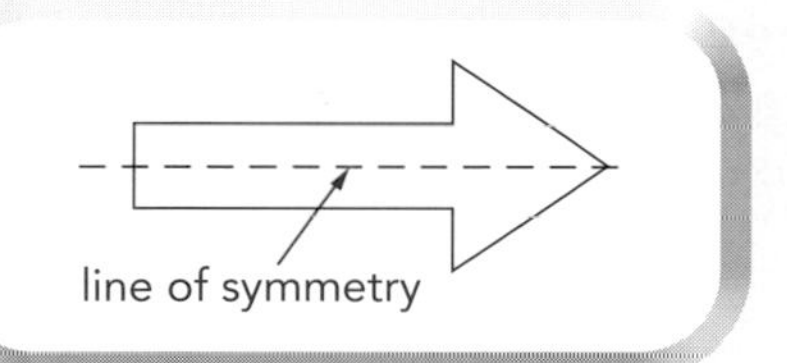

line plot: A diagram showing frequency of data on a number line. **(282)**

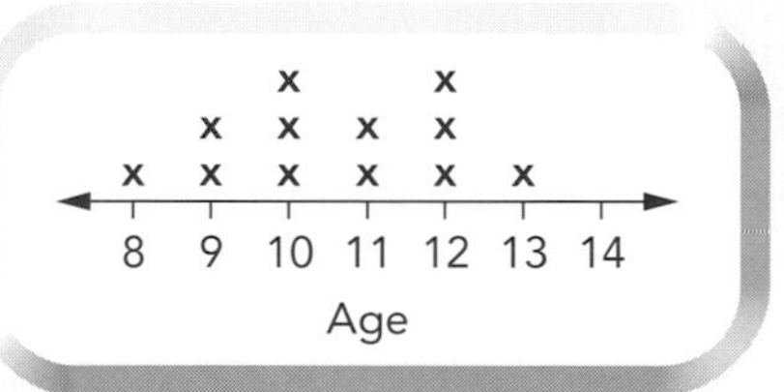

line segment (—): A part of a line defined by two endpoints. **(339)**

line symmetry: A geometric property. If a figure can be folded along a line so that the two halves match exactly, then the figure has line symmetry. **(387)**

logic: The mathematical study of ways to reason through problems.

longitude: Distance around the earth from a line running north and south through Greenwich, England. Measured in degrees from 0° at Greenwich east and west to 180°. **(490)**

lowest terms: *See simplest form*

M

mass: The amount of matter in an object. Usually measured by balancing against an object of known mass. While gravity influences weight, it does not affect mass. **(316)**

mean: A number found by dividing the sum of two or more addends by the number of addends. The mean is often referred to as the average. **(260)**

measure: (1) The dimensions, quantity, length, or capacity of something. (2) To find the measure of something.

median: When the numbers are arranged from least to greatest, the middle number of a set of numbers, or the mean of two middle numbers when the set has two middle numbers. **(261)**

mental math: Computing an exact answer without using paper and pencil or other physical aids. **(071)**

metric system: A system of measurement based on tens. The basic unit of capacity is the liter. The basic unit of length is the meter. The basic unit of mass is the gram. **(485)**

midpoint: The point on a line segment that divides it into two congruent segments.

minuend: In subtraction, the minuend is the number you subtract from.

$$\begin{array}{r} 1496 \\ -\ 647 \\ \hline 849 \end{array}\begin{array}{l} \leftarrow \text{minuend} \\ \leftarrow \text{subtrahend} \\ \leftarrow \text{difference} \end{array}$$

(127)

minute ('): (1) One-sixtieth of an hour. (2) One-sixtieth of a degree of angle measure.

mirror image (flip): *See reflection*

mixed decimal: A decimal number with an integer part and a decimal part. 3.9

mixed fraction: A number with an integer part and a fraction part. $4\frac{2}{3}$ **(034)**

mixed number: *See mixed decimal and mixed fraction*

mode: The number that appears most frequently in a set of numbers. There may be one, more than one, or no mode. **(262)**

multiple: The product of a whole number and any other whole number. **(050)**

multiplicand: In multiplication, the multiplicand is the factor being multiplied.

$$\begin{array}{r} 75 \\ \times\ 3 \\ \hline 225 \end{array}\begin{array}{l} \leftarrow \text{multiplicand} \\ \leftarrow \text{multiplier} \\ \leftarrow \text{product} \end{array}$$

multiplication: The operation of repeated addition. 4×3 is the same as $4 + 4 + 4$. **(136)**

multiplier: In multiplication, the multiplier is the factor being multiplied by

$$\begin{array}{r} 75 \\ \times\ 3 \\ \hline 225 \end{array}\begin{array}{l} \leftarrow \text{multiplicand} \\ \leftarrow \text{multiplier} \\ \leftarrow \text{product} \end{array}$$

multiply (× or ·): *See multiplication*

natural numbers: The counting numbers: 1, 2, 3, 4, 5, . . .

negative numbers: Numbers less than zero. **(045)**

net: A 2-dimensional shape that can be folded into a 3-dimensional figure is a net of that figure. **(384)**

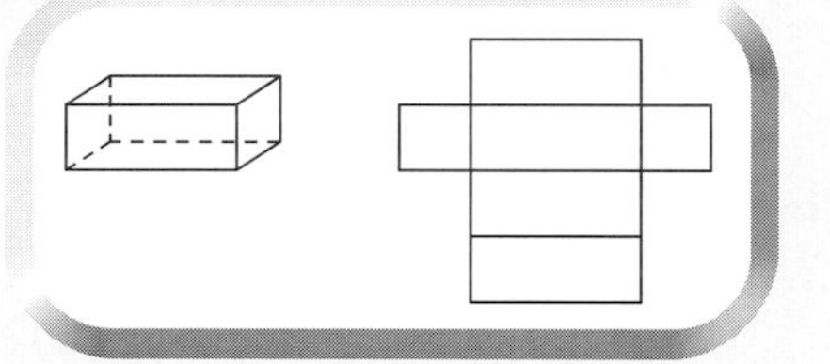

number line: A diagram that represents numbers as points on a line. **(046)**

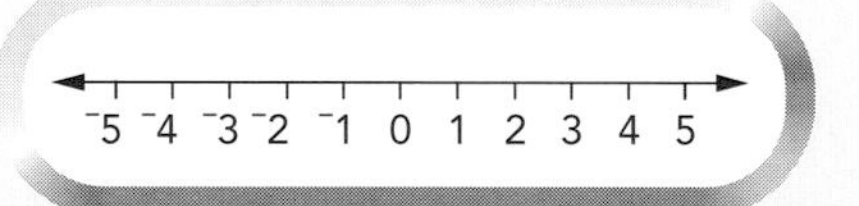

number sentence: An equation or inequality with numbers. $6 + 3 = 9 \quad 8 + 1 < 12$

numeral: A symbol (not a variable) used to represent a number.

numerator: The number that tells how many equal parts are described by the fraction. In $\frac{3}{5}$, the numerator is 3. **(028)**

O

obtuse angle: An angle with a measure greater than 90° and less than 180°. **(347)**

obtuse triangle: A triangle whose largest angle measures greater than 90°. **(361)**

octagon: A polygon with 8 sides.

odd number: A whole number that is not divisible by 2. All odd numbers have 1, 3, 5, 7, or 9 in the ones place. **(063)**

odds: The ratio of favorable outcomes to unfavorable outcomes.

operation: Addition (+), subtraction (−), multiplication (×), division (÷), raising to a power, and taking a root ($\sqrt{\ }$) are mathematical operations.

opposite: (1) Directly across from. (2) Having a different sign but the same numeral. $^{-}6$ is the opposite of $^{+}6$.

opposite angles: (1) Angles in a quadrilateral that have no common sides. ∠*A* and ∠*C* are opposite angles.

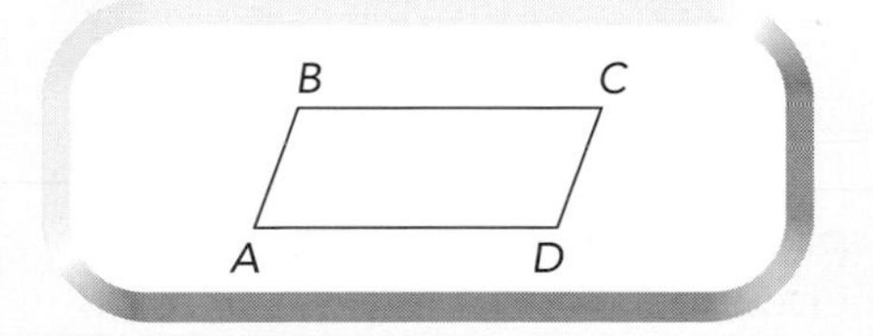

(2) *See also vertical angles*

order of operations: Rules describing what sequence to use in evaluating expressions:
(1) Evaluate within grouping symbols.
(2) Do powers or roots.
(3) Multiply or divide left to right.
(4) Add or subtract left to right. **(212)**

ordered pair: A pair of numbers that gives the coordinates of a point on a grid in this order (horizontal coordinate, vertical coordinate). Point *A* is at (3, 2). Point *B* is at (2, 3). **(265)**

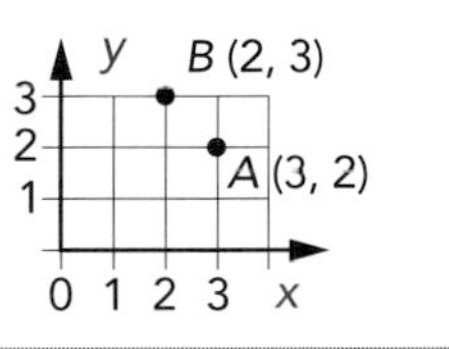

ordinal number: A whole number that names the position of an object in sequence. First, second, and third are ordinal numbers.

origin: The intersection of the x- and y-axes in a coordinate plane. It is described by the ordered pair (0, 0). **(265)**

outcome: One of the possible events in a probability situation. **(286)**

outlier: A number in a set of data that is much larger or smaller than most of the other numbers in the set. 1, 52, 55, 55, 57, 59, 125 **(258)**

overestimate: An estimate greater than the exact answer. **(107)**

P

parallel (||): Always the same distance apart. **(340)**

parallelogram: A quadrilateral with two pairs of parallel and congruent sides. **(365)**

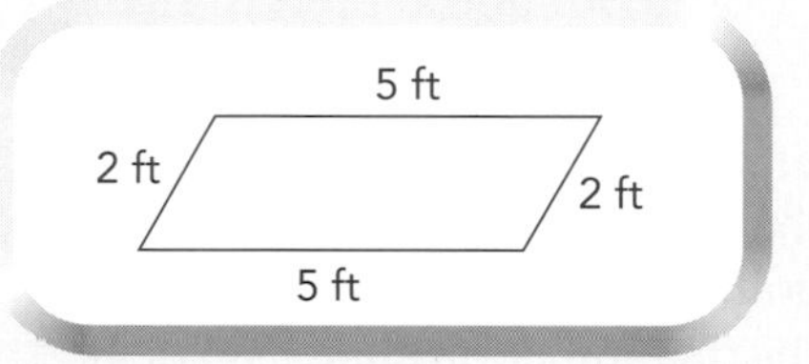

pentagon: A polygon that has five sides.

pentomino: A plane figure made up of five congruent squares, with each square having at least one side shared with another square. There are 12 different pentominoes.

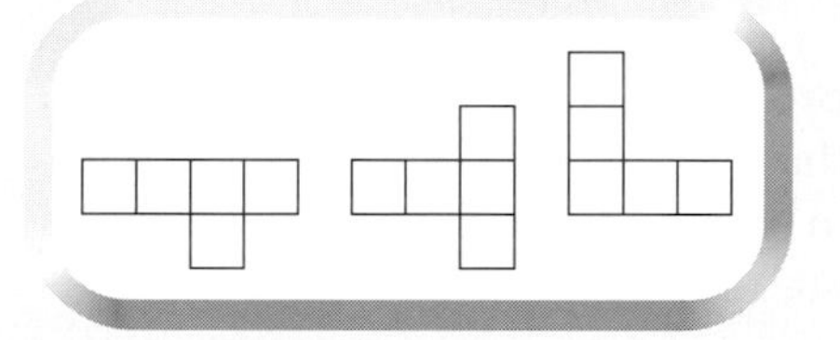

per: For each.

percent (%): A special ratio that compares a number to 100 using the symbol %. The word *percent* means *hundredths* or *out of 100.* 40% of 200 is 80. **(020)**

percentage: A number that is a given percent of another number. In 25% of 60 is 15, the percentage is 15.

perfect number: A whole number that is equal to the sum of its factors (excluding the number itself). 6 is a perfect number because $1 + 2 + 3 = 6$.

perfect square: The product of an integer multiplied by itself. 36 is a perfect square because $6 \times 6 = 36$ **(070)**

perimeter (*P*): The distance around a figure. **(295)**

permutations: Possible orders or arrangements of a set of events or items. If you put them into a different order, you have a different permutation. RAT, TAR, and ART are three of the possible permutations of the letters A, R, and T.

perpendicular (⊥): Forming right angles. **(341)**

perpendicular bisector: A line that divides a line segment in half and meets the segment at right angles. $\overline{CD}$ is the perpendicular bisector of $\overline{AB}$. **(471)**

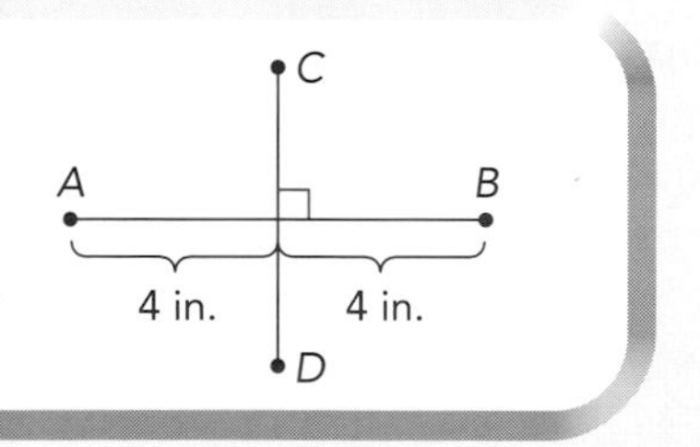

pi (π): The ratio of the circumference of any circle to its diameter, approximately equal to 3.14, or $\frac{22}{7}$. **(298)**

place value: The value of the position of a digit in a number. In 7863, the 8 is in the hundreds place, so it stands for 800. **(004)**

plane: A flat surface that extends infinitely in all directions. **(336)**

plane figure: Any 2-dimensional figure. Circles, polygons, and angles are all plane figures. **(356)**

point: An exact location in space. **(335)**

point symmetry: A geometric property. A figure that can be turned exactly 180° about a point and fit exactly on itself has point symmetry. A parallelogram has point symmetry. **(378)**

polygon: A closed plane figure formed from line segments that meet only at their endpoints. **(357)**

polyhedron: A 3-dimensional figure in which all the surfaces are polygons.

population: A group of people (or objects or events) that fit a particular description. **(250)**

positive numbers: Numbers that are greater than zero. **(045)**

power: An exponent. 8 to the second power is 8^2. **(064)**

power of 10: A number with 10 as a base and a whole-number exponent. 10^3 and 10^5 are powers of 10. **(071)**

prime factorization: A way to write a number as the product of prime factors. The prime factorization of 12 is $2 \times 2 \times 3$. **(056)**

prime number: A number that has exactly two different positive factors, itself and 1. 7 is a prime number. 1 is not a prime number. **(053)**

prism: A 3-dimensional figure that has two congruent and parallel faces that are polygons. The rest of the faces are parallelograms. **(383)**

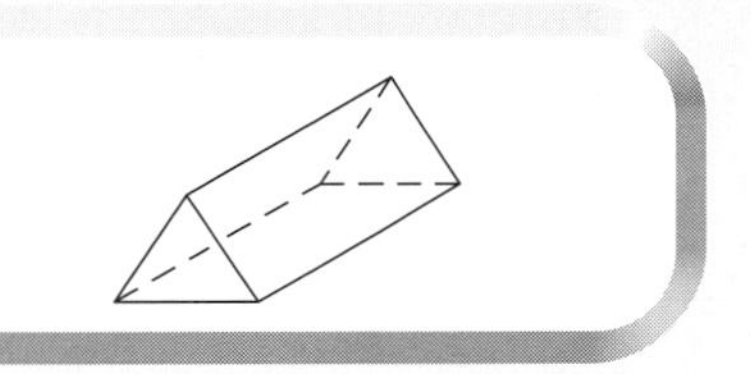

probability: The chance of an event occurring. If all outcomes of an event are equally likely, the probability of an event is equal to the number of favorable outcomes divided by the number of possible outcomes. **(285)**

$$P(\text{event}) = \frac{\text{number of favorable outcomes}}{\text{number of possible outcomes}}$$

product: The result of multiplication.

$$\begin{array}{r} 4 \leftarrow \textbf{factor} \\ \times\ 3 \leftarrow \textbf{factor} \\ 12 \leftarrow \textbf{product} \end{array}$$

(106)

proper fraction: A fraction whose numerator is an integer smaller than its integer denominator.

proportion: An equation showing two equivalent ratios. **(181)**

proportional: Having equivalent ratios.

pyramid: A polyhedron whose base is a polygon and whose other faces are triangles that share a common vertex. **(385)**

Pythagorean theorem: The sum of the squares of the lengths of the two legs of a right triangle is equal to the square of the length of the hypotenuse. $3^2 + 4^2 = 5^2$

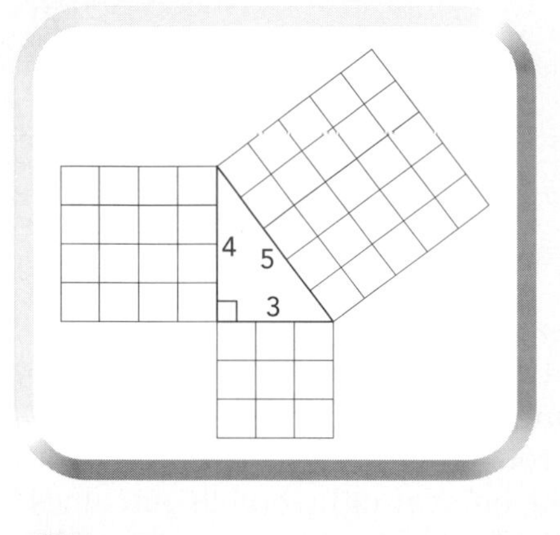

Q

quadrants: The four sections of a coordinate grid that are separated by the axes.

quadrilateral: A four-sided polygon. **(364)**

quantity: An amount.

quarter turn: One-fourth of a revolution (90°) about a given point (turn center). **(378)**

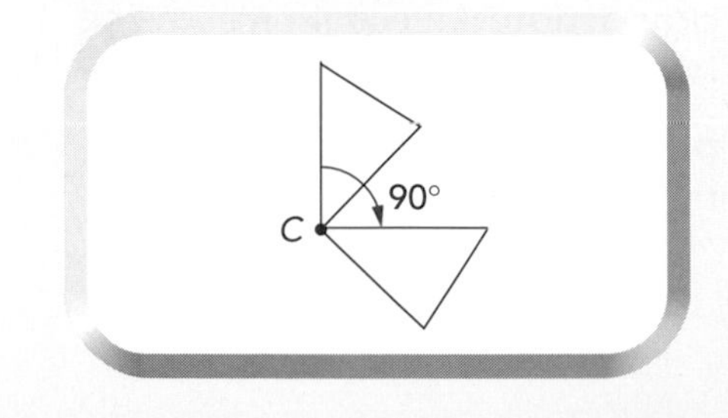

quotient: The result of division.
dividend ÷ divisor = quotient

$$\text{divisor}\overline{)\text{dividend}}^{\text{quotient}}$$ **(144)**

R

radii: Plural of *radius.*

radius (*r*): The segment, or the length of the segment, from the center of a circle to any point on the circle. **(367)**

random: By chance, with no outcome any more likely than another.

random sample: A sample in which every person, object, or event in the population has an equal chance of being included. **(252)**

range: The difference between the greatest and the least value in a set of data. **(257)**

rate: A ratio comparing two different units. Miles per hour is a rate. **(185)**

ratio: A comparison of two numbers or measures using division. The ratio of vowels to consonants in the word *number* is 2 to 4, 2:4, or $\frac{2}{4}$. **(178)**

rational number: A number that can be expressed as a ratio of two non-zero integers.

ray (⟶): A part of a line that has one endpoint and extends indefinitely in one direction. **(338)**

real numbers: The combined set of the rational and irrational numbers.

reciprocals: Two numbers that have a product of 1. $\frac{3}{4}$ and $\frac{4}{3}$ are reciprocals because $\frac{3}{4} \times \frac{4}{3} = 1$ **(171)**

rectangle: A quadrilateral with two pairs of congruent, parallel sides and four right angles. **(365)**

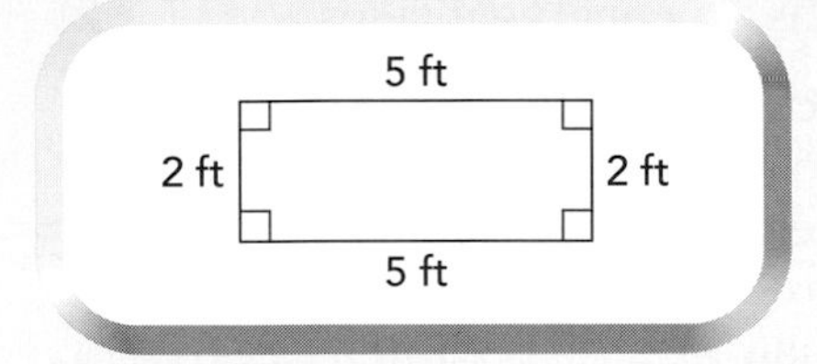

rectangular prism: A prism with six rectangular faces. **(383)**

reduce: Put a fraction into simplest form. $\frac{4}{8} = \frac{1}{2}$

reflection (flip): A transformation creating a mirror image of a figure on the opposite side of a line. **(379)**

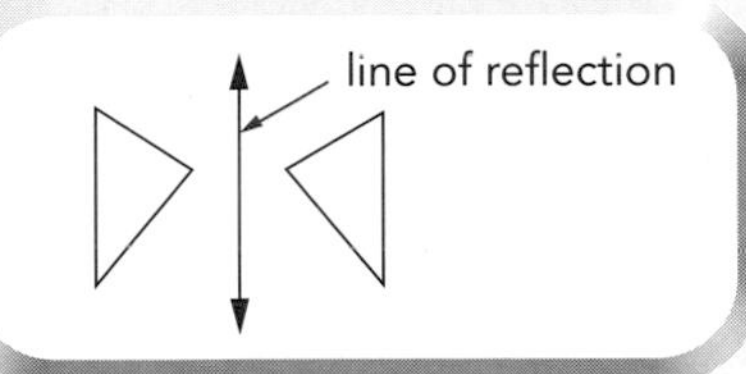

reflex angle: An angle that measures more than 180°. **(347)**

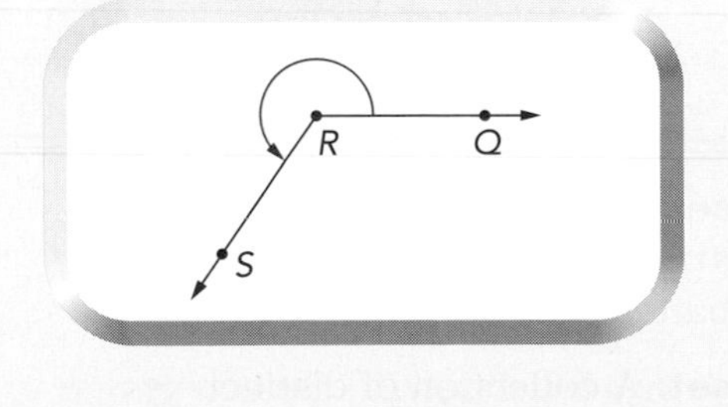

region: A part of a plane.

regroup: Use place value to think of a number in a different way to make addition and subtraction easier. You can think of

$$\begin{array}{r} 43 \\ -\ 27 \\ \hline \end{array} \quad \text{as} \quad \begin{array}{r} 30 + 13 \\ -\ (20 + \ 7) \\ \hline \end{array}$$

Either way, the difference is 16. **(121)**

regular polygon: A polygon with all sides the same length and all angles the same measure. A square is a regular polygon. **(357)**

regular polyhedron: A solid figure with congruent regular polygons for all faces.

regular solid: *See regular polyhedron*

relatively prime: Two numbers are relatively prime if they have no common factors other than 1. 5 and 12 are relatively prime. **(058)**

remainder: In whole-number division, when you have divided as far as you can without using decimals, what has not been divided yet is the remainder. **(148)**

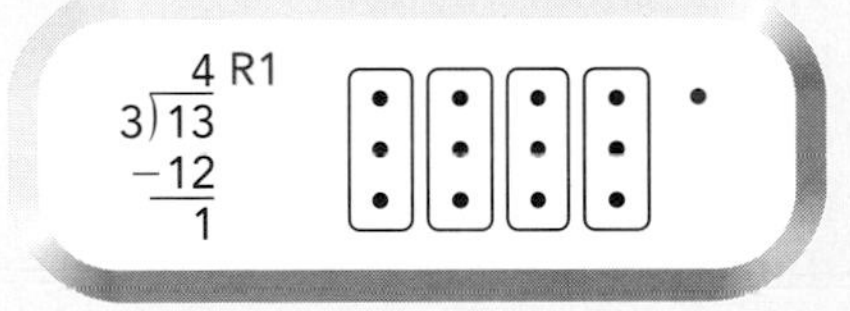

rename: *See regroup*

repeating decimal: A decimal that has an infinitely repeating sequence of digits.
$5.2424\ldots = 5.\overline{24}$ **(021)**

revolution: One turn of 360° about a point.

rhombus: A parallelogram with all four sides equal in length. **(365)**

right angle (∟): An angle that measures exactly 90°. **(347)**

right triangle: A triangle that has one 90° angle. **(361)**

Roman numerals: The symbols used in the ancient Roman number system. **(502)**

rotation (turn): A transformation in which a figure is turned a given angle and direction around a point. **(377)**

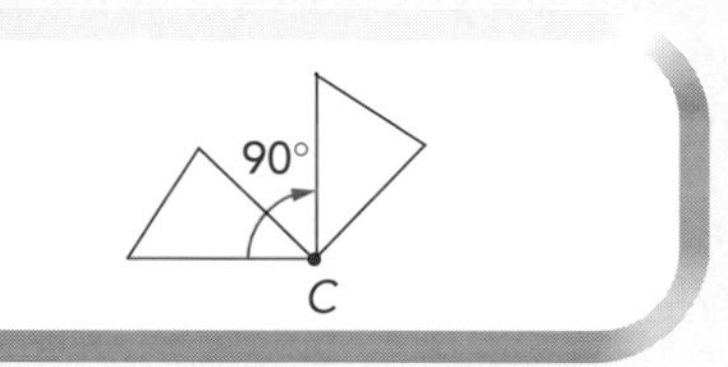

round: To find the nearest ten, hundred, thousand or tenth, hundredth, thousandth, and so on, using these rules:

- Look one place to the right of the digit you want to round to.
- If the digit is 5 or more, add 1 to the digit in the rounding place. This is rounding up.
- If the digit is less than 5, do not change the number in the rounding place. This is rounding down. **(094)**

rubric: A rule, direction, or explanation. **(431)**

S

sample: A number of people, objects, or events chosen from a given population to represent the entire group. **(249)**

scale: (1) The ratio of length used in a drawing, map, or model to its length in reality. (2) A system of marks at fixed intervals used in measurement or graphing. (3) An instrument used for weighing. **(037)**

scale factor: The ratio of the lengths of corresponding sides of two similar figures.

scalene triangle: A triangle that has no congruent sides. **(362)**

scientific notation: A form of writing numbers as the product of a power of 10 and a decimal number greater than or equal to 1 and less than 10. 2600 is written as 2.6×10^3 in scientific notation.

second ("): (1) One-sixtieth of an angle minute (one three-hundred-sixtieth of a degree). (2) One-sixtieth of a minute of time (one three-hundred-sixtieth of an hour). (3) The number two position in a line.

segment: (1) *See line segment* (2) A part of a circle bounded by a chord and the arc it creates.

semicircle: An arc that is exactly half of a circle. A diameter intersects a circle at the endpoints of two semicircles.

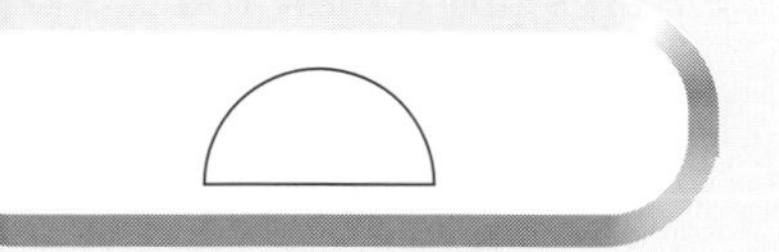

sequence: A set of numbers arranged in a special order or pattern.

set: A collection of distinct elements or items.

side: (1) A line segment connected to other segments to form a polygon. (2) An edge of a polyhedron.

Sieve of Eratosthenes: A way of finding prime numbers in a sequential list of whole numbers.

sign: A symbol ($^+$) or ($^-$) that indicates whether a number is greater than or less than zero.

signed number: Positive or negative number. $^+25$ and $^-30$ are signed numbers.

similar figures (~): Figures that have the **same shape**, but not necessarily the same size. **(369)**

simplest form: A fraction whose numerator and denominator have no common factor greater than 1. The simplest form of $\frac{4}{8}$ is $\frac{1}{2}$. **(037)**

simplify: Combine like terms and apply properties to an expression to make computation easier. $3n + 12 + 2n - 5$ simplifies to $5n + 7$.

simplify a fraction: To divide the numerator and denominator of a fraction by a common factor. *See also simplest form*

slant height: The height of one of the triangular faces of a pyramid.

slide: *See translation*

solid: *See solid figure*

solid figure: A geometric figure with 3 dimensions. **(382)**

solution: Any value for a variable that makes an equation true. A solution of $2x = 24$ is $x = 12$.

space figure: A 3-dimensional figure.

sphere: A 3-dimensional figure made up of all points that are equally distant from a point called the center. **(392)**

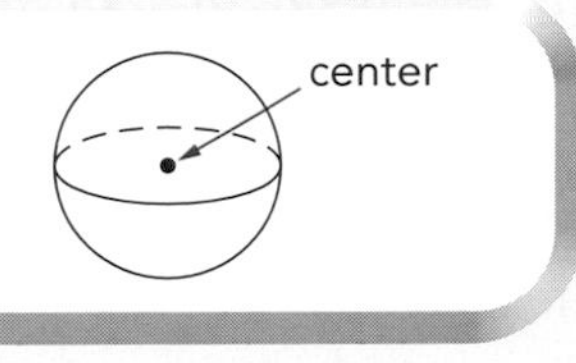

square: A parallelogram with four congruent sides and four right angles. **(365)**

square measure: A unit, such as a square meter, or a system of units used to measure area. **(299)**

square number: The number of dots in a square array. The first two square numbers are 4 and 9. *See also perfect square* **(495)**

square root ($\sqrt{}$): The length of one side of a square with an area equal to a given number. $\sqrt{81} = 9$ **(067)**

square unit: *See square measure*

standard form: A number written with one digit for each place value. The standard form for the number three thousand three is 3003. **(006)**

straight angle: An angle with a measure of 180°. **(347)**

subtract (−): *See subtraction*

subtraction: An operation that gives the difference between two numbers. Subtraction is also used to compare two numbers. **(127)**

subtrahend: In subtraction, the subtrahend is the number being subtracted.

$$\begin{array}{r l} 1496 & \leftarrow \text{minuend} \\ -\ 647 & \leftarrow \text{subtrahend} \\ \hline 849 & \leftarrow \text{difference} \end{array}$$

(127)

sum: The result of addition. The sum of 32 and 46 is 78. **(100)**

supplementary angles: Two angles that have measures whose sum is 180°. **(350)**

surface area: The total area of the faces (including bases) and curved surfaces of a solid figure. **(306)**

symbol: Something that represents something else. + means add, − means subtract, and < means less than. **(514)**

symmetry: *See line symmetry and point symmetry*

T

tangent: Touching at exactly one point. **(367)**

tangent line: A line that touches a circle at just one point. **(367)**

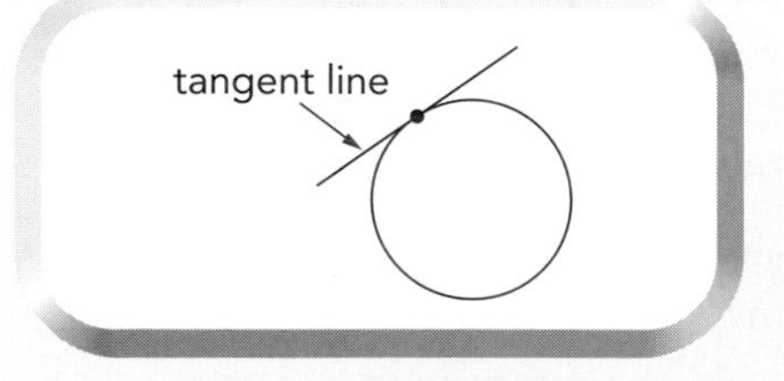

term: A number, variable, product, or quotient in an expression. A term is not a sum or difference. In $6x^2 + 5x + 3$, there are three terms, $6x^2$, $5x$, and 3.

terminating decimal: A decimal with a finite number of digits.

terms of a fraction: Each element of a fraction is a term. In $\frac{4}{5}$, 4 and 5 are both terms.

tessellation: A covering of a plane without overlaps or gaps using combinations of congruent figures. **(381)**

theoretical probability: The ratio of the number of ways an event can happen to the total number of outcomes.

three-dimensional: Existing in 3 dimensions; having length, width, and height. **(382)**

topology: The study of those properties of a figure that don't change when it changes shape.

transformation: A rule for moving every point in a plane figure to a new location. **(375)**

translation (slide): A transformation that slides a figure a given distance in a given direction. **(376)**

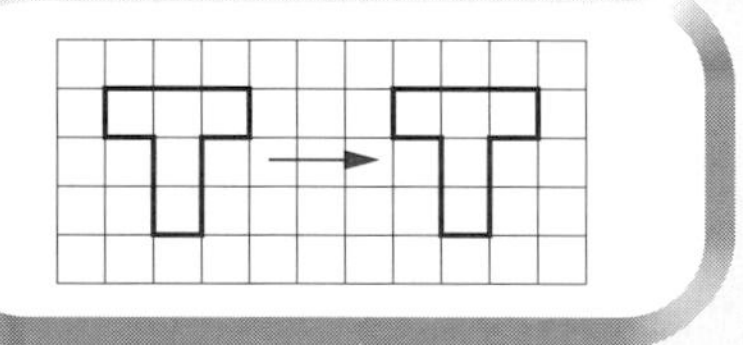

transversal: A line that intersects two or more other lines.

trapezium: A quadrilateral with no parallel sides.

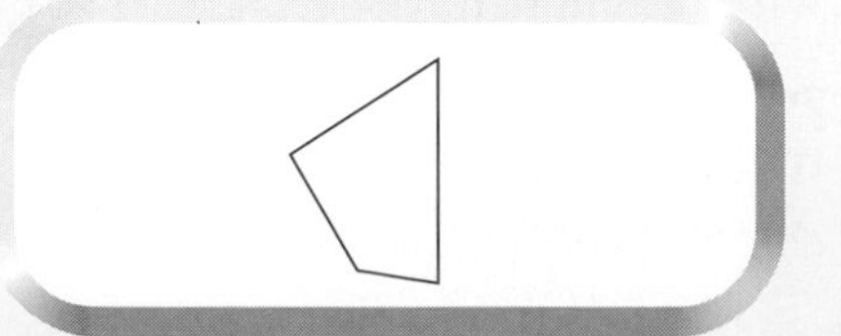

trapezoid: A quadrilateral with exactly two parallel sides. **(365)**

triangle: A polygon with three angles and three sides. **(358)**

Triangle Inequality: In any triangle, no side can be longer than the sum of the lengths of the other two sides. **(360)**

triangular number: The number of dots in a triangular arrangement. The first three triangular numbers are 1, 3, and 6. **(494)**

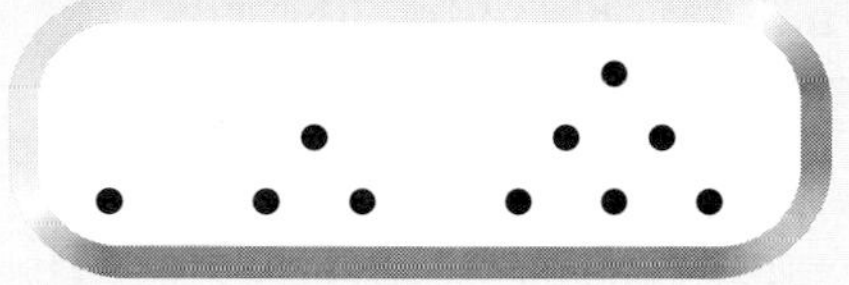

triangular pyramid: A pyramid with a triangular base. **(386)**

truncate: (1) To ignore all digits to the right of a chosen place. 7.0668 ⟶ 7.06 (2) To cut off part of a geometric figure. This is a truncated pyramid.

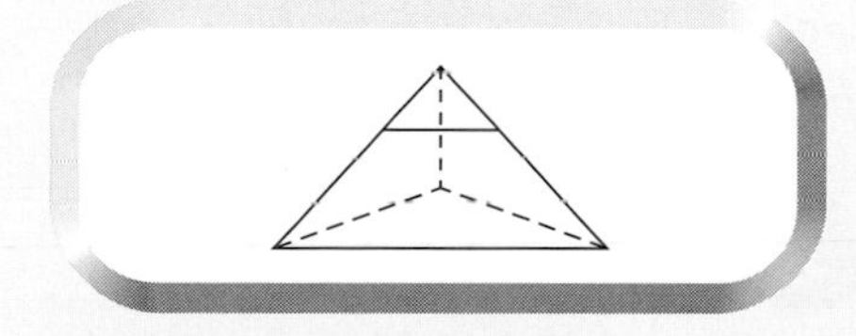

turn: *See rotation*

turn center: The point around which a figure is rotated. **(377)**

turn symmetry: A figure that can be turned less than 360° about a point and fit exactly on itself has turn (or rotational) symmetry. A square has turn symmetry. **(378)**

twin primes: Two prime numbers that are also consecutive odd numbers. 3 and 5 are twin primes but 2 and 3 are not.

two-dimensional: Having length and width. **(356)**

U

underestimate: An estimate less than the actual answer.

unit: A precisely fixed quantity used to measure.

unit fraction: A fraction with a numerator of 1.

unit rate: A rate with a denominator of 1. A rate of 50 miles per hour is a unit rate. **(185)**

V

variable: (1) A quantity that can have different values. (2) A symbol that can stand for a variable. In $5n$, the variable is n. **(236)**

variable expression: An expression that represents an amount that can have different values. $8r$ has a different value for every value assigned to r.

Venn diagram: A drawing that shows relationships among sets of objects. **(283)**

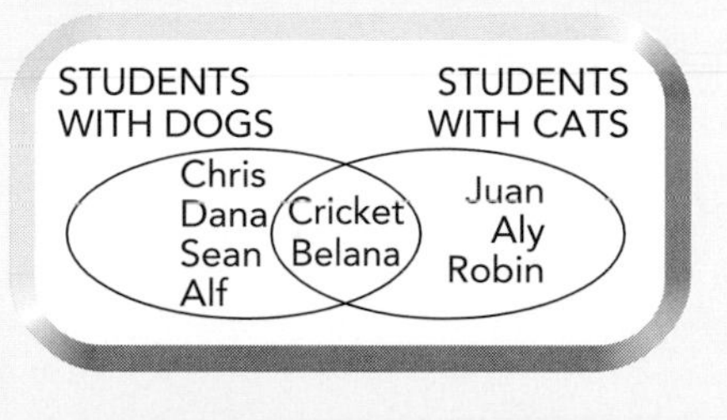

vertex: (1) The point at which two line segments, lines, or rays meet to form an angle. (2) A point on a polyhedron where three or more faces intersect. **(345)**

vertical: At right angles to the horizon. A vertical line is straight up and down.

vertical angles: The congruent angles formed when two lines intersect. Angles 1 and 3 are vertical angles. So are angles 2 and 4. **(353)**

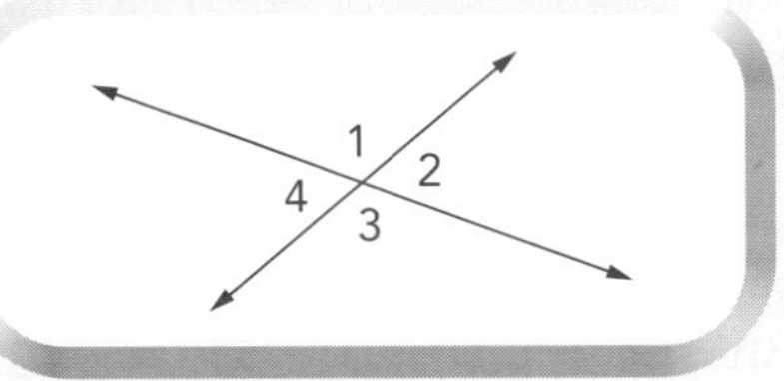

vertices: Plural of *vertex*.

volume (*V*): The number of cubic units it takes to fill a solid. **(309)**

weight: A measure of the heaviness of an object. **(316)**

whole number: Any of the numbers 0, 1, 2, and so on. **(003)**

width (*w*): One dimension of a 2- or 3-dimensional figure.

X

x-axis: On a coordinate grid, the horizontal axis. **(265)**

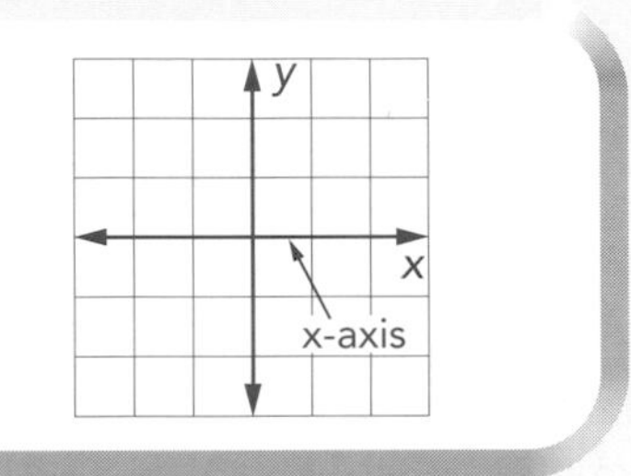

x-coordinate: In an ordered pair, the value that is written first. In (2, 3), 2 is the x-coordinate. **(265)**

y-axis: On a coordinate grid, the vertical axis. **(265)**

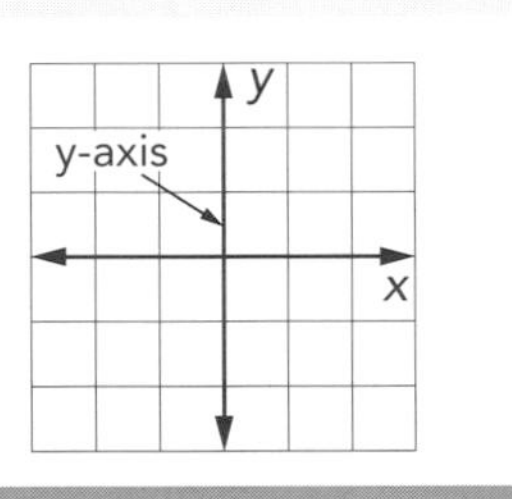

y-coordinate: In an ordered pair, the value that is always written second. In (2, 3), 3 is the y-coordinate. **(265)**

zero pair: Two numbers whose sum is zero. $^{-}4 + 4 = 0$ **(201)**

Zero Property: The product of any number and zero is zero. $6 \times 0 = 0$ and $0 \times 6 = 0$ **(230)**

Index

This index contains topic numbers, not page numbers. You will find topic numbers at the top of each page and next to each new piece of information in the book.

A

B

C

D

E

F

G

N

O

P

Q

R

S

Illustration Credits

Teresa Anderko: 007, 008, 009, 013, 014, 028, 030, 033, 041, 042, 053, 054, 055, 062, 063, 065, 075, 079, 080, 084, 086, 088, 093, 098, 099, 113, 120, 130, 133, 139, 143, 148, 149, 150, 159, 160, 165, 168, 173, 178, 181, 182, 184, 185, 187, 188, 193, 194, 195, 196, 197, 213, 218, 233, 234, 238, 241, 242, 244, 252, 254, 255, 257, 259, 261, 275, 276, 281, 282, 292, 295, 296, 297, 298, 314, 315, 316, 317, 318, 319, 321, 369, 375, 379, 394, 398, 402, 403, 404, 405, 408, 409, 411, 416, 417, 428, 430, 465, 468, 480, 498

Michael Gelen: 001, 012, 024, 025, 026, 027, 032, 034, 046, 048, 072, 076, 077, 087, 101, 107, 111, 113, 115, 123, 136, 149, 168, 177, 179, 186, 190, 193, 200, 246, 251, 253, 260, 265, 268, 280, 286, 287, 290, 294, 307, 310, 315, 323, 325, 360, 370, 378, 380, 383, 388, 389, 397, 398, 399, 401, 410, 425, 426, 427, 430, 440, 456

Joe Spooner: 001, 049, 071, 117, 156, 177, 198, 247, 293, 333, 335, 393

Robot Characters: Michael Gelen

Technical Art: Burmar Technical Corporation

Map Art: The Write Source development group

Cover Illustration: Dan Hubig